The Last Hunter-Gatherer-Fishermen in Sandy Flanders (NW Belgium)

The Verrebroek and Doel Excavation Projects

Volume 1: Palaeo-environment, chronology and features

Archaeological Reports Ghent University 3

Archaeological Reports Ghent University 3

Official abbreviation: ARGU

This new series is the successor of the Dissertationes Archaeologicae Gandenses, edited at Ghent University from 1953 to 1984 (XXII volumes)

GHENT UNIVERSITY

ISBN 90 382 0679 0

Philippe Crombé (ed.)

The Last Hunter-Gatherer-Fishermen in Sandy Flanders (NW Belgium)

The Verrebroek and Doel Excavation Projects

Volume 1: Palaeo-environment, chronology and features

ACADEMIA
PRESS

ARCHAEOLOGICAL REPORTS GHENT UNIVERSITY 3
Official abbreviation: ARGU

This new series is the successor of the Dissertationes Archaeologicae Gandenses, edited at Ghent University from 1953 to 1984 (XXII volumes)

Published by
Academia Press
Eeckhout 2
B-9000 Ghent, Belgium
Info@academiapress.be www.academiapress.be

In co-operation with
Ghent University
Department of Archaeology and Ancient History of Europe
Blandijnberg 2
B-9000 Ghent, Belgium

GHENT UNIVERSITY

Academia Press is part of
J. Story-Scientia
P. Van Duyseplein 8
B-9000 Gent
info@story.be www.story.be

Series editors:
Arjen Bosman, Jean Bourgeois, Philippe Crombé, Johnny De Meulemeester, Johnny Devreker, Roald Docter, Anton Ervynck, Ernie Haerinck and Frank Vermeulen

Editor of this volume:
Philippe Crombé

The Last Hunter-Gatherer-Fishermen in Sandy Flanders (NW Belgium)
The Verrebroek and Doel Excavation Projects
Volume 1: Palaeo-environment, chronology and features

ISBN 90 382 0679 0
D/2005/4804/33
U 651
NUR 682

Contents

PART III: FEATURES

I.

INTRODUCTION

1.1. | General introduction

Philippe Crombé

This volume in the series "Archaeological Reports Ghent University" is a first publication of the results of a major research project financed between 1998 and 2002 mainly by the Research Council of Ghent University[1]. It focuses on the presentation and discussion of all environmental and chronological data gathered during this project. A second volume, to be published later, will be dedicated entirely to the discussion of the archaeological information (spatial analysis, typo-chronological analysis, technological analysis, microwear analysis, etc).

The aim of the research project was the interdisciplinary investigation of Stone Age sites in one of the most promising, and unfortunately also one of the most threatened, areas of northern Belgium, the so called *Waasland Scheldepolders*. In terms of prehistoric research, in casu Final Palaeolithic, Mesolithic and Neolithic research, this particular area constitutes a very interesting one, mainly due to the presence of sealed sites and landscapes. Whereas in most of the Belgian territory the majority of prehistoric sites currently lie at the surface and have therefore been severely disturbed by erosion and ploughing, the prehistoric heritage in the *Waasland Scheldepolders* is generally well-preserved. It results from the deposition of sediments covering the prehistoric surfaces. Unfortunately since 1970 the area has been under constant threat, in particular from Antwerp harbour, one of the largest in Europe. Indeed, large parts of the *Waasland Scheldepolders* (± 6,000 ha), situated on the left bank of the river Schelde, are constantly threatened by the expansion of the harbour, i.e. by the digging of new docks and the installation of large industrial complexes.

1 The project (GOA n° 12050298) was entitled "Large-scale palaeo-ethnographic and palaeo-environmental (salvage) research at the Early Mesolithic settlement of Verrebroek (Province of East Flanders, Belgium)". Promoters J. Bourgeois (archaeology), P. De Paepe (petrography), R. Langohr (soil science), G. Stoops (soil science) and C. Verbruggen (palaeobotany); project manager Ph. Crombé.
Additional funding was given by the Flemish Ministry of Public Works, the municipality of Temse, the FWO (post-doctoral mandate) and the ADW.

Since, in the mid 1990s, the Convention of Malta had not yet been signed by Belgium, no financial support was available to organise large-scale salvage operations in this area. Several attempts were undertaken to obtain funding e.g. by submitting research project proposals to various scientific and governmental institutions such as the Fund for Scientific Research-Flanders (FWO), the European Union (Interreg), the Flemish Ministry of Culture and the *Vlaamse Vereniging voor Archeologisch Onderzoek* (V.V.A.O.). None of these met with success. A final call for research money was then made, this time addressed to Ghent University. After severe screening, this last application was approved, with substantial funding for 5 years. This funding allowed us to organise large-scale salvage excavations and surveys in and around two harbour docks which were in construction, i.e. the *Verrebroekdok* (ca. 500ha) and the *Deurganckdok* (ca. 600ha). In the course of the project, a total of 5 sealed prehistoric settlement locations were investigated by a multidisciplinary team consisting of specialists not only from Ghent University, but also from various other scientific and archaeological institutes, such as the *Archeologische Dienst Waasland* (ADW, Sint-Niklaas), the *Vlaams Instituut voor het Onroerend Erfgoed* (VIOE, Brussel), the *Koninklijk Instituut voor het Kunstpatrimonium* (KIK, Brussel), the *Katholieke Universiteit Leuven* (KUL, Leuven), the *Koninklijk Museum voor Midden Afrika* (KMMA, Tervuren) and the *Koninklijk Belgisch Instituut voor Natuurwetenschappen* (KBIN, Brussel).

1.2. | Situation

Philippe Crombé

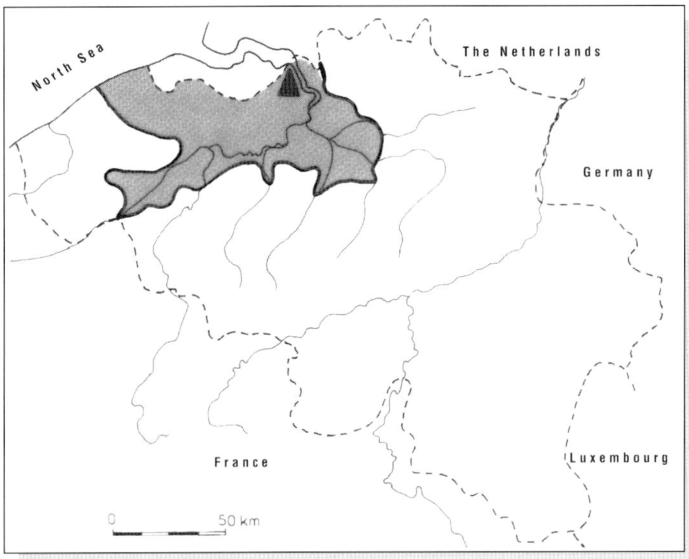

Figure 1: map of Belgium, marking the area of Sandy Flanders (grey) and the *Waasland Scheldepolders* (triangle).

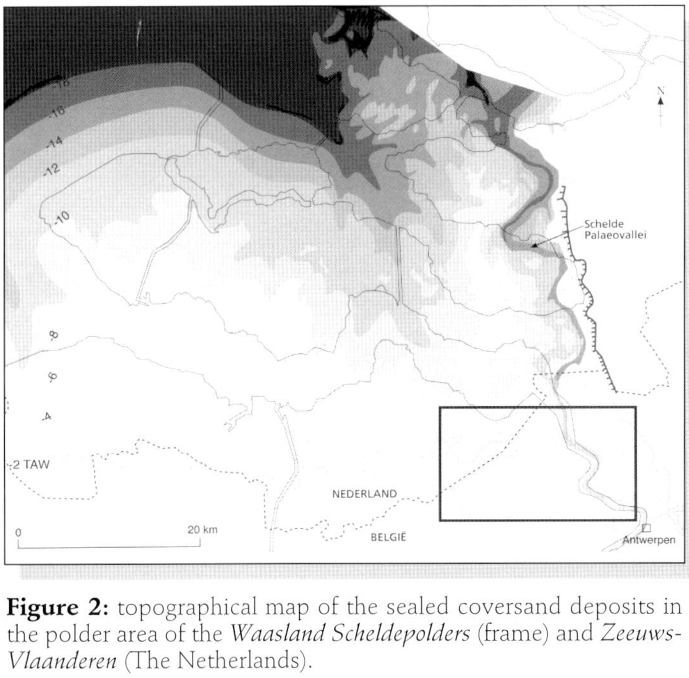

Figure 2: topographical map of the sealed coversand deposits in the polder area of the *Waasland Scheldepolders* (frame) and *Zeeuws-Vlaanderen* (The Netherlands).

1.2.1. | Introduction

The *Waasland Scheldepolders* (province of East Flanders) are part of an area known as *Zandig-Vlaanderen* (Sandy Flanders), a sandy lowland situated in the northwest of Belgium (fig. 1). They lie in the north-eastern district of Sandy Flanders, on the left bank of the River Schelde, covering a surface of approximately 10,650ha which in the north-northwest reaches up to the Dutch state border. The present elevation ranges between two and three metres above present sea-level (TAW[2]). The prehistoric surface (fig. 2), which mainly consists of coversand deposits, however, lies at a deeper level underneath peat (so called Holland peat) and/or (peri-)marine clayey deposits. As the top of the sealed coversand landscape gently slopes in a north-easterly direction towards the valley of the palaeo-Schelde, the thickness of the Holocene cover varies from a few decimetres along the (west)southern limit of the *Waasland Scheldepolders* to several meters (maximum 4 to 5m) at its (north)eastern limit.

1.2.2. | The Verrebroekdok-sites

In and around the *Verrebroekdok* (fig. 3), situated in the municipality of Verrebroek (Beveren), three prehistoric areas were located and subsequently investigated. All three locations, named Verrebroek "Dok 1" (N 51°15'53"/ E 4°12'40"), "Dok 2" (N 49°18'43"/E 4°35'19") and "Dok 3" (N 51°15'39"/ E 4°12'32"), are situated within close proximity of each other (fig. 4). They were found on a large and wide coversand dune, which forms part of an even larger coversand complex, known as the cover-

2 TAW = *Tweede Algemene Waterpassing* = second general levelling

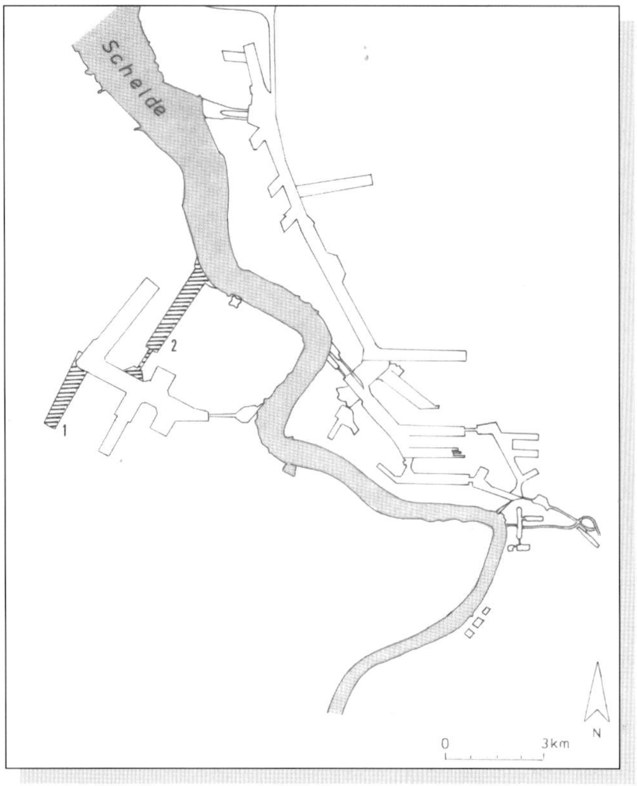

Figure 3: plan of the Antwerp habour, marking the *Verrebroekdok* (1) and the *Deurganckdok* (2).

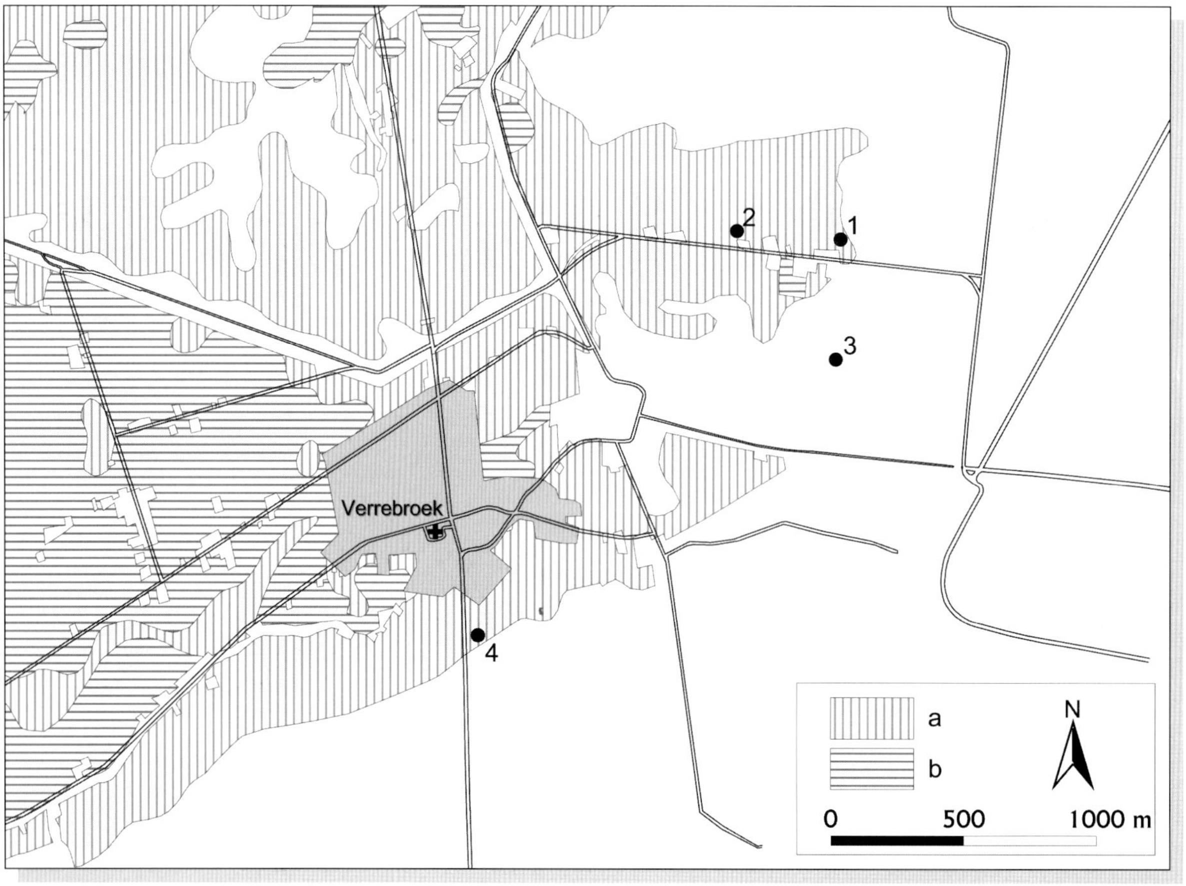

Figure 4: simplified map of the coversand ridge at Verrebroek (based on the soil map Beveren-Waas 27E, 1961), marking the investigated sites of Verrebroek "Dok 1" (1), Verrebroek "Dok 2" (2), Verrebroek "Dok 3" (3) and Verrebroek "Ambachtelijke Zone" (4). Key: (a) coversands sealed by (peri)marine deposits; (b) unsealed coversands.

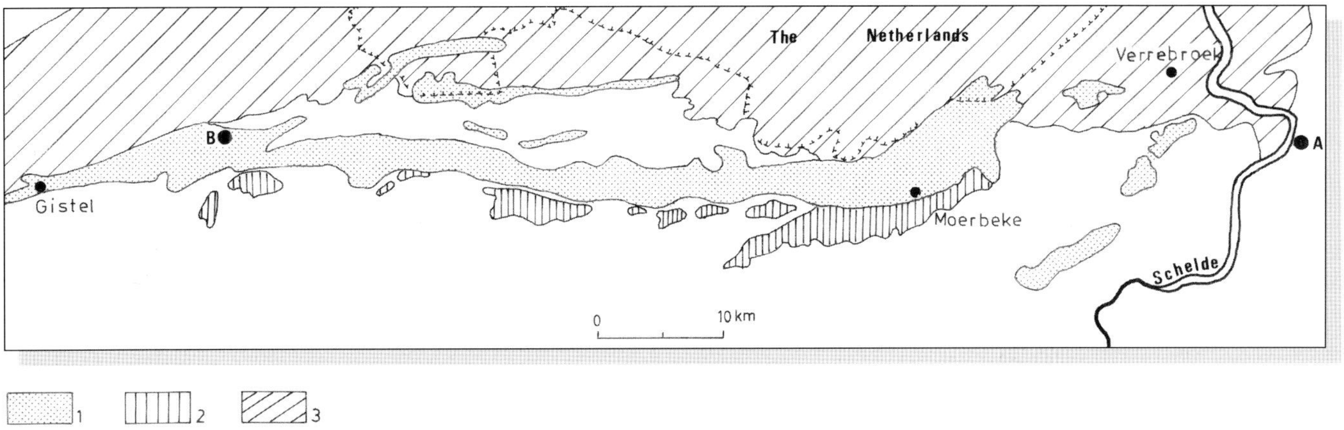

Figure 5: simplified geomorphological map of the northern part of Sandy Flanders (after De Moor, Heyse 1978). Key: (1) coversand ridge or "great ridge" of Verrebroek (Stekene) - (Maldegem) Gistel and smaller parallel oriented sand ridges, (2) Late Glacial lakes formerly lying at the southern edge of the "great ridge",(3) area covered by (peri)marine deposits, (A) Antwerp, (B) Bruges.

sand ridge of Verrebroek (Stekene) - (Maldegem) Gistel (Heyse 1979) or "the Great Ridge" (Crombé, Verbruggen 2002), a massive Pleni Glacial coversand formation which runs over the entire northern part of Sandy Flanders beginning on the North Sea coast in the west and ending at Verrebroek a few kilometres from the Schelde river (fig. 5). Its total length amounts to about 80km, its width being 1.5 - 3km.

Morphologically this sand ridge presents a rather steep south(-east)ern slope (1-4%) opposite a gentle northern slope (1-1.5%). The higher parts display a marked micro-relief with numerous shallow depressions, dating back to different stages of the Late Glacial. During excavations in the *Verrebroekdok* at least one Bølling and one Allerød lake, represented by humiferous to peaty sub-aquatic sediments sealed by aeolian sand, were found. The site of Verrebroek "Dok 2", dating back to the Final Palaeolithic, was found at the border of the former lake. Depressions of clearly younger age – presumably Younger Dryas - were found near the Mesolithic sites of Verrebroek "Dok 1" and "Dok 3" situated along the (south)eastern edge of the sand dune. In contrast to the other lakes, these depressions were formed at the top of the coversand deposits and do not contain organic or aeolian deposits.

1.2.3. | The Deurganckdok-sites

The *Deurganckdok*, situated in the municipality of Doel (Beveren), lies a few kilometres to the northeast of the *Verrebroekdok* and nearby the actual course of the River Schelde (fig. 3). Two locations[3] with prehistoric remains were discovered in the foundation trench of the western dock wall (fig. 6). Both were situated on a narrow, ca. 30-40m wide sand ridge running more or less N-S, which could be followed over a length of ca. 220m.

3 In 2003 two new prehistoric zones (sector J/L and M), both with Early Mesolithic and Swifterbant remains, were discovered in the *Deurganckdok* (Bats *et alii* 2003).

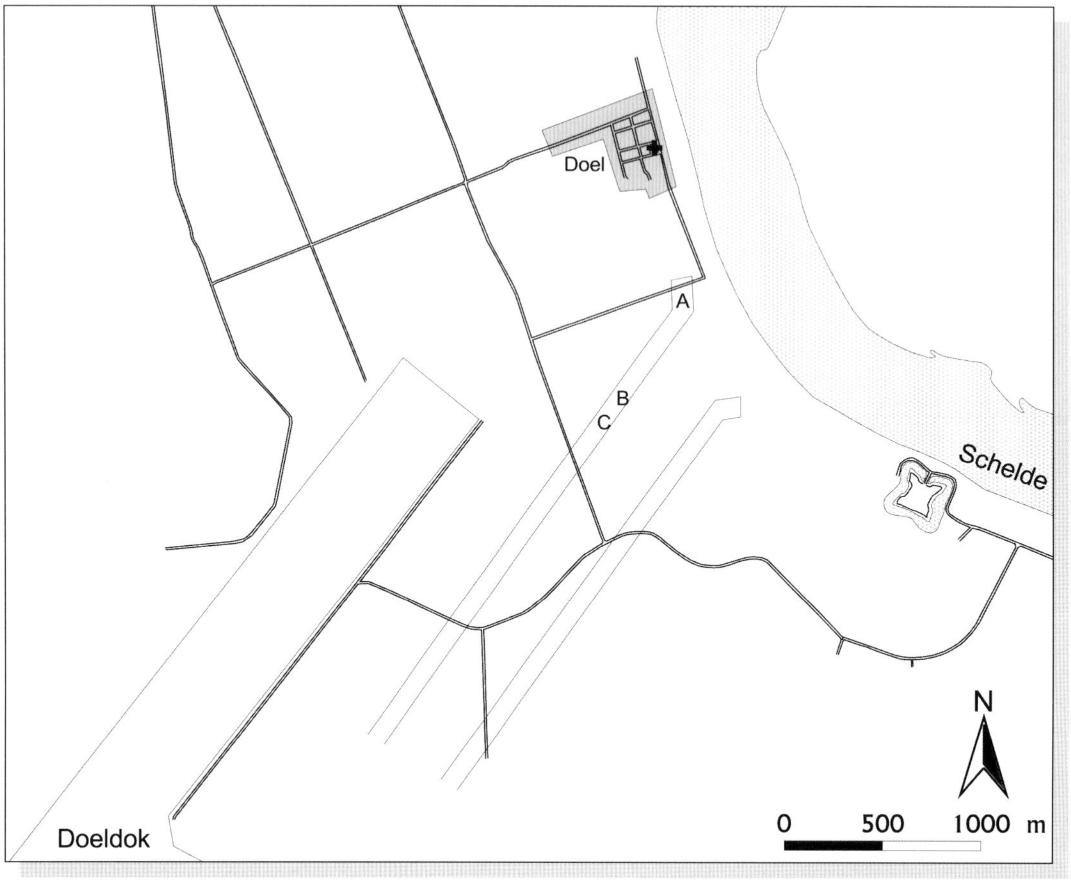

Figure 6: simplified map of the *Deurganckdok* under construction, marking the investigated sectors A, B and C.

A first site, called Doel "Deurganckdok-sector B" (N 51°18'00"/E 4°15'59") was situated in the northern part of this dune at a minimum depth of 0.5m below the actual sea level. At this spot remains of different occupation episodes were recovered, amongst which a Final Palaeolithic (*Federmesser* tradition) and a Final Mesolithic (*Swifterbant* tradition) occupation.

A second prehistoric location, called Doel "Deurganckdok-sector C" (N 51°17'57"/ E 4°15'58") was found in the southern part of the same sand ridge (fig. 16), situated at a minimum depth of 0.10m below the actual sea level. Here the recovered remains mainly pertain to a Middle Neolithic occupation which can be correlated with the *Michelsberg* culture.

To the north of both prehistoric locations, several Holocene soil sections were studied palynologically. One particular section, called Doel "Deurganckdok-sector A" (N 51°18'15"/E 4°16'4"), is discussed in this volume.

1.2.4. | Moerbeke "Moervaart"

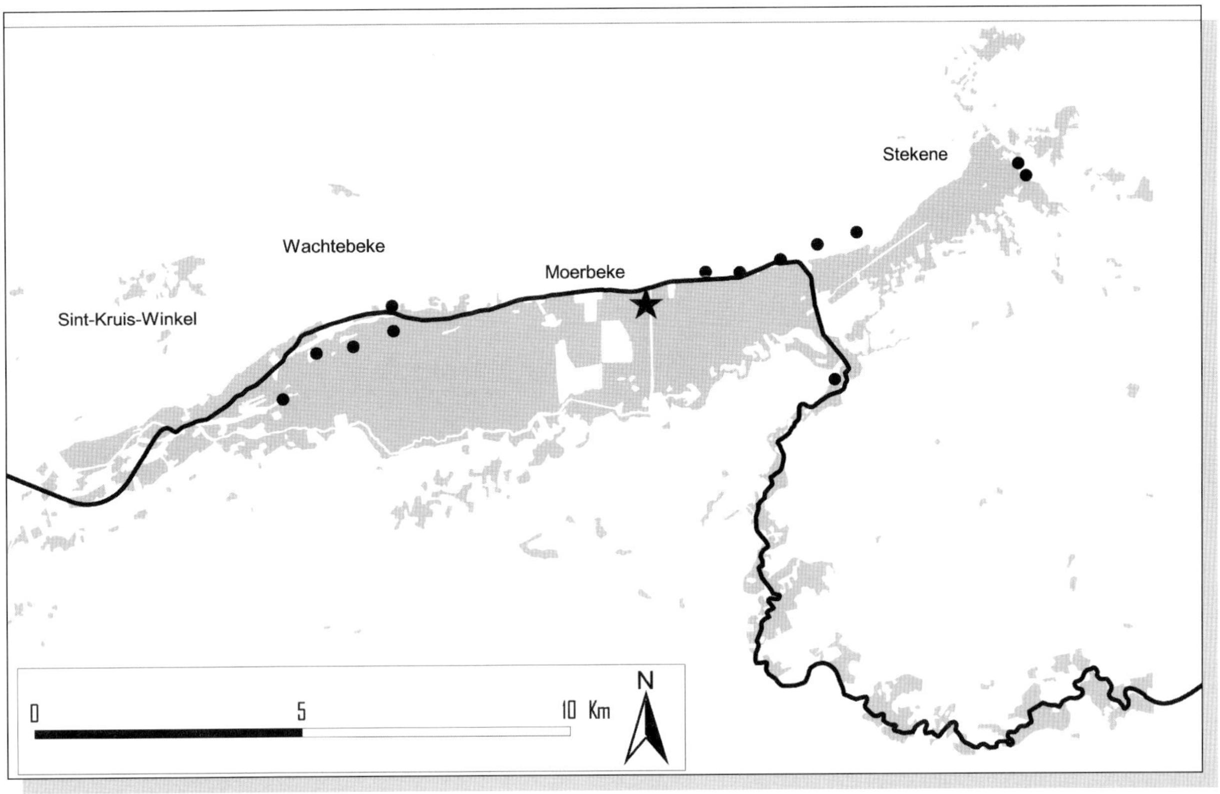

Figure 7: simplified map of the "Moervaart" lake at Moerbeke, marking the *Federmesser* sites (dots) and the drilling location (star).

About 25km to the southwest of the *Waasland Scheldepolders* lies a former Late Glacial lake landscape, known as the "Moervaart" lake or "Moervaart" depression. It is situated at the southern edge of the coversand ridge of Verrebroek (Stekene) - (Maldegem) Gistel mentioned above (fig. 5), and is by far the most extensive palaeo-lake of Sandy Flanders. Its total length amounts to ca. 15km, its maximum width to ca. 2.5km (fig. 7). From a palaeo-environmental point of view, this large but shallow lake constitutes one of the most important sites because of the presence of a well-developed Late Glacial organic sequence. The latter has already been analysed palynologically in detail by C. Verbruggen (1971, 1979) in the context of the geomorphological and palaeoecological investigation of Sandy Flanders. In pursuit of a better understanding of the Late Glacial horizons excavated at Verrebroek it was decided in 2001 to re-investigate the "Moervaart" sequence, especially in terms of its absolute chronology. Manual drilling was conducted in the eastern and deepest part of the palaeo-lake (N 51°10'17"/E 3°56'53"), in the vicinity of two earlier drillings performed in the seventies.

1.3. | Research history and method

Philippe Crombé

1.3.1. | Verrebroek "Dok 2"

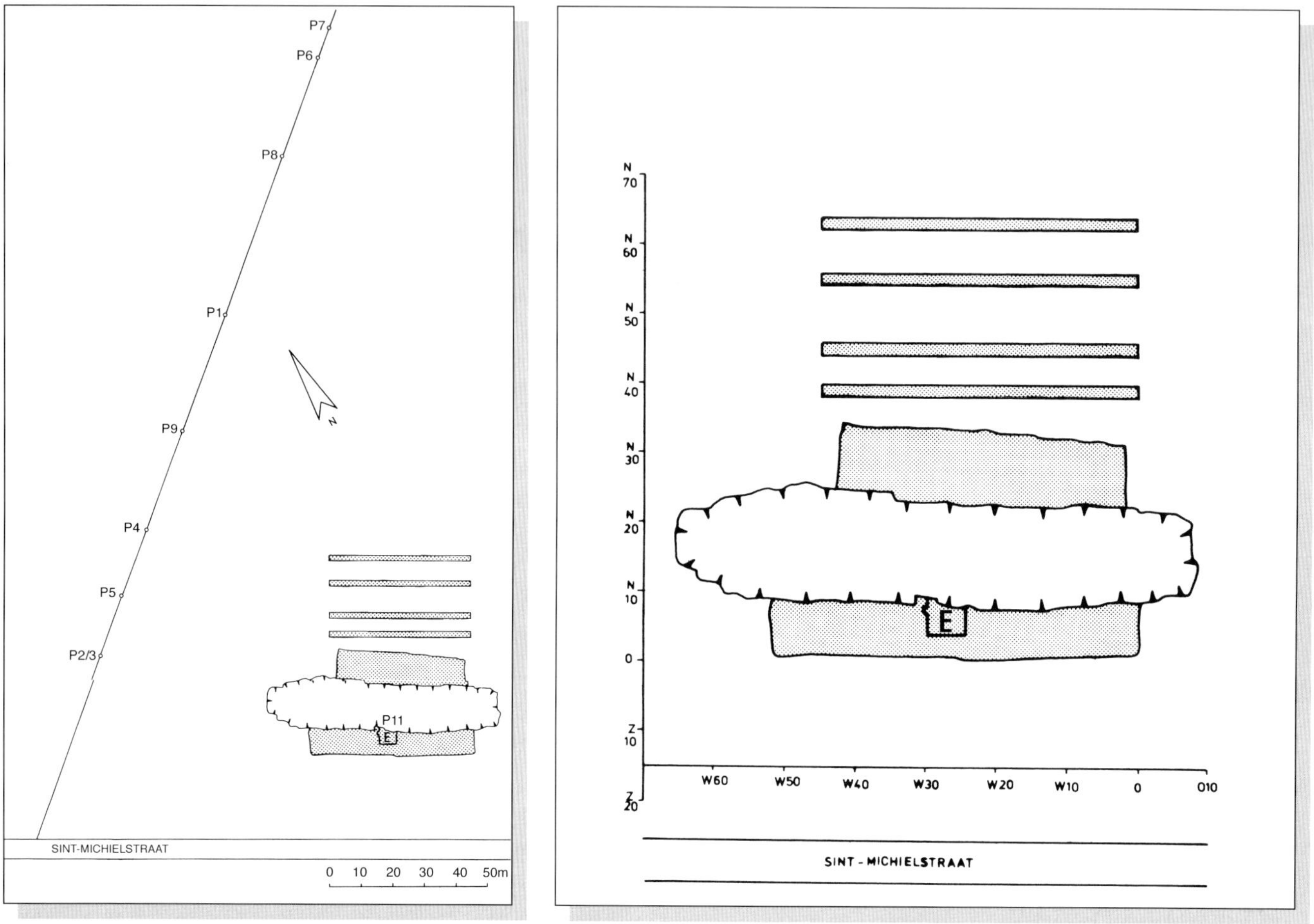

Figure 8: Verrebroek "Dok 2"; (a) location of the excavated trenches (grey shaded) and the studied profiles (P). The letter E indicates the area that has been excavated in detail; (b) detail of the excavated trenches situated to the north and south of a construction pit.

In April of 1999 archaeological survey of dike construction works to the west of the *Verrebroekdok* led to the discovery of some flint artefacts which were apparently situated in a Late Glacial, slightly humiferous soil about 1m beneath the present surface (Crombé *et alii* 1999c). Subsequent excavations were organised in the immediate vicinity of these first discoveries (fig. 8).

In May two trenches immediately to the south (ca. 55 x 7/8m) and the north (ca. 41 x 8/10m) of the pit in which the first artefacts were found were stripped of mechanically down to the humiferous find-layer. Both trenches were consequently sampled by means of $^1/_4\,m^2$ test-pits placed at a distance of 1.5m, in order to locate possible artefact concentrations. In the southern trench at one spot, just before the place where the humiferous layer seemed to have been truncated, a small artefact-rich area was located. This restricted area of ca. $24m^2$ was carefully excavated in units of 1/4 m^2 by means of trowels and three-dimensional registration of the finds. Soil dirt was subsequently wet-sieved through meshes of 2mm width. Artefacts were found in a small concentration of ca. 0.80m in diameter, which was partially destroyed by the dike construction along its northern limit. The northern trench, on the other hand, revealed no finds.

In addition four trial-trenches of ca. 45 x 2m situated further north were excavated before the whole area was raised with a sand-in-water slurry. Here too, not a single artefact was found. All the trenches dug for the construction of the dikes were also inspected for additional archaeological finds. This inspection revealed that the humiferous horizon was preserved over several hundreds of metres to the west and northwest of the site. It could also be noted that further away from the site the layer gradually changed into a very humiferous-to-peaty horizon. At several places this horizon was studied and sampled for various palaeo-environmental analyses.

1.3.2. | Verrebroek "Dok 1"

Figure 9: Verrebroek "Dok 1"; aerial view of the *Verrebroekdok* under construction and the archaeological excavations situated along the road in the foreground (photo J. Semey, Ghent University).

The site of Verrebroek "Dok 1" had already been discovered by the ADW in 1981 during the demolition of a farmstead, which was being removed for the planned construction of the *Verrebroekdok*. The first excavations, however, did not start until 1992 and were conducted by the Ghent University in collaboration with the ADW (Crombé 1993a). Between 1992 and 1994 (Crombé 1994; Crombé, Van Strydonck 1994) seven trenches covering about $880m^2$ were excavated, the results of which were subsequently analysed and published in a PhD thesis (Crombé 1998). After an interruption of a year, the investigation of the site was continued. In 1996 a systematic drilling project was conducted in order to define the limits and extent of the site with a view to better and more efficient planning of future excavations (Crombé, Meganck 1996). Meanwhile the building of the *Verrebroekdok* had started, making the excavation of this vast Mesolithic site very imminent (fig. 9). Starting in 1997, excavations were conducted over four seasons in the northern sector of the site, leading to a total excavated surface of approximately $6{,}210m^2$ (Crombé *et alii* 1997, 1998, 1999b). Finally, during the summer of 2000 the site became inaccessible for further research, as the whole area was submerged under several metres of water (fig. 10).

Excavation period	Excavated surface (m^2)	Excavated trenches	Drilled surface (ha)
1992	±300	1-4	
1993	±410	5-6	
1994	±170	7	
1996	-		±4.5
1997	±800	8	
1998	±1,920	9-10	
1999	±2,130	11-13	±4
2000	±480	13-14	±2
Total	±6,210		±10.5

Table 1: overview of the excavated and drilled area at the Mesolithic sites of Verrebroek "Dok 1" and "Dok 3".

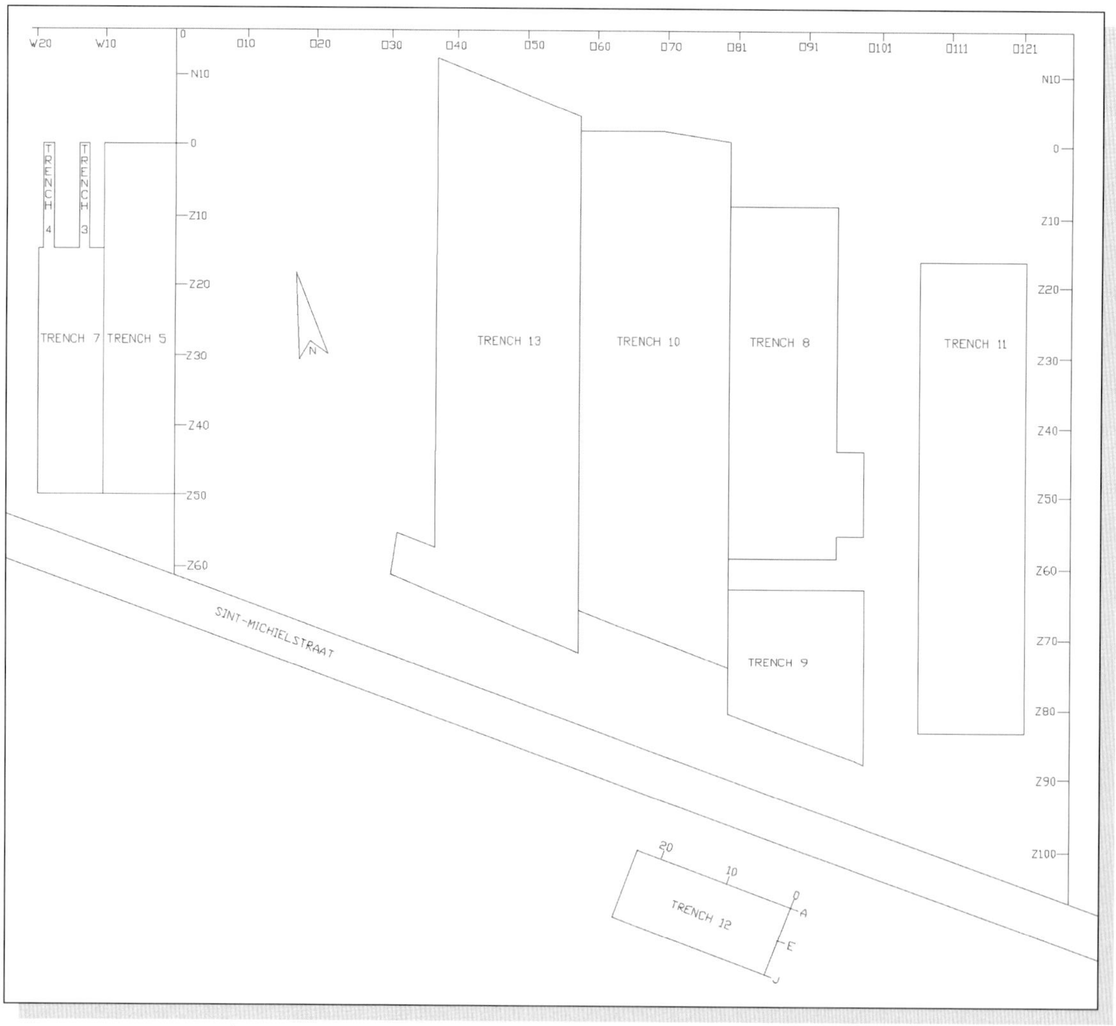

Figure 10: Verrebroek "Dok 1"; plan of the trenches, excavated between 1992 and 2000.

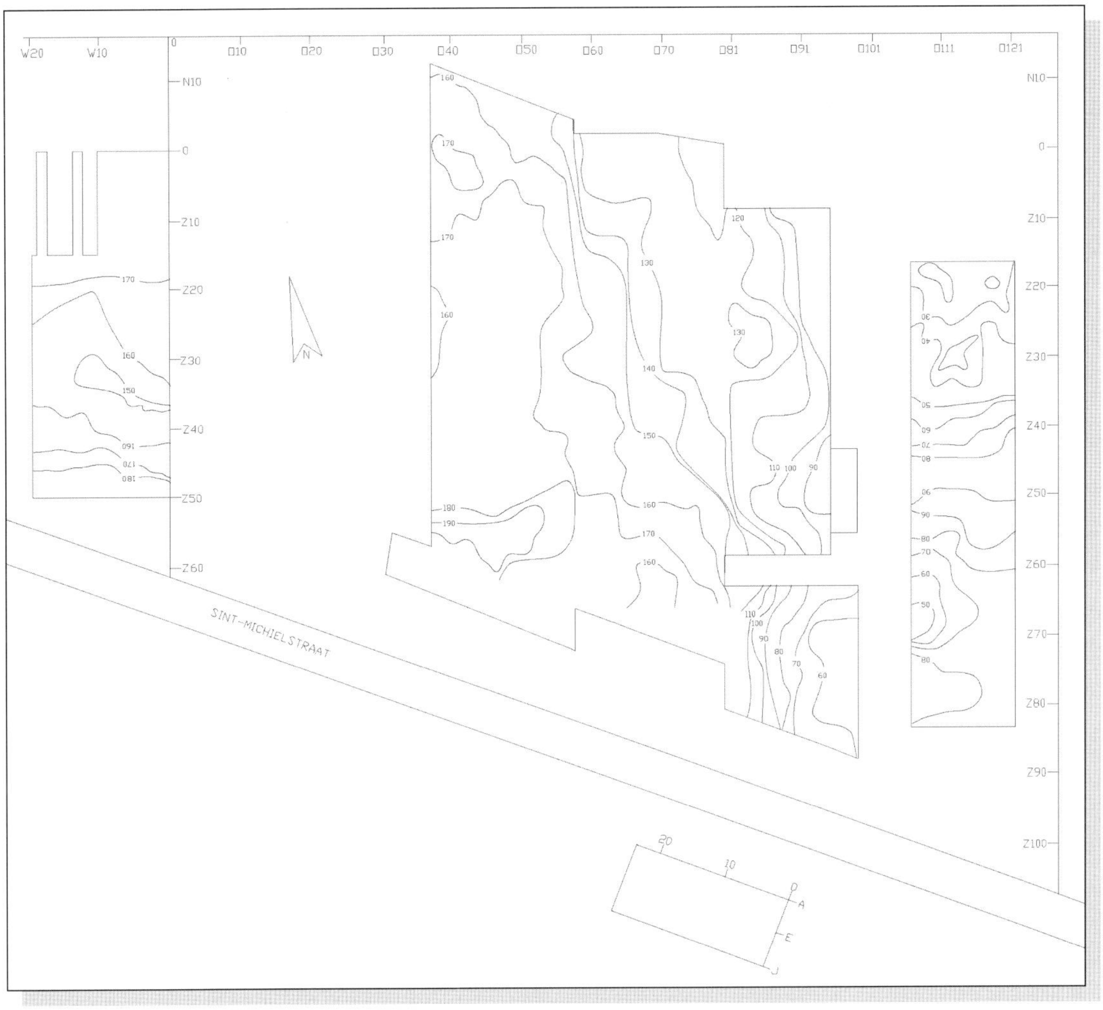

Figure 11: Verrebroek "Dok 1"; topographical map of the Pleistocene (coversand) surface (TAW).

Except for the area excavated in 1992, the whole Mesolithic surface was investigated by means of systematic wet-sieving through meshes of 2mm. After mechanical removal of the alluvial and peaty layers, the Pleistocene surface was excavated in squares of $^1/_4$ m^2 and artificial layers of 0.10m. First the Pleistocene surface was levelled (fig. 11) and sampled in order to locate concentrations of lithic artefacts to be excavated in more detail. This sampling was done by means of $^1/_4$ m^2 test-pits which were set out in a staggered grid at 1.5m intervals (fig. 12). Based on a rough counting of the retrieved artefacts and ecofacts, the areas with high to medium find-density were identified. In some trenches (n° 1-7, 8, 9 & 11) only these areas were subsequently excavated and sieved systematically, leaving the intermediate zones unexcavated. In other trenches (10, 12 & 13), in particular those yielding a very high density of artefact units, it was decided to excavate the entire surface, low-density areas included. The depth to which the soil was sieved depended on the number of finds, and was consequently not constant over the entire site-surface. The limit to which each individual $^1/_4$ m^2 was sieved was set at five artefacts (chips included). Once less than five artefacts were found, the excavation of the square was stopped. Occasionally after sieving, the area was further stripped mechanically in search of possible deeper features.

Figure 12: Verrebroek "Dok 1"; view of the sampling method (trench 11).

Features, either of natural or anthropogenic origin, were mapped in detail, either by archaeologists or by specialists (pedologists, palaeoecologists, etc.). Transects were made of all clearly anthropogenic features, the entire filling being collected in plastic bags for further laboratory analyses (^{14}C-dating, anthracological analysis, botanical analysis, etc.). Features of clear or presumed natural origin (e.g. tree-fall features, charcoal-pits, etc.) were not sampled systematically but only selectively. Latent features, such as spots with a high incidence of carbonized organic residue (hazelnut shells, bone, etc.) and/or burnt lithic artefacts, were also sampled systematically, taking two to five litre bulk samples at different levels. It was assumed that most of these spots might coincide with former fireplaces, in particular with so called surface-hearths.

1.3.3. | Verrebroek "Dok 3"

This site was first detected in 1999 through field-walking performed by the ADW over large parts of the grounds to be destroyed by the future *Verrebroekdok*. Less than 500m south of the site of Verrebroek "Dok 1" (fig. 4), hundreds of lithic artefacts were collected from the ground surface, indicating the presence of a strongly perturbed Mesolithic site. In anticipation of possible future excavations, this new site was drilled and sampled in 1999 and 2000 (Bats 2000-2001, Cordemans *et alii* 2001; cf appendix A). The borings pointed to the existence of a vast site, of possible Early Mesolithic age, which extended over an area of at least four to five hectares along the south-eastern border of the large coversand ridge (cf 1.2.2.). Contrary to what was initially thought, large parts of this site seemed to be relatively well-preserved as they were covered by clayey and locally organic sediments. Only those parts of the site situated on the top of the sand ridge had been extensively affected by modern ploughing. Although it was not possible to extend the drilling project to the limits of the site, it could be observed that the distance to the adjacent site of Verrebroek "Dok 1" was considerably smaller than previously thought. It may even be assumed that both sites originally formed one large complex, consisting of a continuous spread of Mesolithic settlement waste. Unfortunately this new area became inaccessible too soon for further fieldwork to be conducted. The spatial and chronological relationship between the two sites therefore remains unexplored.

Figure 13: Doel "Deurganckdok-sector B"; view of the excavations.

1.3.4. | Doel "Deurganckdok-sector B"

This site was discovered by ADW-members in May 2000 during archaeological survey in the western foundation trench of the *Deurganckdok* (fig. 13). At the time of discovery the north-eastern part of the site had already been destroyed and/or perturbed to a certain degree by the harbour works. Salvage excavations were allowed only for 5 weeks (2nd April to 14th May), and had to be conducted in very unfavourable circumstances (Crombé *et alii* 2000). As it was fully clear that the granted time was much too short for a full and systematic excavation, certain choices had to be made. After careful consideration it was decided to focus mainly on the Final Mesolithic remains and to investigate the Final Palaeolithic site only superficially. This decision was mainly based on stratigraphical arguments. Contrary to the Final Palaeolithic artefacts, which occur vertically dispersed over ca. 0.3 to 0.4m, the Mesolithic finds concentrate within the upper 0.05 to 0.1m of the Pleistocene deposits. Hence, the latter could be relatively easily excavated over a large surface. Furthermore the Final Mesolithic site also yielded a considerable number of features, mainly hearth-pits, which had to be registered and examined. In addition it could be assumed that large parts of the underlying *Federmesser* site had already been perturbed by the construction of these pits.

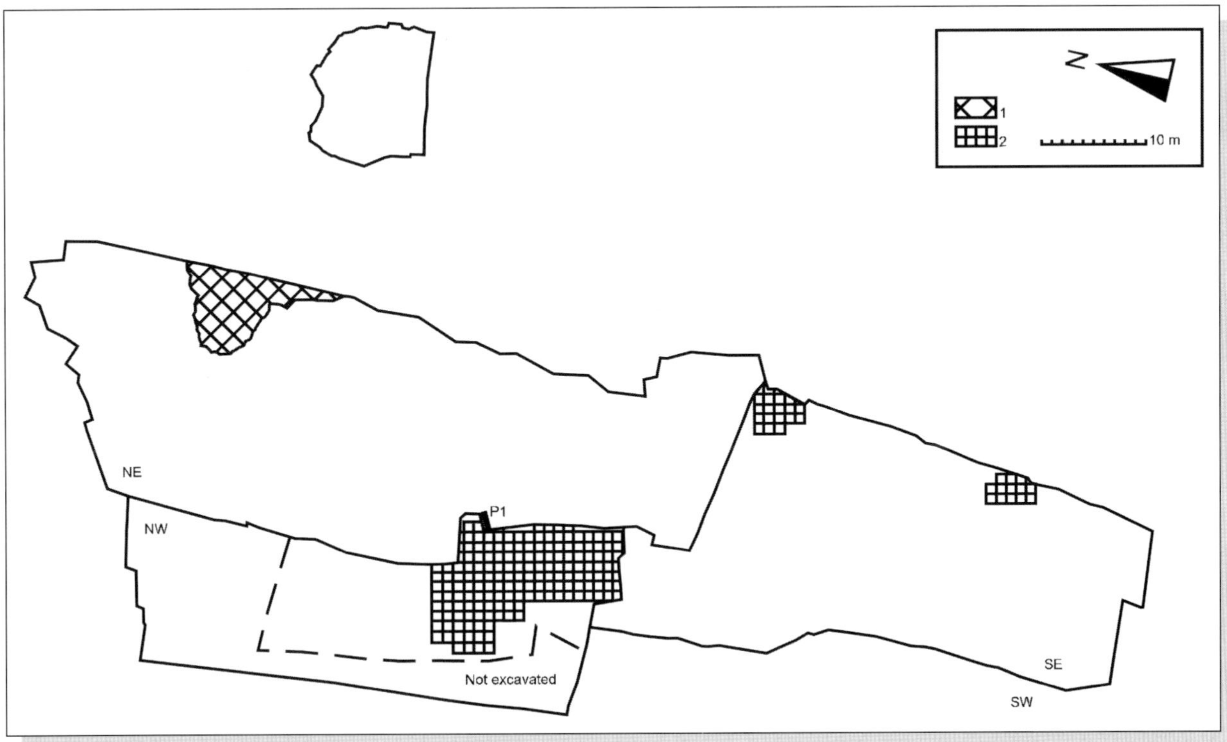

Figure 14: Doel "Deurganckdok-sector B"; plan of the sub-zones, marking the reference profile for pedological and palynological analyses (P1). Key: (1) disturbed area, (2) sieved area.

At the start of the investigations, the site-surface was divided artificially into 4 sub-zones or sectors, to be excavated successively (fig. 14). The excavations started in the north-eastern sector by careful shovelling and three-dimensional registration of individual finds using a total station. In this sub-zone the topsoil and large parts of the underlying A-horizon had already been destroyed by the harbour works. Hence, Final Mesolithic finds could only be retrieved from some local depressions. On the other hand this sub-zone yielded the largest number of hearth-pits. All these features were registered in horizontal and vertical plan. The filling was systematically sampled for botanical analyses and absolute dating purposes. About half of the feature fill was put in plastic bags for subsequent wet-sieving.

The south-eastern sector was investigated more thoroughly. Here too the soil was shovelled systematically and finds three-dimensionally registered, except for two small but high-density areas of $18m^2$ and $14m^2$. These were investigated in more detail with trowels instead of shovels and subsequent sieving of the soil dirt.

As time was pressing during the last week of excavations it was decided to excavate the north-western part of the site by means of wet-sieving. The soil was excavated in a grid of $^1/_4$ m^2 and put in large plastic bags to be transported to another location where it could be wet sieved later. Bulk samples were taken from each $^1/_4$ m^2 with a view to palaeo-botanical analyses. Unfortunately the hearth-pits found within this sector could not be sampled properly due to lack of time. Likewise it was no longer possible to excavate the south-western sector at all. It could only be noted that this part of the site corresponded to a former depression. In the end a total surface of approximately 3,500 m^2 was excavated.

Figure 15: Doel "Deurganckdok-sector C"; view of the excavations.

1.3.5. | Doel "Deurganckdok-sector C"

In this southern sector excavations had to be done in similarly unfavourable circumstances (fig. 15). Here too, the major part of the site had already been removed by digging machines, leaving only a relatively small sector (ca. 570m^2) to be excavated (fig. 16). Moreover, archaeological investigation could only last for two weeks (16th-24th August; 4th-7th September). The area was investigated by shovelling the top of the sandy deposits and three-dimensional registration of the finds. Afterwards the underlying sand was removed mechanically up to ca. 0.20-0.30m in order to verify the presence of archaeological features, such as hearth-pits.

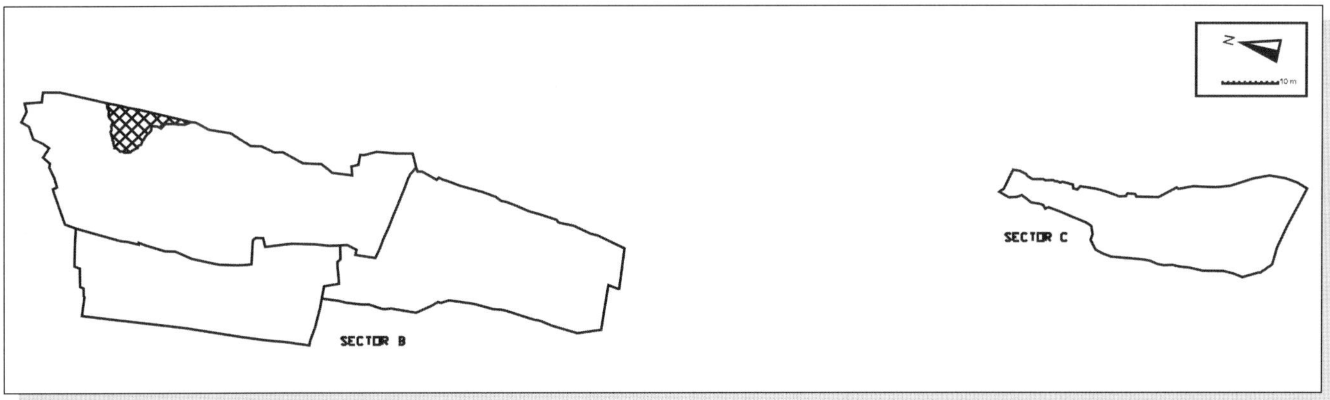

Figure 16: Doel "Deurganckdok-sector C"; location in relation to sector B.

1.3.6. | Doel "Deurganckdok-sector A"

To the north of both prehistoric locations, several Holocene soil sections have been cleaned up by members of the ADW and Ghent University, with a view to detailed description and analysis. One of these sections (sector A, section I; fig. 14), situated at about 500m away, was sampled intensively for palaeo-environmental and radiometric purposes. This section included clayey deposits so called "Calais-deposits" intercalated in peat layers, all of which date to the same period as the prehistoric occupations excavated at the nearby sand dune.

II.

PALAEO-ENVIRONMENT

2.1. | Pedo-lithostratigraphical analyses

Geertrui Louwagie & Roger Langohr
(With a contribution from Cyriel Verbruggen)

2.1.1. | General introduction to the Waasland Scheldepolders

2.1.1.1. | Soil forming factors

Substratum and parent material

The upper layers of the Tertiary substratum at Verrebroek "Dok" (Geological Map of Belgium, Antwerpen (15) 2002) and Doel "Deurganckdok" (Geological Map of Belgium, Kapellen-Essen 7-1 2002) consist of marine, fine glauconiferous shelly Pliocene sediments: the 5 to 10m thick Formation of Kattendijk, buried by the Formation of Kallo that has a thickness of about 5m and 20m in Verrebroek and Doel respectively.

The Quaternary cover is at both sites up to 5m thick (Geological Map of Belgium, Antwerpen (15) 2002; Geological Map of Belgium, Kapellen-Essen 7-1 2002) and consists of Upper Pleniglacial coversands (Maréchal, Maarleveld 1955; Pissart 1987b; Verbruggen *et alii* 1991; Van Vliet-Lanoë 1996). The first phases of the Upper Pleniglacial (25000 to 15000 BP) were dry and cold and allowed coversand sedimentation (Pissart 1987b; Verbruggen *et alii* 1991). These coversands are dominantly of local origin, i.e. they are the result of aeolian reworking of Lower and Upper Pleniglacial fluviatile sediments (Verbruggen *et alii* 1991). Macromorphological observations reveal that they are dominantly horizontally laminated. These coversands are remobilised and redeposited in the Late Glacial Dryas stadials.

Natural and strategic inundations coming from the Schelde River regularly affected the area in Subatlantic Late Medieval times, with subsequent sedimentation of calcareous flood clay and sand.

Topography and hydrology
The studied area has a very flat topography. Under normal conditions, these soils need to be drained when used for crop production purposes. During the archaeological excavations however, the pumping for the harbour construction works considerably lowered the current water table level.

Climate
Present-day climatic data (Poncelet, Martin 1947) are given for the climatic station of Doel, situated at 8.5km northeast of Verrebroek. The mean annual temperature over the period 1901-30 is 10.1°C; the mean January temperature is 3.1°C; the mean July temperature is 18.0°C. Annual mean precipitation and evapotranspiration over the same period are 713mm and 658mm.

Human impact
At Verrebroek "Dok", two main occupation phases have been detected: Final Palaeolithic at Verrebroek "Dok 2" and Early Mesolithic at Verrebroek "Dok 1". Verrebroek "Dok 1" also showed evidence of Medieval land use.

The soil study at Doel "Deurganckdok-sector B" was focussed at the stratigraphic levels coinciding with the Final Palaeolithic (Federmesser) and Final Mesolithic (Swifterbant) human occupation.

More detailed data with respect to type and amount of archaeological evidences revealed are provided during the discussion on palaeo-environment reconstruction.

2.1.1.2. | Soils

Soil map information
On the Soil Map of Belgium (Beveren-Waas 27E 1961), the symbol for the Verrebroek "Dok 1" site is "sEep"; for the Verrebroek "Dok 2" site it is "sUdp". At Doel "Deurganckdok" the soil series symbol (Soil Map of Belgium, Lillo 14E 1960) is "Eep". In the soil map legend, "s" stands for a sandy substratum that starts within the 1.25m deep soil observation. "E" and "U" correspond to light and to heavy clay, "e" and "d" stand for poor and imperfect drainage and "p" for soils without a B-horizon, mostly corresponding to relatively young soils (eroded, colluvium, alluvium). The upper, clayey part of these soils is developed in flood sediments of the Schelde River. The underlying non-calcareous sand deposits correspond most likely to the eastern limit of the Quaternary Upper Pleniglacial coversand ridge of Verrebroek (Stekene) - (Maldegem) Gistel (ca. 80km long, ca. 2-3km wide).

Soil profile information
Figures and macromorphological observations of the studied profiles are mentioned throughout the text, whereas micromorphological, analytical and similarity index data are given in Appendix D.

2.1.2. | Moerbeke "Moervaart"

Cyriel Verbruggen

The northern part of Sandy Flanders is characterised by a distinctive periglacial geomorphology. The prominent feature is the so-called coversand ridge of Verrebroek (Stekene) - (Maldegem) Gistel (cf 1.2.2.), which is a Late Weichselian-Pleniglacial coversand ridge extending over more than 80km between the present day Schelde polders in the East and the coastal plain in the West. During its formation the ridge caused a damming effect on the water discharge from the Flemish Valley and the Tertiary hills from the south to the north. This resulted in the forming of shallow depressions at the southern edge of the coversand ridge. When the braided river discharge changed into a meandering one at the beginning of the Late Glacial more than 15000 years ago, the depressions became isolated mires. The "Moervaart" depression was by far the largest of them. It got gradually filled in with a strongly calcareous gyttja. In present times this gyttja is still within the reach of ploughing activities. The soil map characterises this sediment as marl.

2.1.3. | Verrebroek "Dok 2"

2.1.3.1. | Pedo-lithostratigraphy

The link between the palaeo-geomorphology and soil macromorphology enables a landscape reconstruction. A schematic description of the pedo-lithostratigraphical units at Verrebroek "Dok 2" from bottom upwards gives the following sequence (fig. 8).

1. Upper Pleniglacial coversands. The Upper Pleniglacial coversands are dominantly horizontally laminated. This lamination points towards recurrent cycles of periodic sediment deposition with possible fluctuating energy levels. Any type of bioturbation processes hasn't visibly disturbed the sedimentary sequence afterwards. However, evidences of congeliturbation, such as the presence of few faint isolated ice vein casts with upward bended side laminae, have been observed (fig. 18). Within these Upper Pleniglacial coversands, a number of soil forming processes occurred and gave rise to the development of soil horizons.

2. Late Glacial humiferous soil of 0.10-0.15m thickness, with
   - locally fragmentary charcoal, mainly clustered;
   - local presence of a peat horizon;
   - no, or very little, bioturbation;
   - an abrupt, somewhat wavy lower boundary;
   - at the bottom, local spherical- to ellipsoidal-shaped structures composed of alternating laminae of white quartz sand-size grains and slightly humiferous sand; they are probably post-depositional deformation structures linked to liquefied sediment and rapid sedimentation conditions;
   - local involutions of the parent material into the Late Glacial soil (fig. 17), occasionally following a linear frost crack or fissure;
   - local involutions of the Late Glacial soil in the underlying sediment (fig. 21).

3. Horizon(s), consisting of accumulation bands of very mobile organic matter and/or clay substances, from a few mm up to 0.30m thick, with abrupt upper and lower boundaries (fig. 21). They probably precipitated at the level of a texture jump or (temporary) water table. Most of this material accumulated on top of the Late Glacial soil, forming in most profiles a continuous horizon of about 0.05-0.30m thickness. Several, often discontinuous accumulation bands are observed underneath the Late Glacial soil too.

4. Redistributed aeolian sands, deposited on top of the Late Glacial soil, where present. When deposited above the groundwater table, they are horizontally up to obliquely stratified; otherwise, when deposited in open water, they lack any stratification.

5. Traces of a Brunified Soil with a colour Bw-horizon (fig. 20), only observed in particular positions, with from top:
   - no or an only slightly developed E-horizon;
   - a homogeneous brown (with a higher chroma than the underlying C-horizon), bioturbated Bw.bi-horizon.

6. Imperfectly to poorly drained Podzolised Soil, commented in detail for Verrebroek "Dok 1" (cf 2.1.4.7.).

7. Peat, commented in detail for Verrebroek "Dok 1" (cf 2.1.4.9.).

8. Calcareous flood clay and sand with variable thickness, commented in detail for Verrebroek "Dok 1" (cf 2.1.4.11.).

2.1.3.2. | Upper Pleniglacial coversands

Profile 11 (fig. 17, tab. 2, D.9, D.10, D.39)

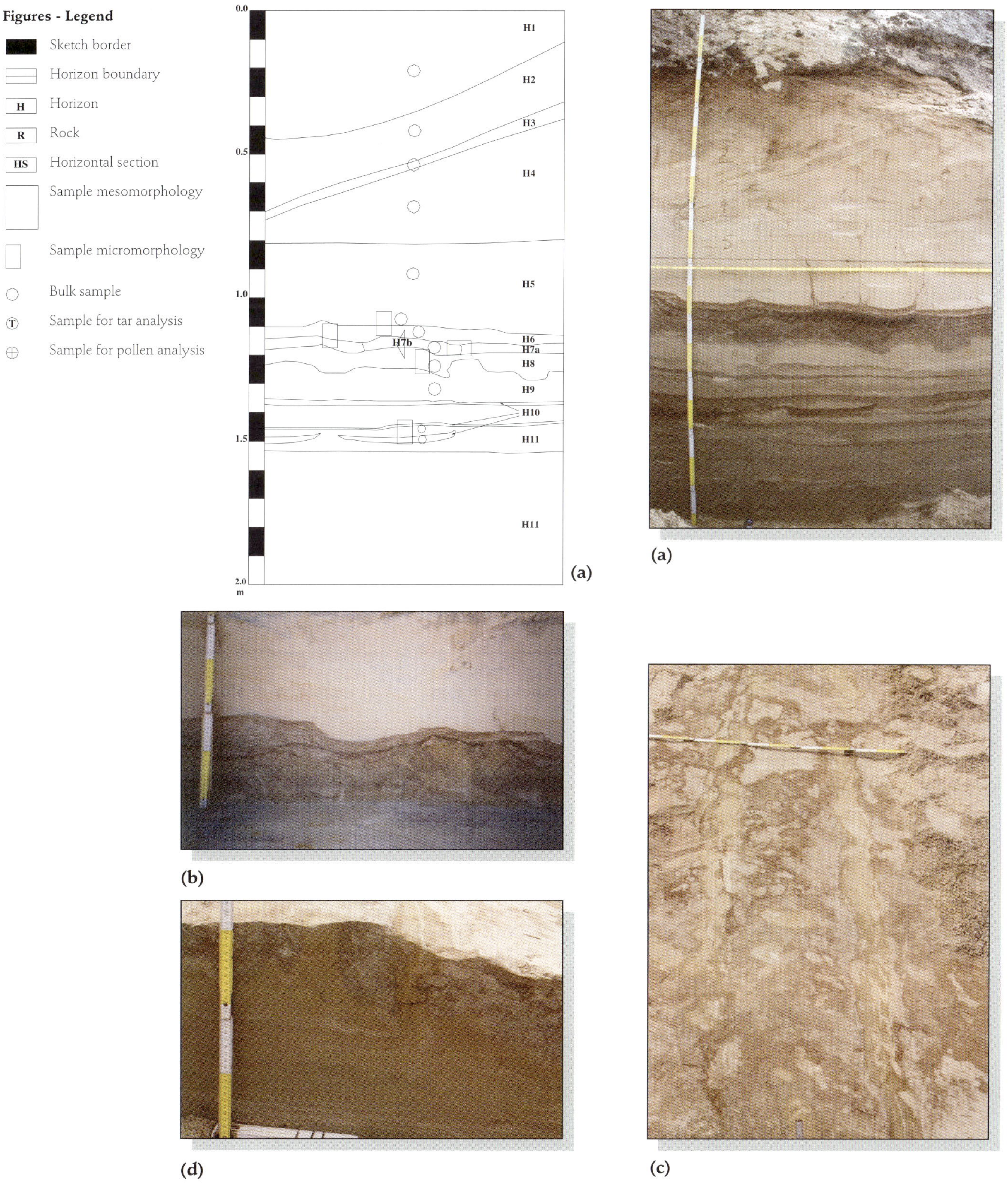

Figure 17: Verrebroek "Dok 2"; (a) profile 11, schematic and photographic representation, (b) sand dyke or diapir, (c)-(d) locally following a bifurcated thermal contraction crack. The bluish colour is a reproduction artefact.

Horizon number	Symbol	Horizon depth (cm)	Profile description
H1	2BC	(0)-11/45	Brown matrix; sand; "B3" of Podzolised Soil
H2	2CB	11/45-32/70	Pinkish white - light grey matrix; admixture of brown sand grains; aeolian sand; obliquely stratified
H3		32/70-38/73	
H4	3C1	38/73-80/81	Pinkish white - light grey matrix; admixture of brown sand grains; aeolian sand; obliquely stratified
H5	3C2	80/81-98/113	Light grey matrix; sand; non-stratified
H6	3BCh.t?	98/113-113/118	Stratified, alternating brown (sand grains coated with organic matter) and pinkish white (sand) matrix; brown matrix connected with above-lying horizon H5 through vertical "veins" (<1cm wide)
H7a	4Ab.a1?	113/117-117/120	Greyish brown - dark greyish brown matrix; sand; few charcoal; few flint Federmesser artefacts with soil capping
H7b	4Ab.a2?	113/118-118/120	Very dark greyish brown - black matrix; sand; common charcoal; common flint Federmesser artefacts with soil capping
H8	4Ab?	117/120-121/130	Dark brown matrix; sand; disturbed morphology; cf Subaqueous Soil?
H9	4C?	121/130-132/135	Light grey matrix; sand
H10	4BCh.t?	135/137-153/155	Local; dark-very dark greyish brown matrix; accumulation strata and laminae of very mobile forms of organic matter
H11	4CBt?	132/135-(200)	Light bluish grey matrix; loamy sand?; stratified; accumulation bands of very mobile forms of organic matter (H10); clay accumulation bands

Table 2: Verrebroek "Dok 2"; profile 11, profile description.

The texture of the Upper Pleniglacial coversands (H9) has a median value of 153µm (tab. D.39). Micromorphological observations show that the quartz grains in the laminated to stratified Upper Pleniglacial sands (H11) are coated with thin (< 5µm) light green (PPL), probably glauconitic clay coatings (tab. D.9, cf tab. D.49). Glauconite clay migration can result of mechanical breakdown of sand-size glauconite (Van Ranst, De Coninck 1983).

2.1.3.3. | Congeliturbation in the Upper Pleniglacial coversands

Evidences of congeliturbation were observed within the Upper Pleniglacial coversands.

Profile 9 (fig. 18)

The top of the Late Glacial soil is situated at 0.91m TAW.

Few faint isolated ice vein casts with upward bended side laminae were observed. Ice veins (< 5cm) are much more probable than ice wedges (≥ 5cm) in seasonally frozen ground and the active layer of permafrost (Washburn 1979, 114). Heyse (1983)

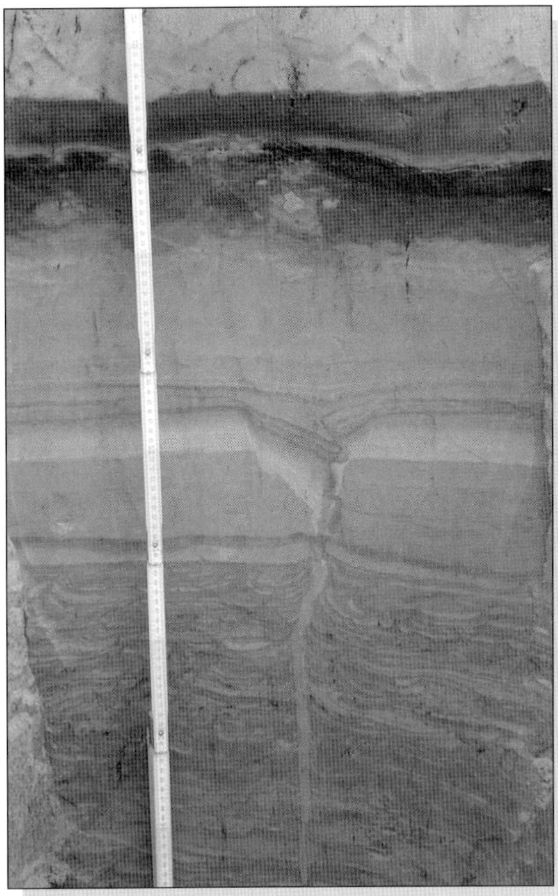

Figure 18: Verrebroek "Dok 2"; profile 9, ice vein cast with upward bended side laminae.

indicates regularly observed ice veins in the dry sandy ridges of the coversand belt of Maldegem - Stekene; he however terms them frost wedges. Van Vliet-Lanoë (1998) on the other hand locates ice wedges in Western Europe, mainly present as small, juvenile forms, essentially at poorly drained sites. Latter position fits indeed better with the relatively low topographic position as measured for this profile. Ice veins are indicators for extreme cold winter temperatures; these seasonal phenomena can develop from -15°C (Dylik, Maarleveld 1967, in Heyse 1983). Pissart (1987b) mentions that permafrost developed at the beginning of the Late Glacial, demonstrated by ice wedge casts.

2.1.3.4. | Late Glacial soil

The redistribution of the Upper Pleniglacial coversands during the Late Glacial occurred in a landscape of dune ridges and dune valleys or slacks. Nearly permanent lakes or ponds were present at the level of at least some dune valleys. Within these ponds or lakes a thin humiferous, locally peaty, soil developed. The interpretation of this soil as subaqueous is based on i) the nearly complete absence of bioturbation by burrowing animals and roots, ii) the absence of tree-fall structures, iii) the abrupt to rather abrupt lower boundary of this soil and iv) the presence of assumed load structures (P4 and P5) probably linked to conditions when water-saturated sediment comes under pressure. Load cast structure formation is driven by an inverse density gradient, which means that heavier coarse-grained sediment sinks into a lighter fine-grained one.

Involutions of the Late Glacial humiferous soil in the underlying sediment concomitantly showing ice segregation eluvial fibres point towards congeliturbation posterior to Late Glacial soil formation (P6). Involutions of the parent material into the overlying Late Glacial soil locally follow a frost crack or fissure (P11). This fits with the fact that flowing soil preferentially moves along pre-existing fissures (Van Vliet-Lanoë *et alii* 1984). This type of involutions seems to correspond to a type of water escape structure (Bjørlykke 1989).

The more humiferous soil is only present in the sections that correspond to dune slack positions. Outside of these, soils are completely, or largely, eroded. In latter positions bioturbations of several decimetres deep and up to a few cm wide root and/or animal galleries are observable below the erosion surface. Possibly these galleries correspond to beetle galleries. Comparable bioturbation was observed in the Upper Pleniglacial coversands at the "Hoge Vaart" sections (Flevoland, The Netherlands), where these galleries were linked with the Bøllng interstadial (13000-12000 BP) (Spek *et alii* 2001a, 13).

Temporary lowering of the groundwater table at least in one area allowed a short human occupation. Over a surface of about 3 over 1.5m, the Late Glacial soil revealed Final Palaeolithic (Federmesser) artefacts embedded in a charcoal-rich horizon (cf 3.1.1.).

The local differences and aspects of the above described soils and soil horizons are represented by different facies, reported in the soil profile descriptions. The different profiles are as much as possible discussed in logical order, mostly following soil genesis chronology, not necessarily the profile numbering sequence.

Profile 11 (fig. 17, tab. 2; D9, D.10, D.39, D.42)

The top of the Late Glacial soil is situated at 1.40m TAW.

This profile represents a facies of the Late Glacial soil (H7 - H8) on Upper Pleniglacial coversands (H11 - H9), subsequently covered by Late Glacial redistributed aeolian sands (H4 - H5) following dynamics of remobilised dunes. The textural change from the parent material (H9) to the Late Glacial soil (H8) can be derived from the change in mean and median particle size diameter (tab. D.39).

The texture of the Upper Pleniglacial coversands (H9) has a median value of 153 µm (tab. D.39). Micromorphological observations show that the quartz grains in the laminated to stratified Upper Pleniglacial sands (H11) are coated with thin (< 5µm) light green (PPL), probably glauconitic clay coatings (tab. D.9, cf tab. D.49). Glauconite clay migration can result of mechanical breakdown of sand-size glauconite (Van Ranst, De Coninck 1983).

Micromorphological observations learn that the greyish dark brown organic matter accumulation bands (H10), as described in the field, correspond to quartz grains coated with parallel oriented, slightly dusty, strong brown (PPL) clay coatings up to 20µm thick and clay infillings. Concomitantly, organic matter, inorganic phosphate and exchangeable cations (Ca, Na, Mg, K) contents as well as cation exchange capacity are twice as high as in the overlying horizon (H9); amorphous and free aluminium and iron contents are equal to slightly higher.

Table D.10 gives an overview of the particle size distribution within and the similarity between the different horizons. From H8 to H4, a reverse graded bedding sedimentary structure, observed via the shift from the fine towards the coarse particle size classes, is detectable.

Boggs (1987, 137) classifies the graded bedding morphology under current- and wave-formed or suspension-settling depositional structures. This phenomenon probably corresponds here to suspension settling in a lake or dune slack environment (Boggs 1987, 142). Absence of stratification or lamination, presence of abrupt lower horizon boundaries and absence of bioturbation from H8 to H5 confirm this hypothesis. Reverse graded bedding, with finer particles at the base grading upward to coarser particles, is however much less common than normal graded bedding. Reverse grading may be due to increasing flow velocity; however, if the velocity is too high, erosion will result (Bjørlykke 1989, 36). It is furthermore known to occur in laminae formed by the migration of wind ripples (Boggs 1987, 144). Ripples are a type of bedform generated by unidirectional currents, both in aeolian and subaqueous environments and develop in shallow water that shows little disturbance at the water surface under slow flow velocity conditions (Boggs 1987, 146). They range in length from about 0.05 to 0.2m and in height from about 0.005 to 0.03m; they characteristically develop in sediment ranging in size from coarse silt to medium sand (0.03 - 0.6mm) (Boggs 1987, 145). The somewhat wavy (pockets less deep than wide) upper boundary of H9, at the base of the reverse graded bedding sedimentary structure, resembles a ripple morphology, albeit more symmetrical than regularly observed (Boggs 1987, 146). As a rule ripples with a symmetrical cross-section are formed by waves eventually caused by wind in very shallow water, rather than by unidirectional currents (Bjørlykke 1989, 38). Furthermore the texture of H9 mainly consists of fine sand (70%) and very fine sand (24%), with minor amounts of clay (2%) and silt (5%), which fits with the above-mentioned size range from coarse silt to medium sand. The observed features are thus more in favour of slow flow velocity conditions in shallow water in which the wind causes tiny waves.

The very high similarity (index > 85) between horizons H8, H7b and H7a indicates these horizons as the core of the Late Glacial soil. Rather than being developed in the Upper Pleniglacial parent material, the graded bedding sedimentary structure, starting at the base of H8, points towards ongoing sedimentary dynamics during the Late Glacial. Moreover, out of the reverse graded bedding structure can be concluded that sedimentation occurred in a dune slack under slow flow velocity, or in stagnant water. The horizons of the Late Glacial soil contain very few (3 - 4%) clay. Micromorphological observations (tab. D.9) show that these horizons have very few (2 - 5%) poorly oriented clay coatings or infillings. It is however questionable whether the clay coatings present in these horizons might fully account for the clay content analysed by laser diffraction, even after pre-treatment with 30%-H_2O_2 in order to remove the organic matter. Mathieu, Pieltain (1997, 21) indeed mention that humin, humified organic matter that remains insoluble after all successive extractions of humic and fulvic acids (Lozet, Mathieu 1997, 202), might resist to H_2O_2-treatment. The Late Glacial soil horizons furthermore contain few (5 -10%) in H7b and H8 to very few (2 - 5%) in H7a organic fine mass coatings and infillings (tab. D.9). Clay can be reworked within the amorphous dusty organic fine mass, lacking any orientation pattern and thus becoming undetectable as mineral clay in thin section. With respect to this, Barratt (1969) indeed mentions different constituents of microscopic organic soil components: (i) humiskel, i.e. mostly undecomposed organic residues, (ii) lithiskel, i.e. mineral grains and fragments, (iii) humicol, i.e. decomposed organic colloids, (iv) mullicol, i.e. a mixture of organic and clay colloids and (v) argillicol, i.e. clay colloids, indeed mentioning clay colloids incorporation in the organic fraction. The amorphous black, possibly carbonised, organic particles, up to 10µm large and the very few to few brown organic fine mass coatings to infillings observed in thin section (2 - 5% in H7a, 5 - 10% in H7b - H8, tab. D.9) might partly correspond to the clay fraction of the particle size analysis by laser diffraction too. These coatings are interpreted as sedimentary organic matter in a dune slack with stagnant or free water and are present throughout the Late Glacial soil (H7 - H8) and even at the base of the overlying H6.

The above-mentioned poorly oriented clay coatings or infillings, up to 0.5 over 1.5mm in size and mainly with horizontal or vertical orientation, could be interpreted as fragments of former surface crusts. Indeed, the Final Palaeolithic (Federmesser) flint artefacts, indicating a former occupation surface, occur in H7.

According to a preliminary analysis (Crombé *et alii* 1999c), around 60% of these artefacts display soil capping, indicating alternating freeze-thaw cycles (Van Vliet-Lanoë 1985). Heating traces occur on 15% of the artefacts. Also few (< 2% in H8 and H7a, 2 - 5% in H7b) charcoal particles up to 5mm in size were observed in thin section. H7 additionally contains few (5 -10% in thin section) black, probably carbonised organic particles up to 10µm large. The relatively high total phosphate content, reaching a peak in H7b (1470mg kg^{-1} P_2O_5t) and decreasing to about half the amount in H8 (674mg kg^{-1} P_2O_5t) (tab. D.42), also sustains human activity. The phosphate is mainly present as inorganic phosphate (Pi/Po = 37.2 in H7a, 7.3 in H7b, 12.0 in H8). Latter fits with the presence of burnt bone on the Final Palaeolithic living floor (cf 3.1.1.) or can furthermore represent phosphate released on carbonising or burning of organic matter and captured in the mainly organic clay-size fraction. Parallel to the relatively high inorganic phosphate content in H7b and H8, exchangeable magnesium (0.15, 0.09cmol(+) kg^{-1} Mg in H7b, H8) and potassium (0.066, 0.087cmol(+) kg^{-1} K in H7b, H8) contents are relatively high too. Release of inorganic phosphate, magnesium and potassium is commonly observed as an effect of fire on the soil (Firmin 1986).

H6 has alternating brown and pinkish white laminae. Micromorphological observations (tab. D.9) show that the brown colour is due to strong brown (PPL) clay coatings on the quartz grains; the pinkish white colour corresponds to uncoated quartz grains.

With regard to particle size distribution, H6 shows higher similarity with the underlying (H7-H10: similarity index from 70 to 80) than with the overlying horizons (H5-H4: similarity index from 65 to 40). However, H6 doesn't contain buried surface crust fragments that could be related to a former occupation surface (H7-H8). Nevertheless, a sample wherein the laminae of H6 were mixed, has a total phosphate amount comparable to that of H7a (P_2O_5t = 459mg kg^{-1} in H7a and 593mg kg^{-1} in H6), again mainly consisting of inorganic phosphate. This probably confirms that H6 is part of the reverse graded bedding sedimentary system with ongoing, though very local, short-distance sediment transport in this dune slack. H6 is furthermore connected to the overlying horizons via vertical brown veins (< 1cm wide) passing through H5, showing the same type of strong brown (PPL) clay coatings as observed in the brown bands of H6. Hence, the clay coatings must be diagenetic; the clay movement might have been blocked on the markedly higher clay and silt contents in the underlying horizons H7a, H7b and H8 and clay accumulation was furthermore favoured in H6 with its prominent presence of very fine sand, silt and clay when compared to the overlying horizons H4 and H5. The latter would indicate that the accumulation only formed once H5 and H4 were already deposited (see below).

A particle size rupture appears between H6 and H5, with a jump in the similarity index from 80 between H7a and H6 to 65 between H6 and H5. It possibly refers to additional external coarse-size sediment input. Indeed the shift in the particle size distribution mainly results out of a strong decrease in very fine sand (from around

40% in H7 - H6 towards 15% in H5 and 4% in H4) in favour of an increase in medium sand in horizons 5 and 4 (from around 5% in H7 - H6 towards 22% in H5 and 47% in H4). Micromorphological observations (tab. D.9) show that the coarse fraction in H5 is moderately sorted whereas in the underlying horizons (H6 - H11) the coarse fraction is only poorly sorted to unsorted. Latter difference in sorting corresponds to fluctuating energy levels (Rapp Jr., in Ellis 2000, 268) and probably reflects different deposition dynamics. In case of additional coarse-size sediment input, sediment transport would have shifted from a small to a large scale, from subaqueous, wind-induced sediment transport in a dune slack towards direct aeolian sediment transport linked to dune dynamics. The fact that H5 is not stratified and only H4 shows clear oblique stratification provides further evidence for this hypothesis. The orientation of the oblique stratification in H4 indicates that western wind carried the sediment. At least for the Younger Dryas stadial (following the Allerød interstadial), De Moor (1963, 348) mentions aeolian remobilisation of the coversands by wind currents coming from the West. Latter could fit within the Bølling/Earlier Dryas age proposed for this soil that is based on pollen analysis (cf 2.2.2.) and ^{14}C-dating (cf 2.3.2.). Parallel, orientation of the coversand ridges in Zeeuws-Vlaanderen (area Sas-van-Gent - Koewacht, The Netherlands, located at about 15km west to southwest of Verrebroek), indicate that dominant wind directions during the Earlier Dryas (following the Bølling interstadial) and Younger Dryas (following the Allerød interstadial) were northwest and southwest respectively (van der Sluijs, Maarleveld 1963).

Profile 1 (fig. 19, tab. 3, D1, D.2, D.39, D.42)

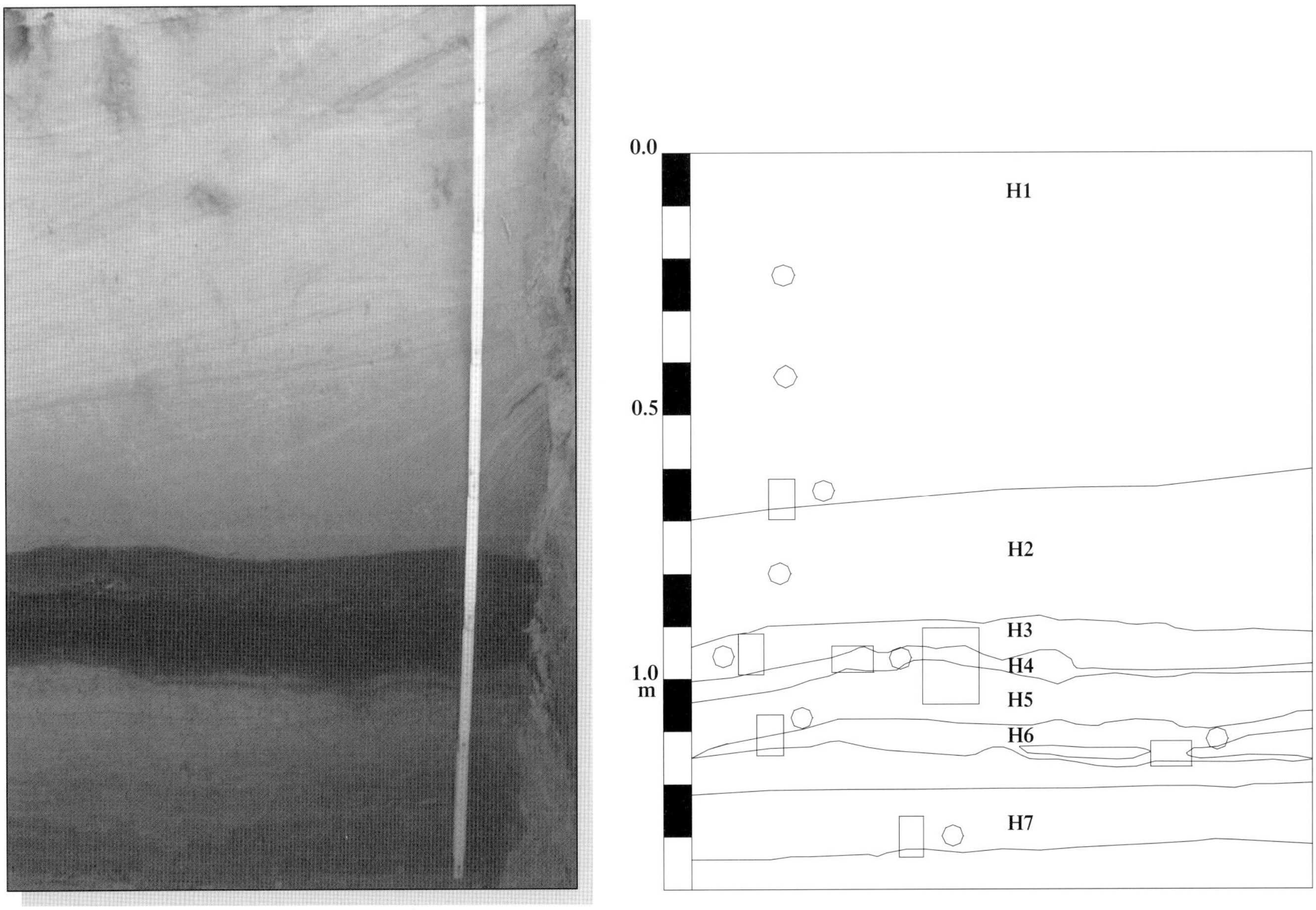

Figure 19: Verrebroek "Dok 2"; profile 1, photographic and schematic representation.

Horizon number	Symbol	Horizon depth (cm)	Profile description
H1		(0)-60/70	Very pale brown - light grey matrix; stratified
H2		60/70-88/94	Light grey matrix; sand; non-stratified
H3		32/70-38/73	Brown - dark greyish brown matrix (sand) with dark brown bands (organic matter)
H4		94/105-108/114	Mixture of pale brown and very dark greyish brown matrix; sand; slightly stratified; liquefied sediment conditions?
H5		108/114-110/117	Dark brown matrix; peat; stratified; with distinct wood fragments
H6		110/117-112/118	Dark grey to dark brown matrix; sand; heterometric; clay-rich, clay accumulation?; presence of micas? (see micromorphology); common excrements, common biological activity
H7		120/122-131/134	Light greenish grey matrix; reduced or presence of glauconite?; stratified; decomposed organic matter (root traces)

Table 3: Verrebroek "Dok 2"; profile 1, profile description.

The top of the Late Glacial soil is situated at 0.81m TAW.

This profile contains a peat horizon (H5) indicating that, at least during the period that the Late Glacial soil formed, P1 was situated in a dune slack depression position saturated with water over long periods. Also Heyse (1983) mentions peat occurrence as restricted to microdepression positions in the coversand ridge of Maldegem - Stekene. The presence of pyrite framboids (cf tab. D.45) in H5 to H7 furthermore points towards continued wet conditions and is linked with the presence of peat (H5).

The similarity index (SI) diagram (tab. D.2) shows that H7 to H3, considered in the field as Upper Pleniglacial coversands, don't form a continuum when it comes to particle size distribution. Indeed, H7, in the field seen as the parent material in which the Late Glacial soil developed, doesn't show important similarity with the overlying horizons (SI < 55, except SI_H7xH3 = 77). The particle size distribution of H6 to H3 again reveals a reverse graded bedding sedimentary structure. The upper boundary of H7 is somewhat wavy, corresponding to the morphology of wind ripples development in shallow water (as formulated higher for P11). Also here, we have to deal with ongoing sedimentation dynamics in a dune slack with stagnating water. The texture of the Upper Pleniglacial coversands (H7) has a median value of 127µm (tab. D.39).

Within the reverse graded bedding sequence, only H6 and H5 have a relatively high similarity (SI_H6xH5 = 80). Nevertheless, H6 has higher amounts in finer particles (clay, fine - medium and coarse silt and very fine sand) than H5.

Micromorphological observations (tab. D.1) only show clay-size fine mass in the organic, not in the mineral fraction; clay pedofeatures, such as clay coatings, are absent as well. The organic fine mass is present in H6 (tab. D.1) mainly as pseudo-particles up to 150µm and possibly coprogenic or as dusty coatings or infillings that eventually can be interpreted as sedimentary, detrital organic matter, i.e. dissolved and particulate organic matter. This is one of the few horizons that shows biological activity evidenced by coprogenic organic fine mass particles or pellets, which let us presume that there were no continuous subaqueous conditions here. Parallel, Van Vliet-Lanoë (1998) only refers to coprogenic humiferous Late Glacial Bølling soils in well-drained landscape positions.

H5, the peat horizon, contains 5 - 10%, locally even 10 - 20% horizontally elongated, yellow to reddish brown (PPL) organ residues (cf tab. D.45), partly showing birefringence. This morphology suggests *in situ*, subaqueous formation of this peat horizon, rather than being organic soil material transported from elsewhere; Exaltus (2001, 18) concludes the same for a comparable morphology in peat horizons at the "Hoge Vaart" (Flevoland, The Netherlands). Part of this organ residue might correspond to the distinct wood fragments observed in the field.

Both H6 and H5 have a high total phosphate content (713 - 783mg kg^{-1} P_2O_5t in H6 - H5), more than half of the amount being present as inorganic phosphate (tab. D.42). As opposed to P11, the relative peak in inorganic phosphate cannot be the result of burning, since there are no other indications for fire activity here. The inorganic phosphate fraction probably results here out of the gradual mineralisation of the organic matter (6.12 - 24.59% OM_500 °C in H6 - H5). Concomitantly, also the cation exchange capacity (CEC) and the exchangeable cations calcium (Ca), sodium (Na), magnesium (Mg) and potassium (K) contents are relatively high when compared to the other horizons. Mineralisation of the organic matter however needs oxidising conditions, which again points towards a discontinuous wet or subaqueous environment or at least occasional aerobic conditions. Leenders *et alii* (1990) indicate retardation or even a break in the peat growth process as possible causes for peat humification and mineralisation.

In the overlying H4, only the base contains this type of horizontal elongated organ residue; the organic matter furthermore mainly consists of amorphous plant fragments and organic fine mass, eventually clustered in particles or deposited as coatings or infillings. The macromorphology of H4 moreover shows a far more disturbed appearance. H4 hence shows more characteristics of reworked, rather than *in situ* deposited organic material.

In the brown to dark greyish brown matrix of H3, very few (2 - 5%), locally few (5 - 10%) horizontal elongated amorphous plant fragments are visible in thin section. This morphology, as in H5, again points towards *in situ* deposition of at least part of the organic matter. Furthermore the quartz grains are coated with limpid organic fine mass of < 15µm thickness (cf tab. D.45); these coatings might however be diagenetic. H3 shows a markedly higher very fine sand, silt and clay content (tab. D.2) than the overlying H2. The finer texture in H3 probably hampered further migration of the organic fine mass. As in P11, this again could be an indication for organic matter migration after sedimentation of the overlying sands (H2 and H1, see below).

The similarity indices of the particle size distribution (tab. D.2) clearly indicate that H2 and H1 belong to the same particle size cluster. These horizons probably correspond to remobilised or redistributed aeolian sands within dune system dynamics. They must have been deposited during one of the Late Glacial colder periods, most likely one of the Dryas stadials. The rupture in particle size distribution is mainly due to (tab. D.2) a shift from clay, silt, very fine sand (from 34 - 25% in H5 and H3 to 3 - 6% in H2 and H1) and fine sand up to 177 µm large towards medium (from 10 - 15% in H5 and H3 to 34 - 48% in H2 and H1) and coarse sand.

<u>Profile 3, H10 - H15</u> (fig. 20, tab. 4, D.3, D.4)

The top of the Late Glacial soil is situated at 0.61m TAW.

The Late Glacial soil horizon sequence was only micromorphologically investigated; neither particle size analyis nor chemical analysis was performed. The interest of this profile lies within the exceptionally preserved colour B-horizon development.

The macro- and micromorphology of the Late Glacial soil horizon sequence is comparable to that of P1, H3 to H6. The poorly sorted parent material (H14) is composed of dominant (> 50%) subrounded (polycrystalline) quartz, very few (2 - 5%) rounded, partly physically weathered glauconite, very few (< 2%) subrounded to rounded chert, feldspars, mica and has additional fragments composed of quartz, glauconite and dusty clay. It has thin (< 5µm) light green (PPL), possibly glauconitic, clay coatings on the coarse grains (cf tab. D.49). As suggested earlier, these coatings might be the result of mechanical breakdown of the sand-size glauconite grains (Van Ranst, De Coninck 1983); part of the glauconite is indeed physically weathered. Throughout the horizon sequence H14 to H10 the sorting of the groundmass evolves from poorly to moderately sorted and the amount of glauconite is slightly reduced. As in P1, H3 to H6, the horizon sequence contains an *in situ* formed peat horizon (H12). The finer-textured underlying horizon (H13) however does have a minor amount of thin

Figure 20: Verrebroek "Dok 2"; profile 3, (a) H0-H9 and (b) H10-H15, photographic and schematic representation.

Horizon number	Symbol	Horizon depth (cm)	Profile description
H0	H+E	+6/+3-+2/2	Black matrix; colloidal organic matter
H1	A	+2/2-2/7	Grey matrix; sand; black decomposed roots
H2	E	2/7-9/11	Light grey - light brown matrix; sand; galleries filled with black decomposed organic matter
H3	ABbi	9/11-12/18	Dark-very dark brown matrix; sand; black mottles consisting of organic matter
H4	Bw.bi1	12/18-18/28	Strong brown - very dark brown matrix; sand; black mottles consisting of organic matter
H5	Bw.bi2	18/28-28/32	Strong brown matrix; sand; black mottles consisting of organic matter; slightly cemented
H6	Bw.bi3	25/32-30/38	Strong brown matrix; pale yellow mottles due to consumption of organic matter; sand
H7	BC	30/38-45/51	Strong brown matrix; abundant pale brown mottles due to consumption of organic matter; sand
H8		45/70-59(70)	Alternating strata of pale brown and bluish matrix, obliquely oriented; stratification or migration?
H9		48/51-(70)	Pale brown matrix; dark brown decomposed roots; laminae of organic matter, obliquely oriented
H10		(110)-114/117	Light grey matrix; sand
H11		114/117-118/120	Brown - dark brown matrix; sand; very dark brown bands of organic matter
H12		118/120-128/131	Dark brown matrix; peat; distinct leaf and wood fragments
H13		128/131-130/133	Light brownish grey matrix; silty - clayey?
H14		130/133-(140)	Light grey-green matrix; sand
H15		134/135-134/136	Pale brown matrix; clay accumulation band?

Table 4: Verrebroek "Dok 2"; profile 3, profile description.

(< 5µm) dusty clay coatings, next to few (5 - 10%) organic fine mass coatings to infillings. However, the disturbed phase, corresponding to H4 in P1 that covered the peat horizon there, is absent here. In P3, the horizon above the peat horizon (H11) has comparable morphology to H3 in P1; it also has limpid organic fine mass coatings up to 15µm (cf tab. D.45), possibly diagenetic.

2.1.3.5. | Involutions from the parent material into the Late Glacial soil

Profile 11 (fig. 17)

Locally involutions from the parent material (H9) into the Late Glacial soil (H8 - H7) and stopping on lateral expansions of H7a were observed (fig. 17b). The involutions thus represent an upward mass flow. In horizontal section, the involutions show a linear, locally continuous pattern, whereby two involutions of about 0.10m wide appear parallel to each other occasionally (fig. 17c-d). One of both follows a narrow (few cm) frost crack or fissure (fig. 17d) indicating that flowing soil preferentially moves along pre-existing fissures (Van Vliet-Lanoë *et alii* 1984). This frost fissure only witnesses frost action, no permafrost conditions. Mackay (1999) observed a similar parallel pattern mentioning thermal contraction cracks (of about 0.10 m wide) in a sand terrace at the exposed lake bottom of North Lake, situated in the Tuktoyaktuk Peninsula area at the western arctic Canadian coast and drained within a few years. He suggests that the two parallel surface cracks resulted from upward cracking with crack bifurcation. Crack bifurcation was observed elsewhere at Garry Island (Canada) in the frozen active layer above an ice wedge and also in the hard packed snow in ice wedge troughs. Mackay (1999) suggests that the active layer sands probably retain little to no memory of the previous winter's crack by the end of the summer, because the sands in the upper layer are dry and loose, and the bottom sands, usually below the water table and saturated, appear unlikely to preserve that memory. As a consequence, downward cracking in the active layer is unlikely, so cracking must usually be upward (Mackay 1999). The mechanism described by Mackay (1999) could also at Verrebroek "Dok 2" explain the observed involutions or at least the observed crack pattern. The involutions from the parent material into the Late Glacial soil then call for permafrost conditions that were active after the Late Glacial soil genesis. As the Late Glacial soil is attributed a Bølling/Earlier Dryas age here (cf 2.2.2., 2.3.2.), climatic conditions during the Earlier or Younger Dryas stadials must have been calling for permafrost. De Moor (1963, 348) mentions permafrost conditions and concomitant congeliturbation in Flanders during the Younger Dryas stadial. Based on different authors from different disciplines, also Pissart (1987b) concludes for discontinuous permafrost in Northern Belgium during Younger Dryas time (11000 to 10000 BP), with a mean July temperature of 10.0 (Van der Hammen *et alii* 1967, in Pissart 1987b) to 11.5°C (Kolstrup 1979, in Pissart 1987b).

Additional mechanisms however must have been responsible for the subsequent infilling of these cracks and hence upward mass movements. Table D.10 shows that H8 has a finer texture than H9, which means that this particular mass displacement can't correspond to load structure formation, whereby coarse-grained, heavier sediment would sink into fine-grained sediment. Other dynamics responsible for upward mass movement, as e.g. water escape structure formation or congeliturbation, must have been active here. The sedimentary sequence nonetheless corresponds to increased frost susceptibility from H9 to the humiferous H8 and H7. Factors and mechanisms responsible for water escape structure formation are discussed in Appendix B. Glossary.

Interpreting the observed structure as sand dykes implies accepting (additional) pressure at some time posterior to the Late Glacial (Bølling/Earlier Dryas) soil formation. It should be questioned whether this pressure is linked to human impact, such as the Final Palaeolithic human occupation, or frost action, such as eventual freezing of the pond or lake water or a combination of both. The above-mentioned permafrost conditions, responsible for thermal contraction crack generation, during one of the Dryas stadial periods, supply a set of pressure producing mechanisms: i) cryostatic pressure in the active layer when both freezing fronts proceed in winter, ii) pore water pressure in the hydrodynamic unstable thawed active layer in summer. Following Mackay's (1999) hypothesis, groundwater under artesian pressure during summer would be sufficient to permit an upward mass movement. This fits with the coincidence of the upward involution base (fig. 17b) and the lake bottom, the latter representing the summer groundwater table level. The mentioned artesian pressure could have been intensified by additional pressure linked to human impact during summer camp occupation of the site. Diapiric upward movement driven by differential bulk density (Swanson *et alii* 1999) or differential frost susceptibility (Van Vliet-Lanoë 1985) within the active layer of a permafrost during the thaw season is an alternative and acceptable mechanism for the observed involutions. Accordingly the involutions stop on lateral expansions of the humiferous surface A-horizon affected by human occupation and containing very few charcoal fragments. Within the dune slack environment as proposed for this site, it is moreover likely that soil material is sufficiently wet to allow the observed mass movements.

Within the context of human impact, French (1988, 43) mentions also the delicate density controlled equilibrium within water-saturated sediments in a seasonally thawed layer. Latter may very easily be upset by very minor disturbances, such as man or animal walking or temporarily, even nearby human occupation in a campsite. Alike disturbances commonly result in volcano-like mudpiles whereby subsurface material reaches the surface under artesian pressure. Swanson *et alii* (1999) mention that disturbance of the surface by fire or animal or human activity can cause thaw of the upper part of the permafrost. When following the hypothesis of diapiric upward movement, driven by suction potential gradient from thawed to frozen soil, human activity thus indirectly enhances development of soil involutions. Alternative deep thaw of the active layer can occur as a result of unusual weather, especially warm summers or deep snow in the preceeding winter (Swanson *et alii* 1999).

Nevertheless, the involutions can hardly be interpreted as trampling features by mammals such as mammoth or moose, as some very typical characteristics for the latter are absent here. Typical characteristics would be occurrence of microbleached horizons, local iron hydroxide staining associated with microcompaction, regular footprints (in horizontal section) and associated absence of ice segregation eluvial fibres (Van Vliet-Lanoë 1998). None of these were clearly identified at the Final Palaeolithic site at Verrebroek "Dok 2".

2.1.3.6. | Congeliturbation structures in the Late Glacial soil - Particular type of flat-based involutions (Drop Soil)

<u>Profile 6</u> (fig. 21, tab. 5, D.8)

The top of the Late Glacial soil is situated at 1.36m TAW. It has a relatively high topographic position when compared to the other soil profiles observed at Verrebroek "Dok 2". It might have been situated on a lower footslope, rather than in a depression; absence of peat enfavours this hypothesis.

This profile is the only illustration of the Drop Soil, a particular type of congeliturbation. For this profile, also featuring a Late Glacial soil, only macro- and micromorphological observations have been performed. The parent material (H8) consists of dominant (> 50%) poorly sorted subrounded (polycrystalline) quartz and very few (< 2%) subrounded to rounded chert, feldspars, partly weathered glauconite, mica and particular to this profile, compound fragments of quartz and dusty parallel oriented clay. The quartz grains have thin (< 5µm) limpid, light green (PPL) clay coatings, probably resulting from physical glauconite weathering (cf tab. D.49); also thin (< 10µm) dusty clay coatings are visible, as observed in the other profiles. This horizon in addition has an incomplete prismatic to subangular microstructure, which could point towards aggregate formation as a result of repeated freeze-thaw processes (Van Vliet-Lanoë *et alii* 1984; Van Vliet-Lanoë 1985; Hartge, Horn 1991, 232-233; Van Vliet-Lanoë 1998). Latter processes are also responsible for the formation of the observed whitish ice segregation eluvial fibres, witnesses of former ice lenses. The observed dusty clay illuviation can in this context be related to conditions

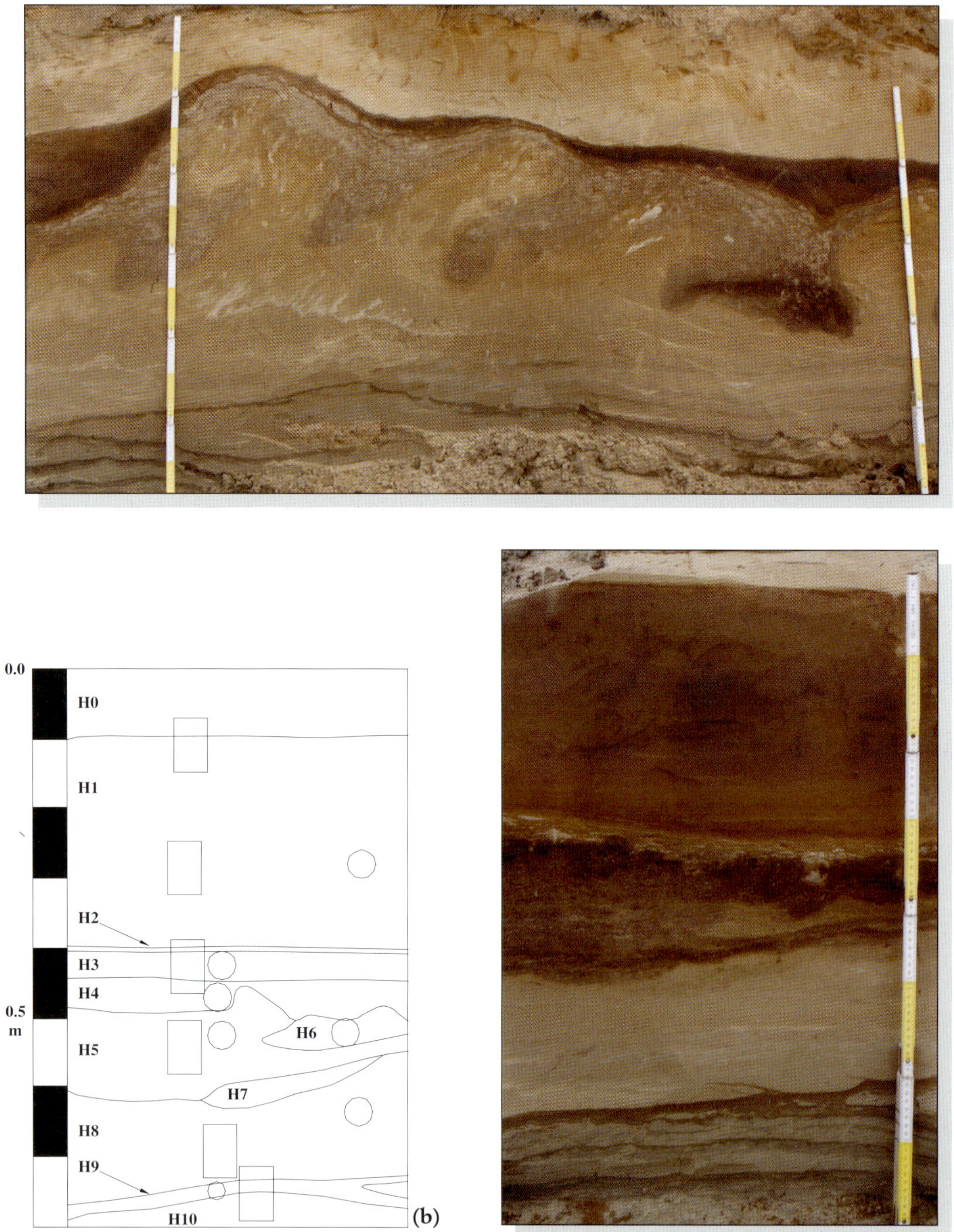

Figure 21: Verrebroek "Dok 2"; (a) Drop Soil, (b)-(c) profile 6, photographic and schematic representation, resembling most to the profile at the left side of the section shown in (a).

Horizon number	Symbol	Horizon depth (cm)	Profile description
H0		(0)-9/10	Light grey matrix; sand
H1		9/10-39/40	Brown - dark brown matrix; sand; very dark brown organic matter bands
H2		39-41	Dark brown matrix; (sand coated with) organic matter?
H3	E+Bt?	40/41-44/45	Alternating laminae of white and brown matrix; sand; disturbed due to influence of congeliturbation (cf Drop Soils to the N)?
H4		44/45-46/50	Very dark-dark brown matrix; sand; disturbed due to influence of congeliturbation (cf Drop Soils to the N)?; few charcoal fragments?
H5	Bh.t?	46/51-55/62	Brown - greyish brown matrix; sand; disturbed due to influence of congeliturbation (cf Drop Soils to the N)?; few charcoal fragments?
H6		48/53-53/55	Light brownish grey - light grey matrix; sand; disturbed due to influence of congeliturbation (cf Drop Soils to the N)?; few charcoal fragments?
H7		55/61-55/63	Laminae of light grey - light greenish grey matrix; sand
H8		55/62-72/77	Light grey - light brown matrix; sand; white ice segregation eluvial fibres?
H9		72/77-75/79	Dark (greyish) brown matrix; (sand coated with) organic matter
H10		75/79-(80)	Light bluish grey matrix; loamy sand?; stratified

Table 5: Verrebroek "Dok 2"; profile 6, profile description.

of thawing, as soils affected by repeated freeze-thaw are sensitive to internal colluviation during thawing, eventually reinforced by snow melting (Van Vliet-Lanoë 1985). This is especially likely to occur when clays are potentially easy to disperse because of their base saturation or their incorporation in organo-mineral complexes (Guillet *et alii* 1981, in Van Vliet-Lanoë 1985). The mechanisms responsible for ice segregation are discussed in detail in Appendix B. Glossary. Micromorphological observations (tab. D.8) show that H5 is somewhat coarser-grained than the underlying H8. Particle size appraisal in the field indicated still a finer texture below H8, in H10. The particle size differences provide important shifts in pore water pressure and are hence favourable for ice segregation development. As stated above, ice segregation eluvial fibres are likely to develop at the freezing front level, eventually coinciding with the permafrost table. The ice segregation eluvial fibres observed within this section probably have been formed under periglacial conditions and even above permafrost present in Upper Pleniglacial or Late Glacial periods.

In the overlying H5, in the field interpreted as a horizon with organic matter accumulation (Bh), few to frequent (5 - 20%) dusty brown organic fine mass coatings or infillings are visible in thin section (tab. D.8). Additionally few (2 - 10%) limpid to slightly dusty, strong brown (PPL) clay coatings up to 0.5mm thick, becoming thinner towards the base, or infillings were observed, as well as thin (< 5μm) slightly dusty light green (PPL), probably glauconitic, clay coatings (cf tab. D.49). Clay accumulation thus occurred next to organic matter accumulation.

In the overlying horizontally laminated, darker coloured H4, only very few (2 - 5%) dusty to very dusty brown organic fine mass coatings or infillings occur. The small quantity of small (< 10µm) amorphous black, possibly carbonised, organic matter particles probably account for the dark colour. Next to the thin light green clay coatings, few limpid to slightly dusty strong brown clay coating fragments up to 350µm in size are present. Latter pedofeatures seem to be reworked, rather than being formed *in situ*.

Flat-based involutions of the Late Glacial humiferous soil horizons H5 and H4 in the underlying sediment (H8) appear a few metres south of P6. They measure up to 0.15m high and 0.50m wide and undulate down to about 0.30m when referred to the base of H3; they appear next to uplifted Late Glacial soil material (fig. 21a). Lamination and ice segregation eluvial fibres in the underlying horizon follow the involution morphology. The concomitant presence of the above-mentioned ice segregation eluvial fibres in the parent material (H8) let us assume congeliturbation as the driving force behind this particular morphology and mass displacement. Factors and mechanisms responsible for congeliturbation are discussed in detail in Appendix B. Glossary.

As mentioned above, relatively coarse-grained material (H5) is on top of relatively fine-grained material (H8) in the original sediment structure; H8 moreover shows ice segregation eluvial fibres. The differential particle size sequence together with the process of ice segregation enabled both differential frost heave and sinking of relatively coarse-grained into fine-grained sediment. H4 and H5 represent in the mean time the organic matter-rich A- and Bh-horizons of the Late Glacial soil respectively, having a relatively lower thermal conductivity than the surrounding mineral soil material and thus less subject to frost heave, but more susceptible to frost and hence to deformation. Additionally, the Late Glacial soil involutions are flattened at the bottom. Latter characterisitic, together with the dune slope topographic position of this profile nevertheless calls for permafrost conditions, providing a perched water table. The bottom then possibly coincides with the upper boundary of the permanently frozen part of the permafrost, called the permafrost table. Considering the lower boundary topography of the soil involutions (fig. 21a), the inferred permafrost table must have had a wavy morphology, also reflected in the topography of the underlying ice segregation eluvial fibres.

Combined (nearby) presence of other congeliturbation features as ice segregation eluvial fibres, relatively higher organic matter content in the intruded Late Glacial soil A- and B-horizons, the reverse graded bedding sedimentary sequence within the Late Glacial soil and the flattened bottom of the Late Glacial soil involutions delivers sufficient arguments to interpret the observed morphology as a "Tropfenboden" or Drop Soil that originated in a periglacial, even permafrost environment, more specifically within the water-saturated and seasonally freezing and thawing active layer. This environment represents conditions favourable for three important mass mouvement mechanisms: i) frost heave, ii) cryostatic pressure within the active layer, and iii) intergranular or pore water pressure within the water-saturated, hydrodynamically unstable, thawed active layer. However, occurrence of ice lenses parallel to the soil surface proves that liquefaction did not occur during thawing (cf Van Vliet-Lanoë 1985), thus counteracting the last-mentioned mass movement mechanism typical for load casting. As a consequence, the observed involutions cannot be attributed to load casting.

De Moor (1963, 348) mentions permafrost conditions and concomitant congeliturbation in Flanders during the Younger Dryas stadial, post-dating the Late Glacial Bølling and Allerød interstadials during which (weak) soil development took place. This fits with the proposed genesis of these involutions in permafrost conditions. Based on different authors from different disciplines, also Pissart (1987b) concludes for discontinuous permafrost in Northern Belgium during Younger Dryas time (11000 to 10000 BP), with a mean July temperature of 10.0 (Van der Hammen *et alii* 1967, in Pissart 1987b) to 11.5°C (Kolstrup 1979, in Pissart 1987b).

In the overlying H3 and H1, thin (< 20µm) limpid, strong brown organic fine mass coatings (cf tab. D.45) are visible. This type of organic matter accumulation is different from the type featuring in the Late Glacial soil (H4 - H5). Morphologically (colour, limpidity) these accumulations resemble the so-called "waterhard", described as a brown, discontinuous, cemented layer with sharp upper and lower boundaries and occurring in coversands below the contact with a former peat blanket in The Netherlands (Koopman 1988; Dekker *et alii* 1991). "Waterhard" consists of quartz grains surrounded by thin coatings and infillings of organic matter resembling dopplerite. Further details on dopplerite are mentioned in Appendix B. Glossary. Organic matter probably precipitated at the level of a texture jump or (temporary) water table.

As dopplerite is defined basically on its chemical composition and characteristics, additional chemical analyses are required in order to check the exact composition of the organic fine mass coatings at Verrebroek "Dok 2". Nevertheless, it concerns very mobile organic matter components that are not syngenetic, but rather diagenetic to the Late Glacial humiferous soil development. The lower boundary of H1 moreover follows the morphology of the involutions of the Late Glacial soil into the underlying sediment, whereas the top of H1 is rather horizontal. This delivers a further evidence for diachronous genesis of the different types of organic matter accumulation.

Nevertheless, the presence of quartz and dusty brown clay internal crusts within H1 indicate that this horizon must have lain at the surface at some time. The internal crusts that now appear within the horizon were probably formed at the surface, e.g. as surface crusts at the bottom of a dune slack that temporarily became dry. By analogy with the other Late Glacial soil profiles observed at Verrebroek "Dok 2", the mineral material of H3 to H1 was most probably deposited during the Late Glacial interstadial sedimentation. H0 would then correspond to remobilised sands during (a) subsequent Dryas stadial(s).

2.1.3.7. | Subaqueous sedimentary structures in the Late Glacial soil

Profile 5 (fig. 22, tab. 6, D.6, D.7, D.39, D.42)

The top of the Late Glacial soil is situated at 1.00m TAW.

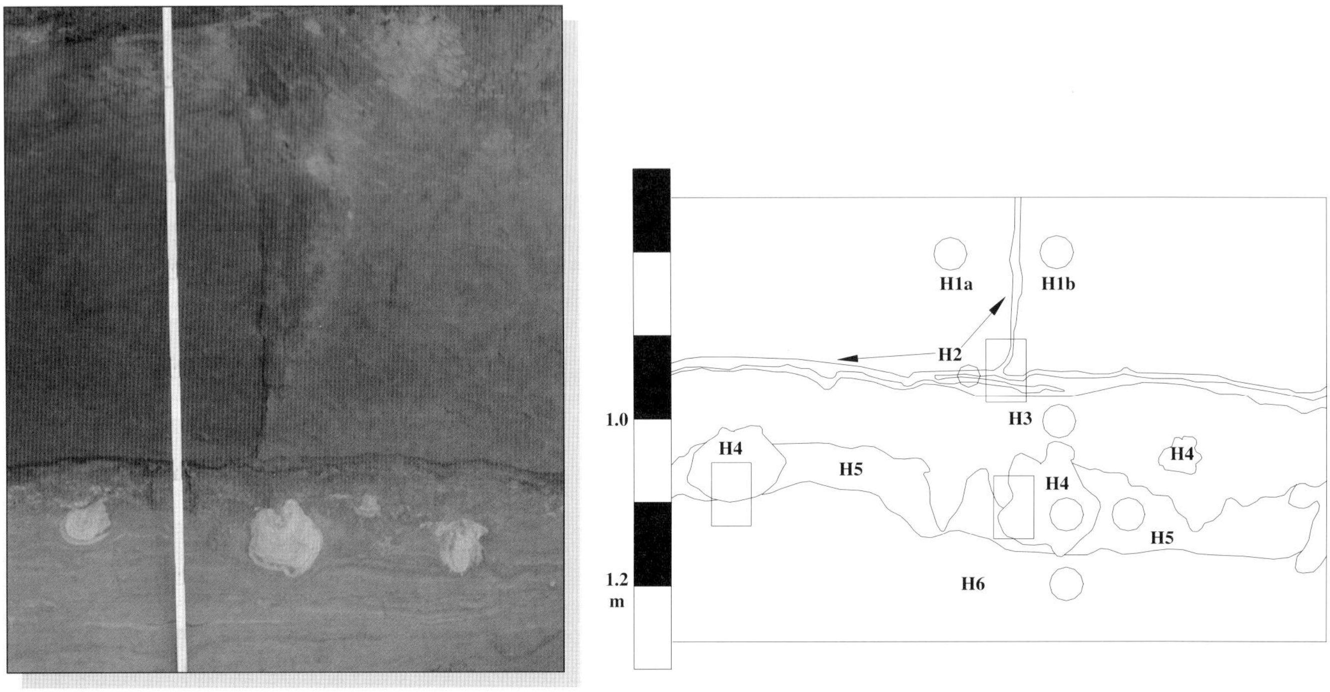

Figure 22: Verrebroek "Dok 2"; profile 5, photographic and schematic representation.

Horizon number	Symbol	Horizon depth (cm)	Profile description
H1a	B	-92/94	Dark brown matrix; sand; B of Podzolised Soil
H1b	B(C)	-94/98	Light brown - brown matrix; sand; B(C) of Podzolised Soil
H2		92/94-94/98	Dark - very dark brown matrix; sand coated with organic matter?
H3		94/99-100/114	Mixture of white and brown - greyish brown matrix; sand; disturbed due to liquefied sediment conditions?, peat fragments?
H4		100/108-107/119	Alternating ellipsoidal laminae of major white and minor brown - light brown matrix; sand
H5		102/113-108/118	Brown - greyish brown matrix; sand; stratification following morphology of H4
H6		108/119-(127)	Pale brown matrix; sand; stratified

Table 6: Verrebroek "Dok 2"; profile 5, profile description.

Below the buried surface horizon (H3), this profile features spherical- to ellipsoidal-shaped structures (H4) up to about 0.20m in diameter, composed of alternating laminae of white quartz sand-size grains and slightly humiferous brown sand (fig. 22) that disappear when scraping less than 0.10m of the section. The laminations in the sandy sediment underneath follow the spherical structure morphology. Stapert and Veenstra (1988) report similar phenomena in lacustrine sediments and aquatic circumstances in the Bølling and Allerød horizons at the type section in Usselo, Twente, Eastern Netherlands. Unlike what Spek *et alii* (2001a, 91-92) suggest for the structures that show a comparable spherical- to ellipsoidal-shaped laminated morphology (dimensions however not mentioned) observed along the "Hoge Vaart-A27" sections in Flevoland (The Netherlands), we cannot agree to relate these features observed here to refilled holes of decomposed tree trunks. In case of decomposed tree trunks, we would expect structures that continued much further in the third dimension (additional to the two dimensions we observed in the section). The particle size distribution analysis (tab. D.7, D.39) moreover shows that H4 is coarser-grained than H5 and H6, which allows us to presume that coarser-grained heavier quartz sank into finer-grained lighter quartz. This aspect corresponds to the genesis of load structures, such as load casts, ball and pillow structures, all driven by an inverse density gradient (Bjørlykke 1989, 37, 46-48). Morphology and formation of load structures are discussed in detail in Appendix B. Glossary.
The observed morphology of the sedimentary structures in this profile fits with the hemispherical or kidney-shaped masses showing eventually internal gently deformed laminations as described for ball and pillow structures (Boggs 1987, 164-165; Cojan, Renard 1997, 125-126). Whether the observed structures should be interpreted as load casts or ball and pillow structures, we can conclude that they are most probably linked to liquefied sediment under rapid sedimentation conditions. Liquefied sediment conditions are likely to occur at the bottom of dune slacks.

Micromorphological observations (tab. D.6) reveal two different types of the sedimentary structures within these deformation structures (H4): Type I, where the brown bands represent quartz coated with amorphous dusty brown to limpid strong brown organic fine mass coatings up to 65µm thick, eventually deposited on dusty brown clay coatings up to 20µm thick, and, Type II, where the brown bands correspond to quartz coated with poorly to well-oriented dusty to slightly dusty strong brown clay coatings up to 65µm thick, or infillings. From the lamination following the sedimentary structure morphology in the underlying H5, we can conclude that pressure, either external or internal, played a crucial role in the development of these structures.

The Late Glacial soil is visibly connected with the overlying Podzolised Soil by a vertical vein (H2_VR). This vein consists of quartz grains coated with, at the top few (5 - 10%), increasing towards the base to frequent (10 - 20%), or even locally common (30 - 40%) limpid (cracked) strong brown organic fine mass coatings up to 50µm thick or infillings (cf tab. D.45). The vein ends in a horizontal deposit (H2_HR) containing about the same type and amount of amorphous organic matter coatings or infillings. The $Al_o+{}^1/_2Fe_o$-content in H2 is 0.22% (tab. D.42), which probably indicates presence of aluminium and iron within these organic fine mass coatings. The amount of free iron is low (Fe_d_H2 < 0.02%), which indicates that migration driven by oxydo-reduc tion processes, as e.g. in case of a placic horizon as suggested by Lapen and Wang (1999), is not probable. As in P6, H1 and H3, the morphology resembles that of dopplerite. Next to this hypothesis, Driessen and Dudal (1991, 65-74) mention signs of beginning podzolisation with accumulation of iron- and aluminium-humus complexes in thin lamellae in weakly developed soils with coarse texture (cf Arenosols) of the temperate zone. Whether the illuviated organic matter is the result of podzolisation or not, it is clear that the latter horizon (H2_HR) is diagenetic and as a consequence doesn't belong to the Late Glacial soil. The similarity index diagram (tab. D.7) indeed indicates H2, eventually together with H3, as a separate unit as opposed to (1) H4 to H6 and (2) H1a to H1b that form other separate soil horizon units. Nevertheless, H3 shows a remarkable increase in clay and silt content with respect to H2 (tab. D.7), probably hampering further migration of the organic matter and aluminium towards deeper horizons. When compared to H1a and H1b, H2 shows an increase in very fine sand, silt and clay content (tab. D.7), which could, as already mentioned for P11 and P1, be favourable for organic fine mass accumulation too. Dosskey and Bertsch (1997) furthermore indicate low organic matter content, presumably maintained by rapid microbial decomposition, combined with a strongly acidic pH, as conditions maximising the sorptive capacity of limited clay surfaces. Interaction between limited clay surfaces and dissolved organic matter in the percolating soil solution can thus be maximised in E-horizons of sandy soils. The coarse, quartz-rich groundmass of H2 seems to be a comparable environment, favourable for organic matter retention.

In the Bh-horizons of the Podzolised Soil (H1a - H1b) above the Late Glacial soil, very few (< 2%) vertically oriented carbonate link coatings and infillings (cf tab. D.48) were locally observed; the coatings are up to 50µm thick and are composed of clay-size carbonate. Seen their vertical orientation, it is likely that these crystalline pedofeatures represent *in situ* secondary carbonate deposits along former roots and their particular rhizosphere microecosystem (Becze-Deák *et alii* 1997). This excludes a primary origin and deposition after aeolian transport. The pH values within these horizons are slightly over neutral (pH_H1a = 7.0, pH_H1b = 7.1), which fits with carbonate formation but not with podzolisation. Carbonate deposition thus must have been diagenetic to podzolisation. Shift from an acid to a neutral pH must have come under influence of the overlying calcareous flood sediments. The necessary ions to form carbonate could have been provided by leaching from the overlying calcareous flood sediments or eventually from the underlying marine, calcareous substratum (Formations of Kattendijk and Kallo).

Profile 4 (fig. 23, tab. 7, D.5)

The top of the Late Glacial soil is situated at 1.22m TAW.

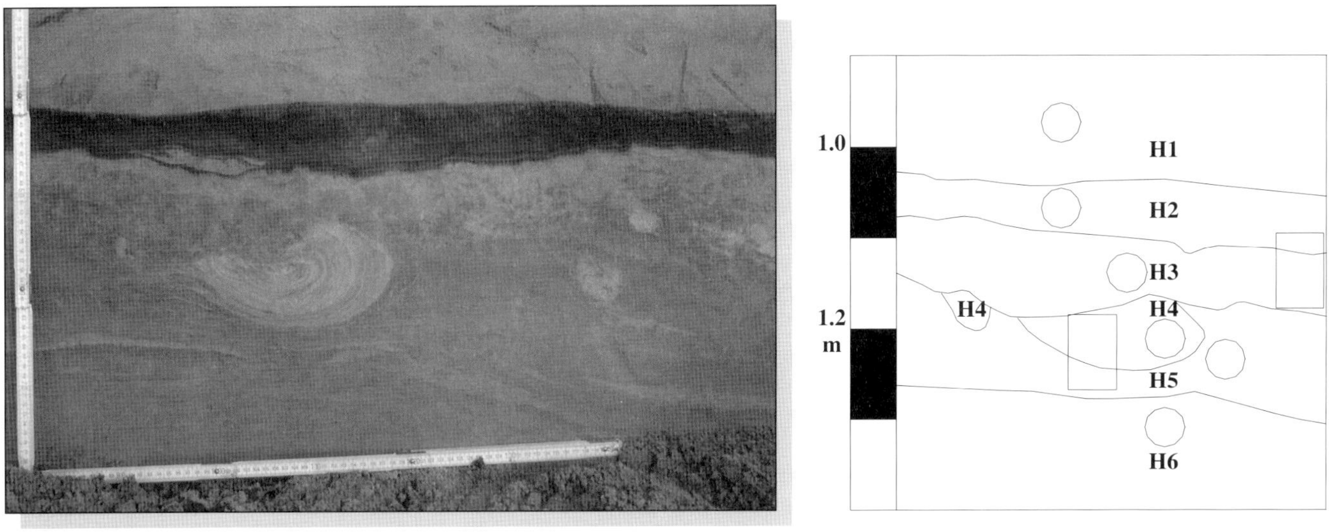

Figure 23: Verrebroek "Dok 2"; profile 4, photographic and schematic representation.

Horizon number	Symbol	Horizon depth (cm)	Profile description
H1		(90)-103/105	Light grey matrix; sand
H2		103/105-108/112	Dark - very dark brown matrix; sand coated with organic matter?
H3		108/112-113/129	Mixture of white and brown - greyish brown matrix; sand; disturbed due to liquefied sediment conditions?
H4		115/119-120/124	Alternating ellipsoidal laminae of major white and minor brown - light brown matrix; sand
H5		113/124-126/130	Brown - greyish brown matrix; sand; stratification following morphology of H4
H6		126/130-(140)	Pale brown matrix; sand; stratified

Table 7: Verrebroek "Dok 2"; profile 4, profile description.

For this profile, only macro- and micromorphological observations have been performed. As in P5, also this profile features spherical- to ellipsoidal-shaped structures below the Late Glacial soil. These structures are composed of alternating laminae of white quartz sand-size grains and slightly humiferous sand (H4) (fig. 23) that disappear when scraping less than 0.10m of the section. The laminations in the sandy sediment underneath follow the spherical structure morphology. As P5 showed comparable structures discussed in detail previously, it is referred to as the reference profile for P4. Micromorphological observations (tab. D.5) in P4 reveal that the brown bands in H4 have dusty to very dusty brown organic fine mass coatings. They are of the same type as those observed in the overlying H3, and differ from the two types observed in P5. Macro- as well as micromorphology point here towards unidirectional flow from the overlying H3 into the underlying H5, however with concomitant sorting of the sediment resulting in the laminated sedimentary structure (H4). The lamination in the underlying H5 is again following the structure morphology, which confirms pressure exertion during structure formation. In accordance with the literature references given for P5 (Boggs 1987; Bjørlykke 1989; Cojan, Renard 1997) on ball and pillow structures and load casts and given the overall structure morphology, we can assume here as well that the structure developed under pressure during rapid sedimentation in water-saturated or liquefied sediment conditions.

2.1.3.8. | Brunified Soil

The soil evolution of the beginning of the Holocene, when Mesolithic hunter-gatherers settled periodically in the area (cf 2.1.4.6. at Verrebroek "Dok 1"), could not be precisely detected. Podzolisation processes active after the Mesolithic occupations erased in the study area the characteristics of the soil types that were present previously. However, evidences of the so-called original profile are preserved in particular spots. It concerns areas with sediments that had a too high clay content in order to enable the podzolisation process and also spots with special features corresponding to tree-falls that were deep enough and of a type (Langohr 1993) that enabled preservation of at least part of the soil horizonation present at the moment of the tree-fall (cf 3.3.1.2.). These profiles show an organic-rich surface horizon, lack an eluvial E-horizon and display a homogeneous brown, bioturbated B-horizon. Environmental conditions must have been in favour of poorly leaching conditions in order to prevent the podzolisation process within the quartzitic sand with low silt and clay content and low in weatherable minerals. Indeed, temperature increased markedly towards the Preboreal (Atkinson *et alii* 1987; Bohncke *et alii* 1987; Verbruggen *et alii* 1996) and the denser vegetation resulted in a relatively high evapotranspiration (Cleveringa *et alii* 1988). Precipitation must have been lower than, or more or less in balance with the evapotranspiration. The latter is confirmed for the first half of the Holocene (10000-5000 BP) by Verbruggen *et alii* (1991).

<u>Profile 3, H0 - H9</u> (fig. 20, tab. 4, D.3, D.4, D.39, D.42)

In this profile, H10 to H15 comprise the Late Glacial soil, whereas H0 to H9 shows Holocene development of a Brunified Soil (H0 - H9) in the redistributed aeolian sands. At this point, we focus on Brunified Soil development. It is important to note that this soil development probably corresponds to the soil type present when Mesolithic hunter-gatherers periodically settled in the vicinity (cf 2.1.4.).

The texture in the parent material has a median value around 230µm (tab. D.39), coarser than in the Upper Pleniglacial coversands in which the Late Glacial soils developed (cf 2.1.3.2.). As the Brunified Soil is located above the Late Glacial soil, it developed in coversands that were remobilised over short distances during the Dryas stadials. The groundmass in H7 to H0 has a better sorting as compared to H14 to H10 and doesn't contain the compound quartz, glauconite and dusty clay fragments observed in H14. The similarity index diagram of H7 to H0 (tab. D.4) reveals two separate units with intra-unit indices above 85 and inter-unit indices between 85 and 60. This pattern indicates an important particle size distribution difference between

(i) H7 to H4 on the one hand and (ii) H3 to H0 on the other hand. H4 occupies an intermediary position but nevertheless rather belongs to the lower unit. H0 to H3 have higher clay to very fine sand contents than the underlying H4 to H7. It is worthwhile finding out whether this particle size distribution difference is due to pedogenesis or lithostratigraphic in origin.

Micromorphological observations (tab. D.3) reveal at first sight that H1 to H3 don't have clay in their groundmass, nor as pedofeatures, but have frequent (2 - 20%) amorphous dark brown, probably humified, organic fine mass coatings to infillings that have the morphology of aged excrements (cf tab. D.46). As mentioned higher, 30%-H_2O_2-treatment prior to particle size analysis possibly didn't destroy resistant forms of organic matter, such as humin, a strongly humified organic matter component (Mathieu, Pieltain 1997, 21). Hence, the organic fine mass coatings to infillings present in these horizons might to some degree account for the higher clay fraction (2.5 - 7% in H0 - H3, versus 0 - 1.7% in H4 - H7) retrieved in the particle size analysis in H0 to H3. Nevertheless, the clay fraction alone cannot be responsible for the observed particle size distribution shift; when comparing both units also the silt and very fine sand fraction increase strongly up to double the value in H7 to H4 (tab. D.39). It should additionally be noted here that the only dispersing agent used during the laser diffraction analysis was filtered tap water, possibly not strong enough to break the strongly humified aggregates apart. The mentioned dusty organic fine mass coatings and infillings observed in thin section indeed measure up to 80, respectively 160µm. Hence, it is quite probable that they represent part of the higher clay (< 3.9µm) and silt (< 62.5µm) fractions present in H0 to H3 as opposed to H4 to H7, causing in the mean time an important shift in the particle size distribution from mainly medium and fine sand in the lower unit towards medium sand up to clay in the higher unit. Last but not least, clay can be reworked within the amorphous dusty organic fine mass, lacking any orientation pattern and thus becoming undetectable as mineral clay in thin section. With respect to this, Barratt (1969) indeed mentions different constituents of microscopic organic soil components: (i) humiskel, i.e. mostly undecomposed organic residues, (ii) lithiskel, i.e. mineral grains and fragments, (iii) humicol, i.e. decomposed organic colloids, (iv) mullicol, i.e. a mixture of organic and clay colloids and (v) argillicol, i.e. clay colloids, thus mentioning clay colloid incorporation in the organic fraction. Duchaufour (1982, 270-305) mentions the importance of hydroxides (Fe and Al) to fix organic matter to clays in the form of salt complexes in which the mineral and organic constituents remain extractable. Humin, the insoluble fraction of organic matter, on the other hand forms adsorption complexes with the iron hydrates in which both constituents are resistant to extraction reagents. Both types of complex cement the aggregates in a mull humus type, whose formation is furthermore favoured by earthworms.

The higher quantities in the clay to very fine sand fractions observed in H0 to H3, recognised as H+E-, A-, E- and ABbi-horizons respectively, are probably evidences of brunification (Duchaufour 1982, 270-305). Clay increases from the bottom to the surface but however shows a drop in H2, in the field recognised as a weakly developed eluvial E-horizon. Active iron, freed by weathering and bound to fine clay is central to the brunification process; latter iron is responsible for the formation of the forest mull humus form, for it forms bridges between clay and humus molecules within clay-humic aggregates. This apparently happened here, where the clay fraction, at its maximum produce at the surface due to weathering from the surface, was intensely mixed by faunal activity with organic fine mass. The formed complexes were observed as dark brown organic fine mass coatings and infillings and recognised as aged excrements in thin section (tab. D.3, tab. D.46). Concluding, there is a fair chance that the observed texture discontinuity between (i) H7 to H4 and (ii) H3 to H0 is pedogenic, rather than lithostratigraphic in origin.

Evidences of alteration as defined for cambic Bw-horizon development (WRB 1998, 28-29) (cf Appendix C. Methods), are observed in H2 and H3 that show neither a redder hue nor a stronger chroma but have a higher clay content than the underlying H4 to H6. The same evidences of alteration can be observed in H5 and H6 of P1 at Doel "Deurganckdok-sector B" when compared to H7, a BC-horizon (cf 2.1.5.2.); they additionally have a redder hue too. The assumed Bw.bi-horizons, H4 to H6, have higher chroma than the overlying horizons H1 to H3, thus showing evidence of alteration as well.

As compared to the Podzolised Soil profiles studied at Verrebroek "Dok 1" (cf 2.1.4.7., P16, P17), the iron (Fe) content, either extracted with dithionite (so-called free iron) or oxalate (so-called amorphous iron) in H2 and H3 of P3 is relatively lower, which fits with the higher formulated absence of appreciable quantities of aluminium and/or iron compounds. The $Al_o+{}^1/_2\,Fe_o$-content is nevertheless below 0.20%, as opposed to the Bh(.s)-horizons of the Podzolised Soils at Verrebroek "Dok 1" (cf 2.1.4.7., P16, P17) that have $Al_o+{}^1/_2\,Fe_o$-contents above 0.20%.

The underlying assumed Bw.bi-horizons, H4 to H6, have very few (< 2%) thin (< 5µm) slightly dusty to dusty clay coatings, indicating clay migration. The overall clay content however is lower than in the overlying E and Bbi-horizons, H2 and H3. H2, recognised as an eluvial horizon, has the lowest clay content; aluminium and exchangeable bases are also at a minimum here. The Fe_o- and Fe_d-contents are also very low in these horizons and are even below the detectable limits of 0.01 and 0.02% respectively. Aluminium (Al) and iron (Fe) possibly leached out which would mean that this profile is in a recharge position with fluctuating groundwater table. Nevertheless we observe a relative peak in Al-content in H4; this peak coincides with the highest quantities of limpid to slightly dusty organic fine mass coatings, possibly metal (Al)-humus compounds indicating weak or beginning podzolisation (cf tab. D.46).

Nevertheless, even though the surface A-, eluvial E-, bioturbated ABbi- and assumed Bw.bi-horizons have evidences of alteration, they don't meet the criteria for a diagnostic cambic horizon proposed by the FAO or USDA classification systems because of a too coarse texture.

Concluding, H1 to H3 seem to correspond to a mull humus type with clay-humic aggregates (aged excrements) as described by Duchaufour (1982, 270-305). Latter complexes are typical for the brunification process and prevent the presence of true mobile complexes that are required for the podzolisation process (Duchaufour 1982, 270-305). These horizons indeed have higher clay content when compared to the underlying horizons (H4 - H6). The assumed Bw.bi-horizons (H4 - H6) show weak evidence of clay illuviation. However, they have lower clay content than the overlying surface (H1), the weak eluvial (H2) and the biologically active Bbi (H3)-horizons. Clay movement must have been slowed down by environmental factors. This can amongst others happen when weathering produces more clay near the surface than in depth (Duchaufour 1982, 270-305). This profile indeed shows increasing clay content from bottom to surface (tab. D.4), except for a drop in H2, recognised as a weakly developed eluvial E-horizon. Duchaufour (1982, 270-305) considers clay migration as a secondary process superposed on the process of brunification. It is questioned whether frost shattering, a particular type of physical weathering due to frost action, or some biological activity, as e.g. mycorrhizal activity, are responsible for the relatively high clay content in the surface horizon. The few to very few limpid to slightly dusty organic fine mass coatings (cf tab. D.46) observed in decreasing quantities from H4 to H6, coinciding with decreasing quantities of Al, possibly represent metal (Al)-humus compounds. Latter probably indicate a weak, beginning tendency of podzolisation. Dimbleby (1962, 44) parallel mentions onset of podzolisation in what previously had been a Brown Forest Soil in British heathlands.

Taking all mentioned characteristics into account, we conclude for a Brunified Soil with weak evidences of clay migration and weak or starting characteristics of podzolisation. From this point of view the evidences of alteration, corresponding to the cambic horizon concept (WRB 1998) and a Brunified Soil development, only represent a first step towards a more strongly expressed soil development. According to Duchaufour's classification system for Temperate Brown Soils (1982, 273-285), this profile shows the typical characteristics of an Ochric Brown Soil, an intergrade between Brunified and Podzolised Soils, with uniformly distributed iron and free aluminium that has begun to migrate. Nevertheless, this profile shows clay illuviation at micromorphological level. Most probably, early eluviation of the little amount of clay and iron present opened the way for podzolisation. Duchaufour (1982, 300) indicates such soils as possible Podzolic Lessived Soils. However, clay migration in P3 is not prominent enough to conclude for this soil type.

The set of characteristics observed in H0 to H6 corresponds quite well to the description of Arenosols in loamy sands of the temperate zone (WRB 1998). The A-, AB- and (A)Bw.bi-horizons in P3 (H0 - H3) have few to frequent (2 - 30%) amorphous dusty dark brown organic fine mass coatings (< 80µm) and infillings (< 160µm), probably coprogenic in origin, resembling aged excrements (cf tab. D.46). The humus type as observed in thin section rather corresponds to a mull-like moder (Kubiena 1970, 224) that shows considerable aggregate formation but with incomplete decomposition of the plant residues. H1 and H2 indeed have very few (< 2%) organ residue of roots and sclerotia, the latter being fungal hyphae of a fungus type living in symbiosis with nearly all trees in forests of boreal and temperate zones (Bal 1973, 53). Mycorrhizal associations between roots and fungi, such as e.g. *Cenococcum sp.*, thus produce sclerotia. Latter are quite frequent in some soils, usually in the lower part or the organic horizons in contact with the mineral material, or they may be preserved in peat (Fitzpatrick 1993, 219-220). The Bw.bi- (H4 - H6) horizons underneath show, next to the above-mentioned very few thin clay coatings, a strong brown matrix and have few (2 - 10%) in H4 to very few (< 2%) in H5 and H6, thin (< 15µm), limpid to slightly dusty organic fine mass coatings and infillings. As mentioned above, iron contents are however too low to assume iron-humus coatings here.

In general, Temperate Brown Soils in equilibrium with their environment have a rapid cycling of nutritive elements (nitrogen, phosphorus and bases) resulting from the presence of a biologically active mull humus type (Duchaufour 1982, 270-305). The neutral to moderately alkaline pH in this profile, probably the result of diagenesis after calcareous flood sediment deposition, however differs from the moderately acid pH commonly measured in Temperate Brown Soils (Duchaufour 1982, 207-305). The C/N-ratio has an almost constant value of 30 from H0 to H6, which is relatively high when compared to C/N-ratios between 12 and 15 commonly analysed in Temperate Brown Soils (Duchaufour 1982, 270-305). C/N-ratios in this profile thus point towards a slower turnover when compared to Temperate Brown Soils in equilibrium with their environment. Apart from the surface H-horizon H0, the highest total phosphate contents occur in H3 and H4 ($P_2O_5t\_H3 = 822 mg\ kg^{-1}$, $P_2O_5t\_H4 = 516 mg\ kg^{-1}$), coinciding with the lowest inorganic over organic phosphorus ratio (Pi/Po_H3 - H4 = 0.3). The highest organic phosphate content occurs in H3, the most pronounced Bbi-horizon. Phosphate is also in the other horizons mainly present as organic phosphate (Pi/Po_H0 - H7 < 1). H3 and H4 moreover have relatively high organic matter content of 2 and 2.5% and low C/Po-ratios (C/Po_H3 - H4 = 54 - 67) as compared to the overlying horizons. The coprogenic organic aggegates observed in H0 to H3 might to a large extent account for their high organic over inorganic phosphates ratio.

2.1.3.9. | Podzolised Soil

Podzolised Soils at Verrebroek "Dok 2" and "Dok 1" have comparable morphology. The topic is extensively discussed under Verrebroek "Dok 1" where development of different Podzolised Soil types is discussed (cf 2.1.4.7., 2.1.4.8.).

2.1.4. | Verrebroek "Dok 1"

2.1.4.1. | Pedo-lithostratigraphy

A schematic description of the pedo-lithostratigraphical units at Verrebroek "Dok 1" from bottom upwards gives the following sequence (fig. 24).

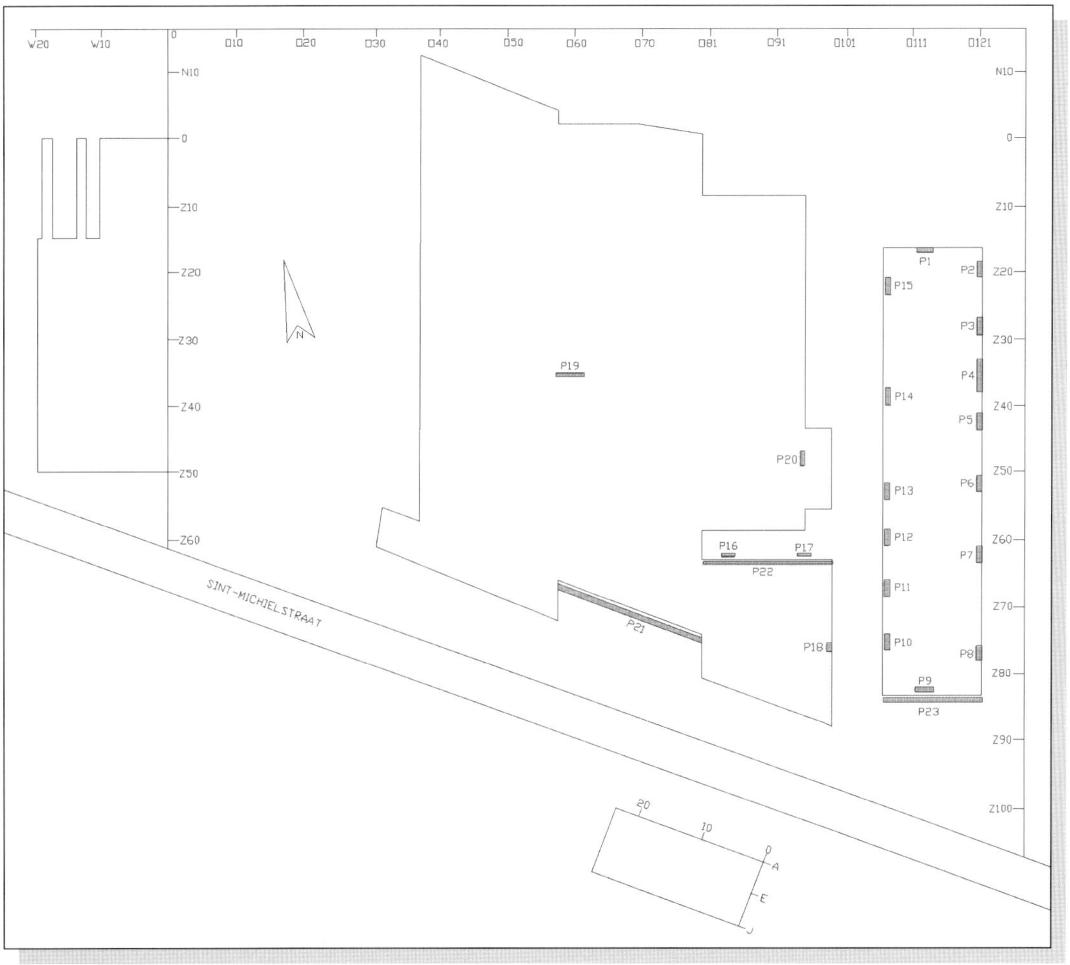

Figure 24: Verrebroek "Dok 1"; general excavation map marking the position of the investigated profiles.

1. Upper Pleniglacial coversands

2. Late Glacial humiferous soil, 0.10-0.15m thick, showing:
   - locally charcoal, mostly powdery;
   - no, or very little bioturbation, except for the presence of roots up to several centimetres diameter, linked to an overlying younger Podzolised Soil (fig. 26);
   - an abrupt somewhat irregular lower boundary;
   - local particular feature synchronous to the Late Glacial soil, probably reflecting an *in situ* fire event linked to human occupation.

3. Horizon(s), consisting of accumulation bands of very mobile organic matter and/or clay substances, with abrupt upper and lower boundaries (fig. 25, 26). Precipitation probably occurred at the level of a texture jump or (temporary) water table. Most of this material accumulated on top of the Late Glacial soil, forming in most profiles a continuous horizon of about 0.05m thickness.

4. Redistributed aeolian sands, deposited on top of the Late Glacial soil, where present. They hey lack any visible stratification here.

5. Traces of a Brunified Soil with a colour Bw-horizon (fig. 144, 145), only observed in particular positions, such as in tree-fall structures, with from top:
   - no E-horizon;
   - a homogeneous brown (with a higher chroma than the underlying C-horizon), bioturbated Bw.bi-horizon.

6. Imperfectly to poorly drained Podzolised Soil, up to 0.5m thick, with from top:
   - a black H-horizon, with a gradual boundary to the overlying peat;
   - an A- or H+E-horizon, weakly developed, composed of black organic pellets (H) and bleached quartz sand grains (E), also called "salt and pepper" (Jabiol *et alii* 1995);
   - a very light-coloured E-horizon, that may be absent in the wetter palaeo-landscape positions;
   - a dark brown Bh.s-horizon with locally patchy strong decomposition of the organic matter due to diagenetic microbiological activity (fig. 31).

7. Peat, up to 0.75m thick, containing non-transported wood fragments and only present in palaeo-depression positions.

8. Cultivation horizon, 0.10-0.20m thick, with 2 facies:
   - on the ridges, without peat cover: humiferous sand (fig. 28), with local spade marks at the base;
   - in the depression positions: peat anthropogenically mixed with sand (fig. 34).

9. Calcareous flood clay and sand (fig. 34) with variable thickness (0.4-1.5m), showing:
   - a weak stratification at the base;
   - a strong bioturbation in the upper 1.0-1.3m;
   - presence of vertically oriented *in situ* shells and shell fragments;
   - presence of few brick fragments;
   - a high porosity and in general a massive pedality, except for the local development of a colour and structure Bw-horizon with subangular blocky structure in the upper part;
   - a present-day plough horizon, which is homogeneous and has a sharp lower boundary, having already some decarbonatation.

The local differences and aspects of the above described soils and soil horizons are represented by different facies, reported in the set of soil profile descriptions. The different soil profiles are commented following the chronostratigraphical sequence.

Throughout the text, regular reference is made to the pedo-lithostratigraphical sequence at the Mesolithic to Early Neolithic site of the "Hoge Vaart" (Flevoland, The Netherlands) (Hogestijn, Peeters 2001), as it has a comparable soil and environment genesis as the Verrebroek "Dok 1" site.

2.1.4.2. | Upper Pleniglacial coversands

Profile 14, H18 - H20 (fig. 25, tab. 8, D.17, D.18, D.40, D.43)

In this profile, the Upper Pleniglacial coversands (H18 - H20) were studied.

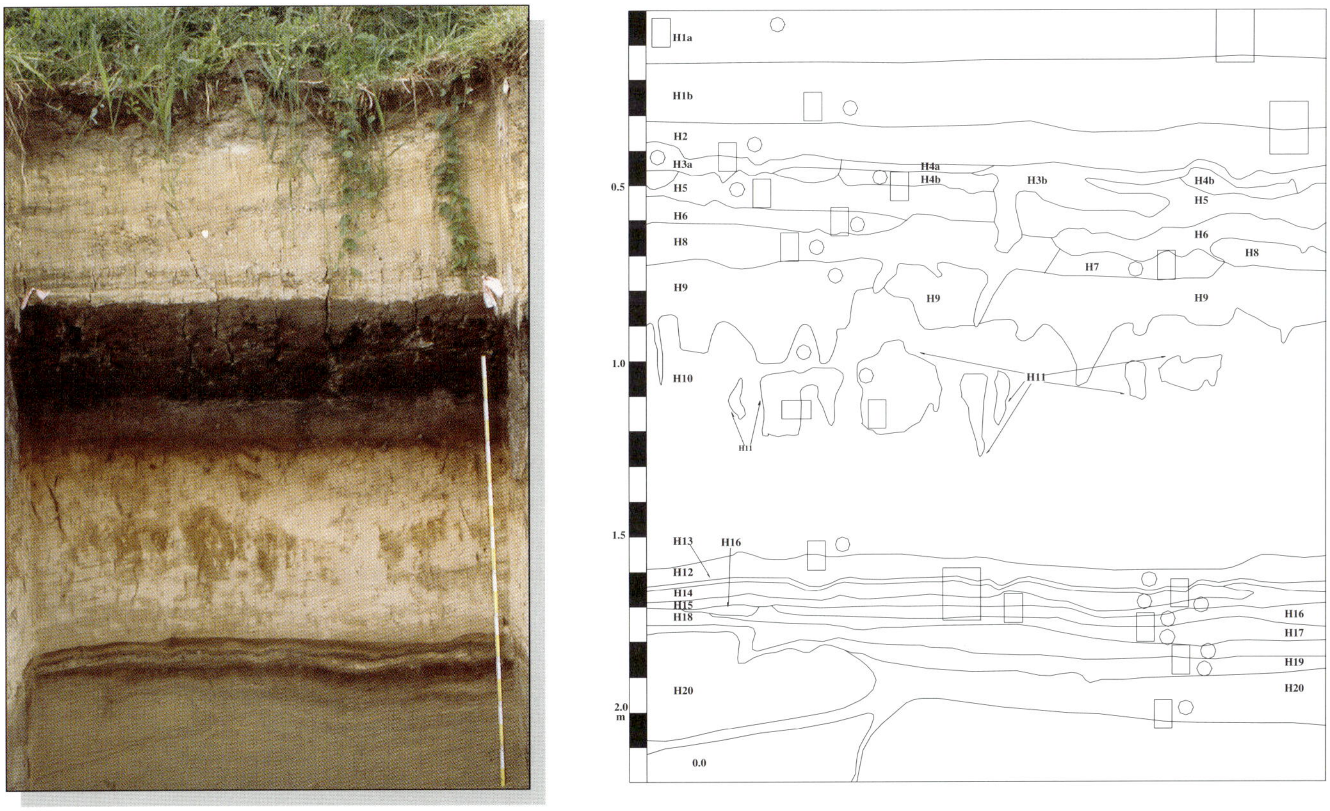

Figure 25: Verrebroek "Dok 1"; profile 14, photographic and schematic representation.

The parent material (C-horizon, H20) shows a horizontal to oblique stratification to lamination and has a wavy morphology. It consists of dominant (> 50%) poorly sorted densely packed subrounded (polycrystalline) quartz, very few (2 - 5%) rounded glauconite that is partly, mainly physically weathered and very few (< 2%) subrounded to rounded chert, feldspars and other mica minerals; some laminae are however well sorted. The standard deviation of the particle size distribution (tab. D.40) (SD_H20 = 88.2µm) accordingly points toward a very poorly sorted sediment. The Upper Pleniglacial coversands at Verrebroek "Dok 1" have a median particle size of 104 µm (tab. D.40), and are thus slightly finer than at Verrebroek "Dok 2" (cf 2.1.3.2.).

Horizon number	Symbol	Horizon depth (cm)	Profile description
H1a	F	0-12/15	Black matrix; decomposed organic matter; admixture of sand; 20-30% granular pedality; no ripening structure?; root rust; most probably in faunal bio-galleries
H1b	F	12/15-32/35	Dark brown - black matrix; peat; distinct wood fragments
H2	H	32/35-37/47	Black matrix; colloidal organic matter; mor or Podzol Soil
H3a	HA?	37/42-45/49	Dark brown - black matrix
H3b	H?	42/56-42/68	Very dark grey matrix; decomposed root
H4a	H+E1	42/43-44	Local presence; transition H2-H4b?
H4b	H+E2	43/49-48/52	Grey - dark grey matrix; "pepper and salt": organic excrements and quartz sand
H5	E+H	47/53-52/65	Greyish brown - brown matrix; "salt and pepper": quartz sand and organic excrements
H6	Bh1	52/62-61/72	Brown - dark brown matrix; sand; best developed Bh of Podzol
H7	Bh2	68/71-61/77	Local presence; light yellowish brown matrix; sand; ice segregation lines
H8	Bh3	60/69-70/90	Dark yellowish brown matrix; sand; common dark brown decomposed roots, probably associated with Podzol
H9	BC	71/76-79/107	Pale-light yellow matrix; sand; common dark brown decomposed roots, probably associated with Podzol
H10	C	79/107-154/159	Light grey matrix; sand
H11	Bfe/s?	93/105-106/128	Olive yellow matrix; sand; firm (consistence); accumulation of Fe or P or organic matter?
H12		154/160-162/164	Brown - dark brown matrix; organic matter coatings?
H13	H?	162/164-163/166	Dark grey matrix; charcoal powder
H14	E+H?	163/166-162/167	Pinkish grey - brown matrix; sand; perturbed (liquefied sediment conditions?)
H15	E?	163/171-169/172	Olive black matrix; sand
H16	BH?	169/172-170/176	Dark brown - black matrix; humiferous; slightly sticky?; charcoal fragments
H17	B	170/176-172/182	Light grey matrix; clay or silt-rich sand?
H18	BC1?	170/181-175/185	Bluish grey matrix; sand
H19	BC2?	175/185-176/190	Bluish grey matrix; clay
H20	C	176/190-(220)	Bluish grey matrix; sand; stratified/laminated; local clay migration/illuvation?

Table 8: Verrebroek "Dok 1"; profile 14, profile description.

The grains have thin (< 5μm) slightly dusty light green (PPL) (cf tab. D.49), possibly glauconitic clay coatings up to link cappings. These cappings are illuvial features as they are oriented parallel to the grain surface. The glauconitic clay is the result of mechanical breakdown of the sand-size glauconite grains (Van Ranst, De Coninck 1983).

Additional very small quantities (2 - 5%) of very dusty to impure dark brown clay coatings or infillings are present. Carbonised organic particles up to 150μm were observed as well. Latter features, together with the wavy lamination morphology, argument for liquefied sediment conditions pointing towards fluviatile sediment input, rather than simple aeolian input. De Moor (1963, 345) indeed mentions two coversand transport and sedimentation mechanisms: niveo-aeolian and niveo-fluviatile. Under periglacial conditions, meltwater on a permanently frozen underground can induce intense runoff and (micro-)mudstream development, giving rise to mentioned dusty to impure infillings within the sediment. The fluviatile nature of the sediment together with its poor sorting might furthermore underline its Upper Pleniglacial origin under periglacial conditions. Alternatively, fluviatile sediment input might also reflect the snow-rich climates of the last ice ages (Gullentops, Wouters 1996, 68-69).

Slightly birefringent organ, probably root residue and amorphous brown plant fragments are present in very small quantities (< 2%). The pH is close to neutral (pH_H20 = 6.92).

The micromorphological characteristics of H18 to H19, in the field considered to be transitional BC-horizons, are comparable to those of H20. The very dusty dark brown clay coatings are however absent here. Few (5 - 10%) amorphous, very dusty organic fine mass is present in these horizons. It possibly concerns sedimentary, detrital organic matter, i.e. particulate and dissolved organic matter, deposited in a still wet, eventually subaquatic system in which the Late Glacial soil developed afterwards. Both the fluviatile sediment character and the presence of a well-developed Late Glacial soil point towards a (local) palaeo-depression position.

2.1.4.3. | Late Glacial soil

The redistribution or remobilisation of the Upper Pleniglacial coversands during the Late Glacial occurred in a landscape of dune ridges and dune slacks. Nearly permanent ponds or lakes were present at the level of the dune valleys. Within these ponds a humiferous, locally peaty, soil developed. The interpretation of this soil as subaqueous is based on the nearly complete absence of bioturbation by burrowing animals and roots and the abrupt to rather abrupt lower boundary of this soil.

<u>Profile 9, H11 - H17, H19</u> (fig. 26, tab. 9, D.13, D.14, D.40, D.43)

In this profile, the Late Glacial soil (H12 - H17) present at Verrebroek "Dok 1" was studied in detail.

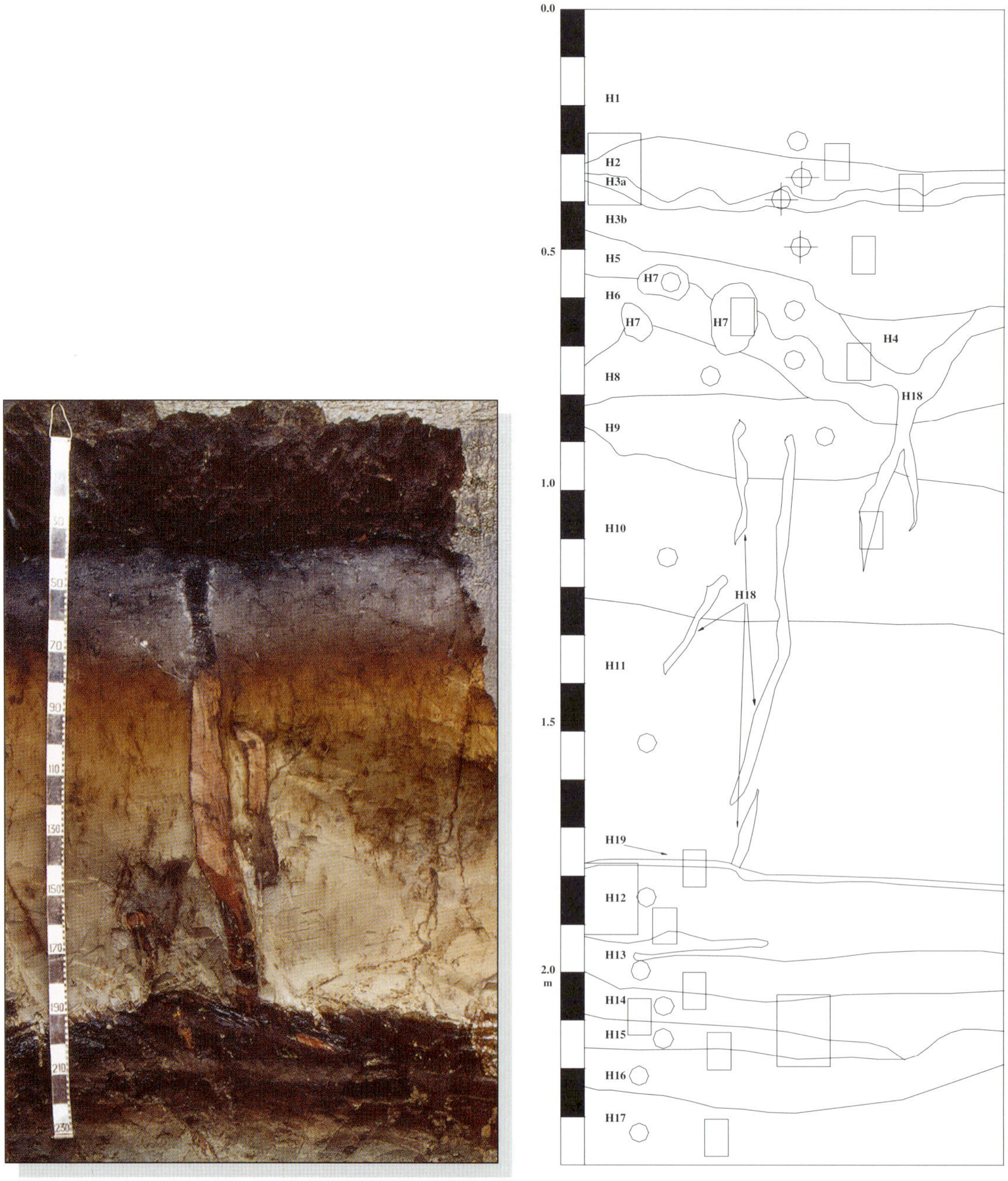

Figure 26: Verrebroek "Dok 1"; profile 9, photographic and schematic representation.

Horizon number	Symbol	Horizon depth (cm)	Profile description
H1	H-F?	0-27/33	Dark brown - black matrix; peat; distinct wood and leaf fragments (fibric?)
H2	H(+E)	27/33-33/40	Black matrix; mor or Podzol Soil
H3a	H+E	33/40-37/42	Dark grey matrix; colloidal organic matter and quartz sand
H3b	E(+H)	37/42-46/65	Brownish grey - grey matrix; sand
H4		62/63-62/76	Local presence; very dark grey - black matrix; sand
H5	Bh1	46/76-52/78	Very dark brown - black matrix; sand; best developed Bh of Podzol
H6	Bh2	55/78-66/87	Dark brown - strong brown matrix; sand
H7	Bh.ir/s?	52/61-60/72	Brownish yellow matrix; sand; Fe-concretions?
H8	Bh3	66/80-79/82	Brown - dark brown matrix; sand
H9	BC	79/82-88/100	Brown matrix; sand
H10	CB	88/100-122/129	Light grey matrix; sand
H11	C	122/129-177/182	Light grey - grey matrix; few strong brown Fe-mottles, linked to root rust
H12	H-F?	176/183-191/198	Dark brown - black matrix; colloidal organic matter; subaqueous deposit?; stratified; distinct vegetative fragments (fibric?)
H13	E+H	191/198-200/206	Pale brown matrix; sand; subaqueous, aeolian deposit?
H14	H or A?	200/206-208/218	Dark brown - very dark brown matrix; abundant colloidal organic matter
H15	Bh?	208/218-215/219	Dark brown matrix; humiferous sand?
H16	BC	215/219-219/228	Olive yellow matrix; sand?
H17	Cr	219/228-(240)	Bluish grey matrix; sand; stratified/laminated
H18	H	87/89-91/118	Root, partly decomposed
H19		177/182-176/183	Very dark grey matrix; dopplerite-like organic matter coatings?

Table 9: Verrebroek "Dok 1"; profile 9, profile description.

The parent material (H17) is stratified to laminated and has a somewhat wavy morphology. It consists of dominant (> 50%) poorly sorted to unsorted densely packed subrounded (polycrystalline) quartz, very few (2 - 10%) rounded glauconite that is partly, mainly physically weathered and very few (< 2%) subrounded to rounded chert, feldspars and other mica minerals. The standard deviation of the particle size distribution (tab. D.40) (SD_H17 = 69.3µm) accordingly points toward a very poorly sorted sediment. The parent material has a median particle size of 113µm (tab. D.40), and is thus slightly finer than at Verrebroek "Dok 2" (cf 2.1.3.2.).

As in P14, the grains have thin (< 5µm) slightly dusty light green (PPL), possibly glauconitic clay coatings (cf tab. D.49), formed after mechanical breakdown of the sand-size glauconite grains (Van Ranst, De Coninck 1983).

Additionally very few (2 - 5%) slightly to very dusty strong to dark brown clay coatings up to 50µm or infillings and infilling fragments are present. Also here, these coatings together with the earlier mentioned wavy stratification to lamination morphology, argument for liquefied sediment conditions pointing towards fluvio (-aeolian) sediment input, rather than pure aeolian input.

Both the fluviatile character of the sediment and the presence of a well-developed Late Glacial soil point towards a (local) palaeo-depression topography.

Special attention should be paid to the presence of very few (2 - 5%) clay-size carbonate coatings and infillings and (spherulitic) carbonate (either calcite $CaCO_3$ or siderite $FeCO_3$) crystals up to 50µm (cf tab. D.48). The crystals probably represent calcified roots, commonly consisting of calcite elements whose mean size varies from 60 to 80µm and which retain the structure of the original cells of the root cortex (Jaillard *et alii* 1991). Bullock *et alii* (1985) and Lozet and Mathieu (1997) link the presence of siderite with reducing conditions in marshy soils, whereas Fitzpatrick (1993, 88-89) mentions that siderite occurs as a replacement substance in peat. Stoops (1983) observed siderite spherulites in bog-ores of the Belgian Campine. The bluish grey matrix of this horizon, as observed during fieldwork indeed points towards reducing conditions. Reduction and its effect on soil characteristics are commented in detail in Appendix B. Glossary. The densely packed mineral groundmass in H17 was a favourable environment for reduction under slightly aerobic to anaerobic conditions, once it was saturated or flooded with water. Organic matter was decomposed while iron (hydr)oxides were concomitantly reduced. The ferrous iron (Fe^{2+}) could subsequently react with carbonate (CO_3^{2-}) ions present in the soil solution to form siderite ($FeCO_3$). Free and amorphous iron contents are indeed relatively high (Fe_o = 0.08%, Fe_d = 0.12%). There is a fair chance that the observed carbonate coatings and crystals in thin section are actually siderite, but further testing on an uncovered thin section with HCl, by staining with alizarine red and potassium ferricyanide producing different colours when applied on calcite (red) and siderite (purple) (Morras 1973) or microprobe analysis (Scanning Electron Microscopy) is needed to confirm this. Given the reducing circumstances, we would expect a high pH here. Surprisingly, this horizon has a very low pH (pH_H17 = 4.0). Cation exchange capacity and exchangeable cations contents are low as well (CEC_H17 = 3.8cmol(+) kg^{-1}). We assume that by exposing the soil section to the air, the artificial oxidation of H17 had an impact on the soil reaction and hence might have resulted in an artificial low pH. Similar chemical reactions occur when bringing vivianite (iron phosphate) from a reducing into an oxidising environment; it turns from colourless or pale yellow to strong blue on oxidation (Fitzpatrick 1993, 92). Alternatively, the intensity of the reduction and the parallel reactions are much stronger in acid than in neutral or alkaline soils (Van Ranst 1991, 167-169). Hence, we can as well assume that reduction went on much further because of originally more acid conditions. Moreover, sandy soil horizons, more specifically the C-horizon containing almost pure quartz, have a low buffering capacity, implying that pH drops or increases relatively fast. Furthermore, Begon *et alii* (1996, 404-405) believe that under anaerobic conditions all sugars are decomposed to by-products such as alcohol and organic acids, which results in a lowering of the pH. However, as organic matter accumulation is only present in minor amounts within this C-horizon, this process must have had minor impact here.

The similarity index diagram (tab. D.14) confirms that H14, H15 and H16, recognised already in the field as a soil horizon sequence with H-, A-, Bh- and BC-horizons, indeed form a unit (similarity indices > 80). Also in these horizons, few (5 - 10% in H16) to frequent (20 - 30% in H14) carbonate clay-size coatings or infillings and silt-size crystals are present (cf tab. D.48). No carbonate coatings were observed in H15; very few (< 2%) pyrite (FeS_2) framboids next to decomposing plant fragments (cf tab. D.45) are however relevant indicators for wet conditions here as well. The pH in these horizons is slightly acid (pH_H14 - H16 = 6.0 - 6.5) which indicates that protons (H^+) released in the soil are mainly taken up and thus neutralised by organic matter and iron and/or aluminium hydroxides (Gisi *et alii* 1997, 128-130). Accordingly calcium is the dominant cation in the cation exchange complex (Gisi *et alii* 1997, 128-130). Furthermore cation exchange capacity and exchangeable cation contents are higher than in the reduced Cr-horizon (H17) and increase from bottom to top (H16 - H14), concomitantly with the organic matter content. An important part of the cation exchange capacity is indeed determined by the organic matter fraction. Furthermore organic matter is subject to mineralisation resulting in nutrient release. Decomposition of organic matter probably occurred here under more or less continued aerobic conditions posterior to Late Glacial Subaqueous Soil Formation. The groundmass has indeed a moderately dense (H16) to loose packing (H15 - H14), which permits more aeration. The fairly high ratio of inorganic over organic phosphate in H14 to H16 (Pi/Po = 2.4, 2.0, 36 in H14, H15, H16) is a further indication for mineralising conditions.

Still, macromorphologically, this soil unit (H16 - H14) shows little evidence of bioturbation and has a rather abrupt lower boundary. As at Verrebroek "Dok 2 ", these characteristics are in favour of subaqueous soil forming conditions.

The Late Glacial soil horizons (H16 - H14), especially H15 (Bh-horizon) and H14 (H- or A-horizon), have few (5 - 10% in H15) to frequent (20 - 30% in H14) amorphous dusty organic fine mass or fine mass particles. These particles might be coprogenic, i.e. remnants of animal excreta, especially when present as coatings or infillings or might correspond to dead plankton, floating biomass. Amorphous limpid cracked strong brown organic fine mass coatings, which remind of the humus gel coatings described in Dy soils, are absent. The chemical characteristics, such as the slightly acid to neutral pH, the very high cation exchange capacity and basic cations (Ca, Na, Mg, K) level, relative to the underlying parent material (H17) and also in absolute figures (tab. D.43), indicate a high nutrient status. Nevertheless, with respect to pH-values and cation contents, it should be mentioned that the Late Glacial soil horizons possibly reflect diagenetic influence of lateral or vertical input of calcium (Ca) cations from the overlying calcareous flood sediments. Soil C/N-ratio commonly varies between 8 and 30 but can reach up to 60 (Gisi *et alii* 1997, 197); in H16 - H14, values are around 20, which indicate a fair portion of nitrogen (N) as compared to carbon (C). All these characteristics, together with the above-mentioned minor presence of carbonate and pyrite, are in favour of a Gyttja soil type.

Furthermore the soil ESP (Exchangeable Sodium Percentage)- and SAR (Sodium Adsorption Ratio)-values in H16 to H14 is above the critical value of 15 and 13 respectively (Brady, Weil 1999; cf Appendix C. Methods) (ESP_H16 = 29%, ESP_H15 =

22%, ESP_H14 = 18%; SAR_H16 = 14, SAR_H15 = 17, SAR_H14 = 22), indicating that sodium (Na) occupies an important place on the exchange complex. Either (i) flooding by salt-containing (brackish) water, or (ii) input/discharge of salts from the underlying marine substratum (Tertiary Kattendijk and Kallo sands) through a fluctuating groundwater table, or (iii) extreme climatic conditions where evapotranspiration largely exceeds precipitation can explain this phenomenon. The second option is the most likely to have occurred within the Late Glacial environmental setting of Verrebroek "Dok".

The low similarity index between H14 and H13 (SI_H14xH13 = 42) indicates a clear texture break. The median particle size shifts from 68 to 147µm when moving from H14 to H13. The textural shift possibly corresponds to incoming new aeolian sediment. H13 has relatively low organic matter content when compared to H14 (OM_H14 = 21.6%, OM_H13 = 4.8%). Micromorphological observations (tab. D.13) show that it contains few (5 - 10%) non-birefringent organ residues and elongated amorphous dark brown plant fragments and very few (2 - 5%), amorphous, brown, organic fine mass. The elongated plant fragments might represent leaves or twigs fallen within standing water.

H12 shows distinct plant fragments, both in macro- and micromorphology, and has over 60% organic matter (OM_H12 = 62.6%); both characteristics let us conclude for a peat horizon.
The matrix of H12 has indeed a dark brown to black colour. The C/N-ratio is about 25, which indicates a fair decomposition of the soil organic matter. The coarse/fine related distribution in H12 is moreover open porphyric, which means that the coarse mass (10 - 25%) is embedded in the fine mass (40 - 50%); the fine mass corresponds to amorphous dusty brown organic fine mass (tab. D.13) (cf tab. D.45). These morphological characteristics correspond well with the strongly humified groundmass that Kubiena (1953, 103-104) mentions for Peat Gyttja. Within H12, very few (2 - 5%) pyrite framboids (FeS_2) (cf tab. D.45) and few (5 - 10%) carbonate, possibly siderite ($FeCO_3$), clay-size coatings or infillings and crystals up to 50µm are visible in thin section (tab. D.13, cf tab. D.48), partly corresponding to the colloidal ferrous carbonate Kubiena (1953) refers to. Free and amorphous iron and aluminium contents in H12 are relatively high, both within the profile and also when compared to the values measured in the Late Glacial peat horizon at Verrebroek "Dok 2" (P1, H5). Total phosphate content is relatively high (P_2O_5t_H12 = 1167mg kg^{-1}) and consists in almost equal quantities of inorganic and organic phosphate.

H19 has few (5 - 10%) limpid (cracked) strong to dark brown organic fine mass coatings up to 30µm thick, or infillings (cf tab. D.45). It also has frequent (5 - 20%) clay-size carbonate coatings, eventually on the organic coatings, or infillings, and very few (2 - 5%) carbonate crystals up to 50µm (cf tab. D48). As in the underlying soil horizons, the crystals probably represent calcified roots, commonly consisting of calcite elements whose mean size varies from 60 to 80µm and which retain the structure of the original cells of the root cortex (Jaillard *et alii* 1991). Amorphous dusty organic fine mass in whatever form (coatings, infillings, particles), commonly pointing towards former organism activity, is however absent.

The illuvial organic pedofeatures probably again represent very soluble organic matter substances that precipitated at the level of a texture jump or (temporary) water table. Chemical analyses for H19 are however absent, which prevents from evaluating the nutrient status or the chemical characteristics of this organic matter accumulation band. The micromorphology learns that (part of the) carbonate illuviation happened after the organic matter illuviation.
Ongoing presence of very few (< 2%) pyrite (FeS_2) and carbonate coatings and crystals throughout the Late Glacial soil and further upwards in the overlying soil indicate continuous wet and anaerobic conditions over a litho-stratigraphic sequence of almost half a metre (H17 - H19). These conditions once more underline the (local)palaeodepression topography and the importance of subsurface water flow in soil genesis. Within this context Verbruggen *et alii* (1991) mention gyttja and peat genesis during infilling of former river gullies (up to 50m wide and 3 to 8m deep). At the "Hoge Vaart" sections (Flevoland, The Netherlands), siderite was observed on the slopes and in the infilling of a flood gully, especially at the contact with the Peat Soil (Spek *et alii* 2001a, 17). Latter authors attribute siderite formation to a chemical bond between ferrous iron (Fe^{2+}) that laterally leached from the coversand ridge and slopes towards the depression with subsequent discharge and the carbonate ions (CO_3^{2-}) that formed *in situ*. A riverine environment could eventually fit with the fluviatile character of the parent material in the studied sections at Verrebroek "Dok 1" (P14, P9).

Based on pollen analysis (cf 2.2.3.1.) and ^{14}C-dating (cf 2.3.3.1.), both units of the Late Glacial soil at Verrebroek "Dok 1" could be attributed an Allerød age. Observations in the Allerød soil developed in the coversand ridges of Zeeuws-Vlaanderen (area Sas-van-Gent - Koewacht, The Netherlands, located at about 15km west to southwest of Verrebroek), mention a comparable, however laterally developed sequence (from bottom to top) of (i) a (Chalk) Gyttja Soil, (ii) a Peat Soil (composed of reed and sedges), (iii) a humiferous silty horizon of about 5cm thickness, covered with (iv) brownish sand (van der Sluijs, Maarleveld 1963). Latter brownish sand horizon is possibly comparable with the diagenetic organic matter accumulation in H19 of P9 at Verrebroek "Dok 1".

2.1.4.4. | Particular feature synchronous to Late Glacial soil

Profile 11 (fig. 27, tab. D.15, D.16, D.40, D.43)

Figure 27: Verrebroek "Dok 1"; profile 11, detail of particular feature synchronous to Late Glacial, Allerød soil, photographic representation.

In this profile, only particular light olive brown and dark grey mottles observed within the upper unit of the Late Glacial soil were studied. Only the light olive brown mottles were chemically analysed.

Within the light olive brown matrix, micromorphological observations (tab. D.15) revealed dusty reddish to strong brown (PPL) fine mass with undifferentiated birefringence-fabric. Few (5 -10%) amorphous cracked orange (PPL) iron infillings were observed; they locally show a radial pattern but do not have any orientation (tab. D.50). Comparison with reference material (oral communictation P. Adderley; Simpson *et alii* 2000) learns that they can eventually be interpreted as calcium iron phosphate. Also few (5 - 10%), poorly oriented, limpid (cracked) yellow (PPL) (calcium) iron phosphate coatings or infillings (tab. D.50), eventually on the amorphous iron infillings, were observed. Some translucent fragments exhibit a morphology that is similar to that of bone. The poorly oriented, yellow (PPL) (calcium) iron phosphate pedofeatures probably represent an intergrade between amorphous and crystalline forms (oral communication P. Adderley; Adderley *et alii*, in press). In this case, they would be diagenetic and possibly reflect alteration of the assumed bone fragments.

Parallel, very few (< 2%) charcoal fragments, few (5 - 10%) carbonised organic fine mass coatings and infillings (cf tab. D.47) as well as few (5 - 10%) clay-size carbonate coatings and infillings and very few (2 - 5%) silt-size carbonate crystals (cf tab. D.48) were observed in thin section in the dark grey matrix. Latter morphological features indicate presence of ash. The carbonised organic fine mass coatings and infillings are in favour of an *in situ* fire event, albeit local.

The pH of these mottles is slightly acid to neutral. The organic matter content (OM = 10.2%) and C/N-ratio (C/N = 64) are high; the nitrogen content is relatively high (N = 0.09%). The C/N-ratio fits with values around 60 commonly reported for wood ash. The organic matter content determined by loss on ignition at 500°C might however be overestimated, since this horizon contains appreciable quantities of clay (17% of the particles < 3.9μm) and sesquioxides, probably losing crystal-lattice water at heating between 105 and 500°C (Rowell 1997, 48).
Exceptionally high quantities of total, mainly inorganic phosphate are present within this accumulation horizon (P_2O_5t = 7236mg kg^{-1}, P_2O_5i = 6628mg kg^{-1}, P_2O_5o = 607mg kg^{-1}). Possibly the phosphate is preferentially associated to the perimeter and internal voids of the carbonised organic matter, as also reported by Macphail (1998) and Macphail *et alii* (2000) in urban context and by Davidson *et alii* (submitted) in Anthrosols. Free iron and aluminium (dithionite-extracted) contents (Fe_d = 9.33%, Al_d = 1.29%), as well as amorphous (oxalate-extracted) iron and aluminium contents (Fe_o = 0.65%, Al_o = 0.26%), are considerable as well when compared to values in Late Glacial soil horizons and B-horizons of the Podzolised Soils. The high free and amorphous iron content could correspond to the amorphous iron infillings observed under the microscope.

As the analysed phosphate contents are quite high, phosphate input might eventually be associated with an occupation event reflecting deposition of animal refuse and/or ashes; the latter were attested in this particular feature. Eventually, remnants of game trapped in an occasional forest fire might be a possible explanation for the high phosphate content too. Bone mainly consists of Ca-phosphate; (partial) replacement of Ca by Fe on decarbonatation can however result in (Ca-)Fe-phosphate (Jansen, Peeters 2001, 19). Excrement input is less probable here, as we would expect mainly organic phosphate in that case, unless post-depositional mineralisation occurred. The radial pattern observed in the iron coatings additionally suggests wet conditions. The hypothesis of human evidence is nevertheless at first sight not supported by archaeological evidences at this particular profile. However, no attempt to observe these accumulations with eventual associated characteristics, ecofacts and/or artefacts in horizontal section has been undertaken either. Seen the particular, albeit very local interest of these phosphate and ash accumulations, we suggest investigating comparable features from an archaeological point of view during future excavations, possibly clarifying eventual human impact.

2.1.4.5. | Particular features posterior to Late Glacial soil but prior to Podzolised Soil

Profile 14, H11 (fig. 25, tab. 8, D.17, D.18, D.40, D.43)

Particular olive yellow zones with a firm consistence (H11) were observed below the poorly developed Podzolised Soil, extending from about 0.50 to 0.95m below the Mesolithic occupation surface. In the field it was questioned whether this patchy horizon is the result of iron (Fe) or phosphate (P_2O_5) or mere organic matter accumulation.

Micromorphological observations revealed that very few (2 - 5%) amorphous, limpid cracked to slightly dusty organic matter coatings up to 30µm thick on or next to limpid cracked olive yellow iron phospate coatings up to 30µm thick (tab. D.50) account for the particular colour. Also here carbonised organic particles up to 100 µm were observed. The pH of this horizon is close to neutral as well (pH_H11 = 6.75). Only minor amounts of organic matter and nitrogen are present (OM_H11 = 0.75%, N_H11 = 0.009%). The total phosphate content (P_2O_5t_H11 = 530mg kg^{-1}) is higher than those commonly observed in Bh(.s)-horizons of Podzolised Soils (cf 2.1.4.7.) where the total phosphate content ranges roughly between 250 and 375mg kg^{-1}, but is comparable to phosphate levels found in organic matter accumulation horizons at Verrebroek "Dok 2" (cf 2.1.3.7.). The phosphate in H11 is mainly present as inorganic phosphate. The Al_o+$^1/_2$ Fe_o-content in H2 is 0.17% (tab. D.43), which possibly indicates presence of aluminium and iron within these coatings. However, the free iron content (Fe_d_H11 = 0.11%) is relatively high when compared to that of Bh(.s)-horizons of Podzolised Soils (cf 2.1.4.7.) or of organic matter accumulation horizons (cf 2.1.3.7.), which indicates that migration driven by oxydo-reduction processes is probable. All mentioned characteristics are in favour of the assumed amorphous iron phosphate accumulations. They might have developed prior to podzolisation. Local concentrated input of phosphate, e.g. as faunal excrements or from bone alteration, and their recharge as inorganic phosphate together with iron are however prerequisites for this local discharge of phosphate. It should be questioned whether we could expect a burial place (single event) here, eventually linked to the Early Mesolithic occupations. Based on microwear analysis, artefact units at Verrebroek "Dok 1" call for recurrent, short-term occupations or visits of Early Mesolithic hunter-gatherers. It is likely that the site has been used for the first phases of processing of hunted game, such as skinning or scraping of fresh hides. Within latter context, high quantities of bones and possible other phosphate-rich material, as e.g. flesh, must have been left at the surface or even dumped towards the depression sites. Posterior migration and accumulation in depth following the groundwater table of these phosphates, eventually in association with iron, might have resulted in the observed olive yellow zones.

2.1.4.6. | Brunified Soil

Occurrence of Brunified Soil horizons at Verrebroek "Dok 1" was restricted to tree-fall features, discussed in detail below (cf 3.3.1.2.).

2.1.4.7. | Podzolised Soil

Environmental conditions during the Boreal and/or Atlantic must have been in favour of podzolisation. Mokma and Buurman (1982) mention that Podzols only develop in regions where precipitation exceeds evapotranspiration, at least seasonally. These conditions are met by the coincidence of stabilising, to even decreasing, temperature and increasing precipitation of the Boreal and Atlantic (Verbruggen *et alii* 1996). It is not excluded that the effect of fire, either natural or anthropogenic, on these already poor sandy soils led to strong leaching conditions and further acidification, thus inducing podzolisation. Latter is confirmed by Little *et alii* (1997) in semi-natural oak woodlands in the marginal uplands of Ireland. Alway and McMiller (1933, in Dimbleby 1962, 27) mention greater acidity, decrease in bases and a lower nitrogen status as a direct effect of fire. Odgaard (1994) indicates the indirect effect of fire, favouring heathland development and as a consequence leading to podzolisation in West Jutland, Denmark.

The intra-site soil variability shows topo-hydrosequences with Podzolised Soil profiles. Well-developed, imperfectly drained Podzolised Soils with a distinct E-horizon appear in the slightly higher palaeo-landscape positions; weakly developed, poorly drained Podzolised Soils with nearly no E-horizon occur in the lower relief positions. In latter areas the later development of the overlying peat horizon is thickest. In such positions, the podzolisation was controlled by a permanent, or almost permanent, acid water table causing by reduction a mobilisation and subsequent lateral and/or vertical migration of iron with groundwater flow. The resulting Hydromorphic Podzolised Soil can be called Groundwater Podzolised Soil, according to the environmental classification of Duchaufour (1982, 307-334).

<u>Profile 16, H5 - H12</u> (fig. 28, tab. 10, D.19, D.20, D.40, D.43)
<u>Profile 17, H6a - H13</u> (fig. 29, tab. 11, D.21, D.22, D.40, D.43)
<u>Profile 18, H7 - H14</u> (fig. 30, tab. 12, D.23, D.24, D.40, D.43)

Profiles 16, H5 to H12, 17, H6a to H13 and 18, H7 to H14 represent a palaeo-topo-hydrosequence of Podzolised Soil development within the redistributed Upper Pleniglacial coversands, going from imperfectly drained to poorly drained soil conditions. The parent material (CB-horizons) has a median particle size between 180 and 200µm (tab. D.40), and is thus coarser than the Upper Pleniglacial coversands in which the Late Glacial soil units at Verrebroek "Dok 1" developed (cf 2.1.4.2., 2.1.4.3.).

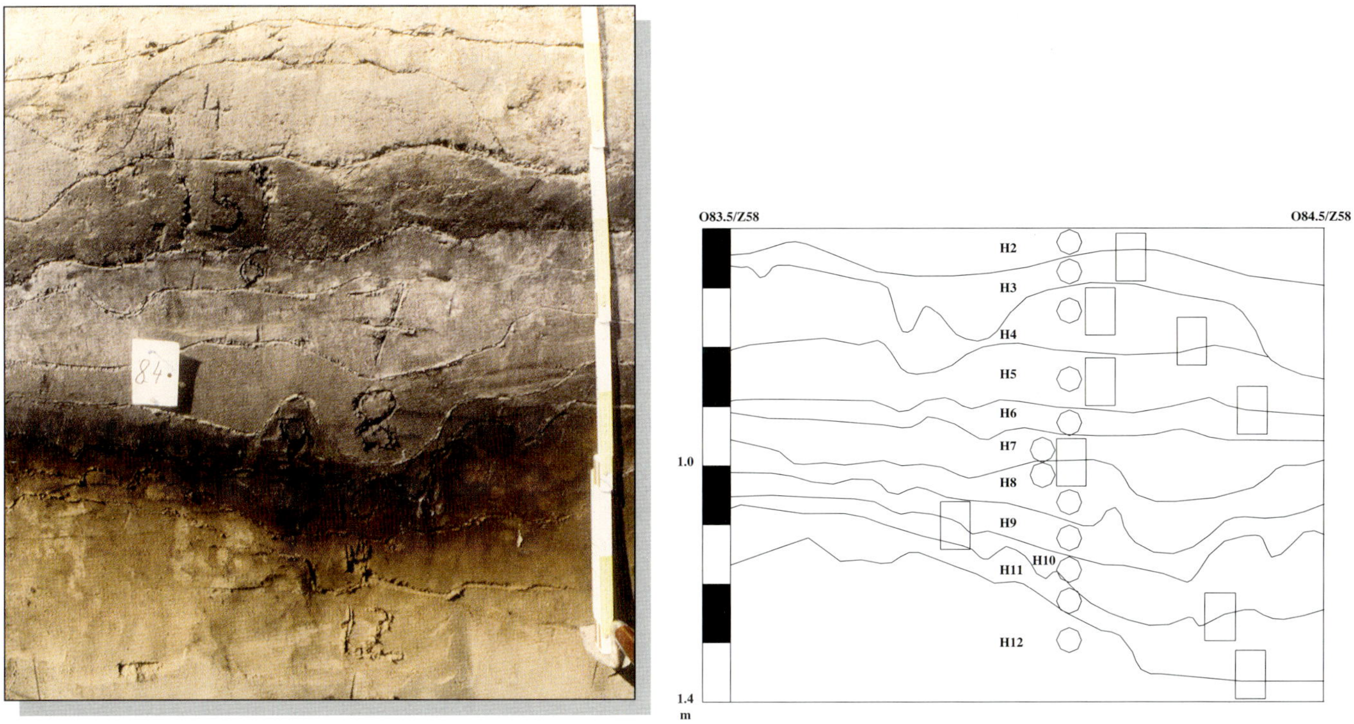

Figure 28: Verrebroek "Dok 1"; profile 16, photographic and schematic representation.

Horizon number	Symbol	Horizon depth (cm)	Profile description
H1			Flood sediments
H2		(60)-62/70	Light grey matrix; abundant strong brown mottles; flood clay?
H3	1CBg	62/70-68/85	Transition
H4	2Ap	68/78-78/83	Very dark brown - grey matrix; sand; strong brown mottles of Fe (root rust?); cultivation horizon
H5	2A	78/85-88/90	Very dark grey - black matrix; sand
H6	2AE	88/90-92/96	Dark grey - very dark grey matrix; sand
H7	2E	92/96-96/107	Grey - dark grey matrix; sand
H8	2EB	96/107-101/115	Transition
H9	2Bh.s?	101/115-105/120	Brownish black matrix; sand
H10	2Bh.s?	105/120-107/127	Dark brown matrix; sand; slightly cemented
H11	2BC or Bh.s?	107/127-112/136	Transition; dark-strong brown matrix; sand
H12	2CB	112/136-(140)	Brown matrix; pale brown mottles due to consumption of organic matter by microbiological activity; sand

Table 10: Verrebroek "Dok 1"; profile 16, profile description.

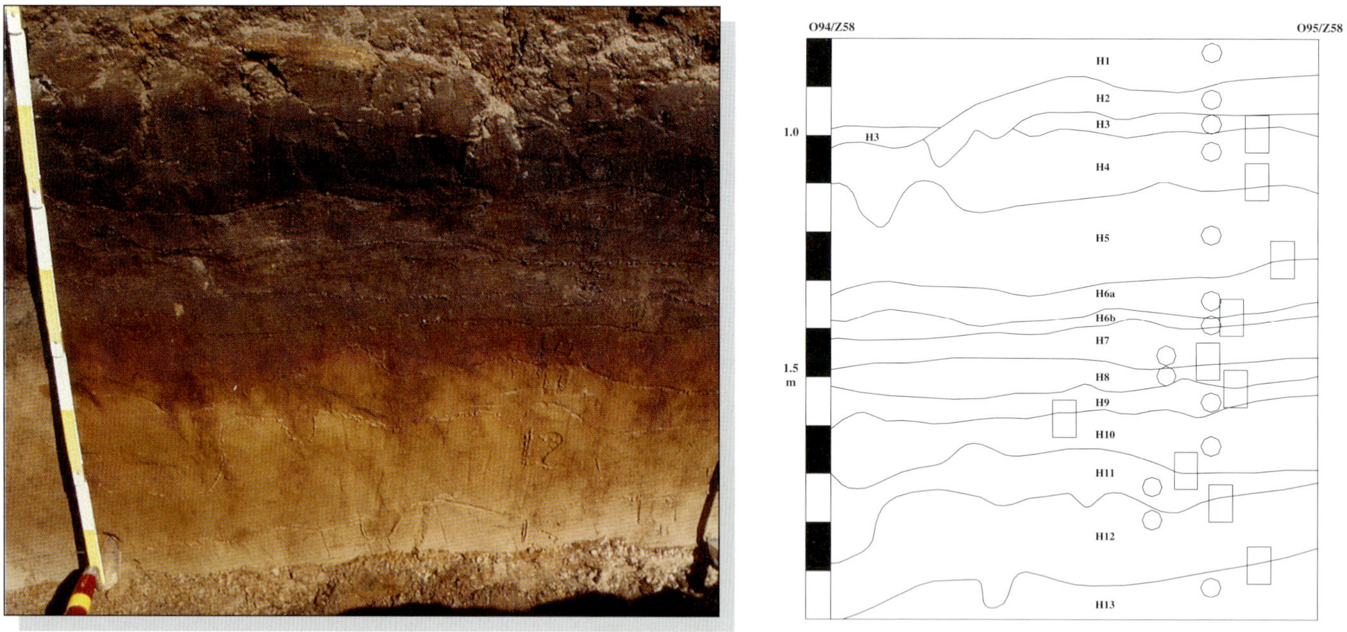

Figure 29: Verrebroek "Dok 1"; profile 17, photographic and schematic representation.

Horizon number	Symbol	Horizon depth (cm)	Profile description
H1		(80)-88/99	Light grey matrix; abundant strong brown mottles; flood clay?
H2		88/101-95/107	Grey - dark grey matrix; abundant strong brown mottles; flood clay?
H3		95/99-99/102	Transition
H4	2Ap	98/102-109/120	Very dark brown - grey matrix; sand; strong brown mottles of Fe (root rust?); cultivation horizon
H5	3F+H	109/120-126/134	Very dark brown - black matrix; peat; distinct wood fragments
H6a	4H	126/134-135/140	Black matrix; colloidal organic matter
H6b	4H+E	135/140-138/142	Black matrix; colloidal organic matter and sand
H7	4AE?	138/142-147/149	Very dark grey - black matrix; mottles of H6
H8	4AE?	147/149-150/155	Grey - dark grey matrix; sand
H9	4Bh.s?	150/155-153/161	Dark grey-brown matrix; dark brown mottles; sand
H10	4Bh.s?	153/161-163/172	Dark-strong brown matrix; sand; well-developed Bh?
H11	4BC	163/173-171/189	Transition
H12	4CB1	171/189-185/200	Light yellowish brown matrix; sand; dark brown mottles of decomposed organic matter (roots)
H13	4CB2	185/200-(200)	Very pale brown matrix; sand; few faint brown mottles of decomposed organic matter (roots)

Table 11: Verrebroek "Dok 1"; profile 17, profile description.

Figure 30: Verrebroek "Dok 1"; profile 18, photographic and schematic representation.

Horizon number	Symbol	Horizon depth (cm)	Profile description
H1		(75)-98/101	Light grey matrix; abundant strong brown mottles; flood clay
H2		98/101-112/119	Grey - dark grey matrix; abundant strong brown mottles; flood clay
H3		112/119-118/123	Alternating whitish grey sand and dark grey clay; stratified; abundant strong brown mottles; flood clay and sand
H4	2Ap1	118/123-120/127	Transition
H5	2Ap2	122/127-127/136	Very dark brown - grey matrix; sand; cultivation horizon
H6	3F+H	135/163-150/168	Very dark brown - black matrix; peat; distinct vegetative (wood and leaves) fragments
H7	4H	150/168-164/173	Black matrix; colloidal organic matter
H8	4A	154/167-167/169	Black matrix; sand
H9	4AE	167/169-168/172	Very dark grey matrix; sand and colloidal organic matter
H10	4Bh1?	168/172-178/195	Dark brown matrix; common black mottles of decomposed organic matter (roots)
H11	4Bh2?	178/180-180/188	Transition
H12	4BC1	180/188-190/195	Brown - light brown matrix; black mottles of decomposed organic matter (roots)
H13	4BC2	190/195-198/206	Transition
H14	4CB	198/206-(215)	Very pale brown - light grey matrix; sand
S1		120/123-123/125	"Intrusion" of flood clay and sands into underlying sediment?
S2		129-135/139	Very dark grey matrix; yellow mottles; fragments of burnt clay?
S3a		126/136-139/162	Tree trunk
S3b		128/129-138/141	Wood

Table 12: Verrebroek "Dok 1"; profile 18, profile description.

Micromorphological observations (tab. D.19) show that the Bh(.s)-horizons in P16, H10 and H11, the assumed Bh- and BC-horizon, have few to frequent (5 - 20%) slightly dusty to limpid cracked, brown organic fine mass coatings up to 25 µm and infillings (cf tab. D.46). Coatings and infillings of the same type, amount and thickness are furthermore observed in the Bh-horizons of P17, H9 and H10. It is questioned whether the limpid organic fine mass coatings are remnants of beginning podzolisation in a Brunified Soil development (cf tab. D.46), as this type of coatings were recognised as such in the Bw.bi-horizons of the Brunified Soil reference profile at Verrebroek "Dok 2" (P3) (cf 2.1.3.8.). Parallel, Spek *et alii* (2001a, 20, 55-56) mention brownish iron (hydro)oxide and organic matter coatings on quartz grains in the B- and BC-horizons of former Brown Forest Soils recognised at several depths in the "Hoge Vaart" sections (Flevoland, The Netherlands). The underlying BC- and CB-horizons in the different profiles have very few (< 2 - 5%), very thin (< 5µm), dominantly slightly dusty organic fine mass coatings. The A(B) - and Bh-horizons in the poorly drained P18 on the contrary have frequent (5 - 10%) to common (10 - 20%) dusty organic fine mass coatings up to 25µm and infillings; the BC- horizon (H12) has very few (< 2%) coatings of the same type. The underlying CB-horizons have very few (< 2%) slightly dusty organic fine mass coatings up to 15µm. De Bakker and Schelling (1989,

33-34) report comparable characteristics in Hydromorphic Podzolised Soils in The Netherlands and accept the presence of illuviated amorphous, iron-poor organic matter as sufficient to call for podzolisation.

All Bh-, BC- and CB-horizons in the three profiles have furthermore very few (< 2 - 5%) distinct amorphous (dark) brown organic fine mass particles up to 100μm in size, commonly called organic pellets. Presence of very few (2 - 5%) to few (5 - 10%) pyrite (FeS_2) framboids (cf tab. D.45) in P17 and P18 throughout the (peat cover and) Podzolised Soil profile confirm the poorer drainage conditions of these profiles as compared to P16.

The Bh(.s)-horizons do not meet all the criteria for a spodic horizon as determined by the World Reference Base for Soil Resources (WRB 1998) (cf Appendix B. Glossary). Although colours are measured on air-dry material, the Bh(.s)-horizons in the three profiles would probably meet the colour requirements as stated for a spodic B-horizon according to the WRB. Even if the colour requirements would not be met, the Bh(.s)-horizons in the considered profiles all have very few (< 2 - 5%) distinct organic pellets. All Bh(.s)-horizons have an organic carbon content of at least 0.6%. The pH requirements are only met in the Bh(.s)-horizons of P18. It should however be mentioned that pH was measured at a ratio 1:5 in stead of the required 1:1 and that only P17 and P18 have a peat cover on top of the Podzolised Soil horizons that prevents from a diagenetic pH increase due to e.g. agricultural activities or the presence of the overlying calcareous flood sediments.

$Al_o+{}^1/_2\,Fe_o$-contents in the Bh(.s)-horizons vary between 0.13% (H11 in P18) and 0.37% (H10 in P16) and are thus lower than the required 0.50%. Either (i) the parent material contained only few aluminium and/or iron, or (ii) most of the iron was leached after reduction out of these profiles that would then be situated within a recharge area. The latter is however less probable under the poorest drainage conditions where the palaeo-depression position is in favour of discharge. Only in P16 and P17, the Bh(.s)-horizons have two times more $Al_o+{}^1/_2\,Fe_o$ than the overlying A-, AE-, or E-horizons; in P18, $Al_o+{}^1/_2\,Fe_o$-content is higher in the A- and AE-horizons than in the Bh(.s)-horizons. Within each separate profile, the summed thickness of the Bh(.s)-horizons reaches at least 2.5cm; only in P16 and P17, the upper limit of the Bh(.s)-horizons is situated below 0.10m of the mineral coversand soil surface.

Podzolised Soils tend to have a typical set of soil chemical characteristics that are reported in Appendix B. Glossary.
The pH in P16 and P17 is sightly acid to neutral due to diagenetic influence of the overlying cultivation horizon and flood sediments. Profile 18 has a slightly to moderately acid pH in the surface H-, A- and AE-horizon, but has a strongly acid pH in the Bh(.s)- and BC-horizons. This profile apparently was better protected from diagenetic influence by the thicker peat cover. C/N-ratios in the Bh.s-horizons in P16 are around 25 and roughly follow the values put forward by Driessen and Dudal (1991, 273-280).

In the three profiles, aluminium (Al) contents are higher than iron (Fe) contents (tab. D.43). In P16 maxima for Fe_o and Fe_d occur in H9, whereas for Al_o and Al_d they are in H10; the Al_o+½ Fe_o-content is concomitantly highest in H10, recognised already in the field as a Bh.s-horizon. This pattern corresponds to Driessen and Dudal's (1991, 273-280) observations that Al-humus compounds tend to accumulate somewhat deeper in the illuviation horizon than Fe-humus compounds. In P17, we observe the reverse pattern, confirming poorer drainage conditions. In P18, only relative peaks in Al_o- and Al_d-contents occur at the level of the assumed Bh-horizons H10 and H11; absolute highest Al- and Fe-contents occur in the surface A-horizon, the only horizon where the the Al_o+½ Fe_o-content exceeds the value of 0.20 and even of 0.50%. In the Bh-horizons of P18, Al-contents are lower than those in the Bh(.s)-horizons in the other two profiles; Fe-contents are higher. This could call for the discharge position of P18 within the topo-hydrosequence.

The Bh(.s)-horizons in the three profiles have comparable total, inorganic and organic phosphate contents. Maximum total phosphate (P_2O_5t) contents coincide in the three profiles with maximum Al and Al_o+½ Fe_o-contents (tab. D.43). In P16 and P18, this peak corresponds as well to the maximum for inorganic phosphate (P_2O_5i); in P17, inorganic phosphate (P_2O_5i) and iron (Fe) peaks coincide. Organic phosphate (P_2O_5o) contents in the Bh(.s)-horizons follow the same pattern as organic matter contents.

Cation exchange capacity (CEC) in the Bh.s-horizons in P16 and P17 varies from 6.5 to 9.5cmol(+) kg^{-1}; in the Bh-horizons of P18, CEC is slightly lower and only about 3.0 to 4.0cmol(+) kg^{-1}, increasing from bottom to top, the individual cations alike. Calcium (Ca) is the dominant cation.

In P18, H10 to H12, the Bh- and BC1-horizons have exchangeable sodium percentages (ESP) that exceed the critical value of 15; sodium adsorption ratios (SAR) however vary between 6 and 8 and are thus below the critical value of 13 for sodic soils. Both ESP- and SAR-values decrease from bottom to top. This would indicate an input of salts from below, eventually from the underlying marine substratum (Tertiary Kattendijk and Kallo sands), through a fluctuating groundwater table.

In P18, the overall nutrient distribution (C, N, P, CEC, Ca, Na, Mg, K, Al, Fe) within the mineral soil below the peat cover shows a general decreasing pattern (tab. D.43) from top to bottom. Except for the above-mentioned relatively high sodium contents, nutrient input must mainly have happened through nutrient (re)cycling from the vegetative cover at the surface as opposed to pedogenic and weathering processes that can happen throughout the profile. The importance of nutrient input by the vegetative cover within the nutrient cycle could explain the relatively high Fe-contents present within this profile. Podzolisation characteristics are however almost lacking. Only iron shows some migration in depth with a slight peak in H10, the assumed Bh1-horizon (tab. D.43). Aluminium doesn't show any mobility at all; the highest values occur in the surface A-horizon, H6 (tab. D.43). Hence, leaching of iron by reduction must have been more prominent than podzolisation. The minor impact of the podzolisation process within this profile can probably be assigned to its poorer drainage conditions, attested by the thicker peat cover (cf 2.1.4.9.) and the constant presence of pyrite throughout the profile. This fits with Juste's (1965, in Duchaufour 1982, 326) statement that iron is more mobile than aluminium in poorly drained Podzolised Soils. Hence, P18 can be designated as a Hydromorphic Podzolised Soil, and according to its particular topographical position in sandy lowlands, a Groundwater Podzolised Soil (after Duchaufour 1982, 306-334).

The three profiles have very few (< 2%) charcoal fragments within their upper, mainly A-, E- and intermediary transition horizons. Only in P18, the wettest profile, the Bh-horizons contain charcoal too. Possible impact of fire, either natural or anthropogenic, on the podzolisation process was discussed earlier. Presence of charcoal is common in soils of boreal forests and is linked to forest fires, commonly caused by summer storm lightning (Bunting 1983). Most burns only affect the L- and F-horizons and some woody components of the F-horizon (Bunting 1983). However, especially in depression and lee sites, added charcoal from fallout (crown fires), deposition from run-off, ash-blow and biologic transfer seemingly deepen that part of the soil profile affected by fire (Bunting 1983). Fire cycles in taiga and boreal forests have a frequency of about one per century (Viereck 1973, in Van Vliet-Lanoë *et alii* 1993). Simmons (1996, 124) especially mentions coniferous forest and heath and moorland habitats as naturally fire-prone. With respect to human-induced occurrence of fire, Mesolithic using fire for hunting to drive game and/or to open up the landscape, eventually to facilitate hazel growth (Dimbleby 1962, 27; Simmons 1996, 138-144) are suggested.

In P16 and P18, the surface A-horizon occupies a particular position with regard to texture; their similarity with the underlying E-, Bh(.s)- and BC-horizons is sometimes lower than 80. The A-horizons, H5 and H8 respectively, however have comparable particle size distributions in the clay to very fine sand fraction: about 5% clay (< 3.9μm), about 10% fine to medium silt (3.9 - 31μm), about 5% coarse silt (31 - 62.5μm) and about 8.5% very fine sand (62.5 - 125μm). Latter surface horizons are rather homogeneous and have been subject to strong bioturbation. This is in contrast with the surface horizon (H6b) in P17 where we can differentiate uncoated quartz grains from colloidal organic pellets. Earthworm activity must have been low or absent here (Jabiol *et alii* 1995, 32) and the corresponding humus type probably is moder. The rather oligotrophic conditions reflected in a neutral pH and a C/N-ratio above 17 (pH_H6b = 6.8, C/N_H6b = 33) are in favour of this humus type as well (Lee 1983). Nevertheless, the pH-value is too high to fit with the observed Podzolised Soil profile; most probably, pH was increased diagenetically by lateral input of more alkaline sediments as e.g. the overlying calcareous flood sediments. The surface A-horizon of the Podzolised Soil in P18 furthermore has an appreciable phosphate content, possibly reflecting accumulation of organic matter in depression position and subsequent nutrient release when affected by fire. As mentioned above, the latter is evidenced by the presence of charcoal throughout this profile.

On the A-horizon of the Podzolised Soil profile P16 (H6), phytolith analysis was performed (cf 2.2.3.3.). In absolute quantities, very few (< 2%) phytoliths were observed. The phytolith identification gives a coherent assemblage of panicoid grasses. Some pollen grains were registered too.

Profile 9, H2 - H11 (fig. 26, tab. 9)

Pollen analysis was performed on H2, H3a and H3b, corresponding with the H(+E)-, H+E- and E(+H)-horizons of the Podzolised Soil (cf 2.2.3.3.). The E-horizon dominantly has *Pinus* and *Corylus* and doesn't contain *Ericaceae*; this spectrum can be attributed a Late Boreal age. This period is characterised by the coincidence of stabilising, to even decreasing temperature and increasing precipitation (Verbruggen *et alii* 1996), providing leaching conditions favourable for the podzolisation process. There is a fair chance that this spectrum corresponds to the vegetation present during podzolisation, as Havinga (1962) mentions gradual pollen preservation in upward direction, whereby the pollen spectrum at the bottom can reflect former, so-called sealed pedogenic stages (e.g. biologically active Brunified Soil development). The H+E- and H-horizon show a wetter spectrum, with limited presence of *Ericaceae* that calls for an Early Subboreal age. Latter spectrum probably fits with the environment wetting, favourable for the subsequent peat growth (cf 2.1.4.9.).

2.1.4.8. | Degraded Podzolised Soil

Another factor responsible for soil variability is the degree of organic matter decomposition by microbiological activity in the Podzolised Soil profiles. This diagenetic process occurs as well in the A- as in the Bh-horizons of Podzolised Soils. It is reflected by the presence of prominent more or less subrounded lighter coloured patches, up to several centimetres in diameter. Locally the Podzolised Soil even evolved to a soil somewhat similar to a "Postpodzol" (Tavernier, Ameryckx 1957). Such soils are considered to be Podzolised Soils characterised by a patchy Bh.s-horizon as a result of the decomposed humus-colloids. This is observed in P19, situated in the immediate vicinity of a former Medieval field road, indicated by the presence of cart tracks. Chemical analyses have been performed on the AE-horizon of the so-called "Postpodzol", on the locally appearing Aa-horizon representing the Medieval field road and on the locally appearing, strongly bioturbated E.Bbi-horizon, probably linked to this field road too. No analysis has been done on the postpodzolised Bpp-horizon. Nevertheless, the association of the postpodzolised Bpp-horizon with a field road, let us assume that, in this particular case, dung of draught-animals or cattle passing over the field road might have led to an intensification of the (micro)biological activity. On other spots the influence of agricultural practices with manuring, related to Medieval land use, might also have induced an activation of the microbiological activity in these soils. An example is amongst others given in the horizons surrounding the assumed tree-fall S90 (cf 3.3.1.2.). Here, micromorphological and chemical analysis have been performed on the Bh-horizon characterised by distinct pale yellow mottles.

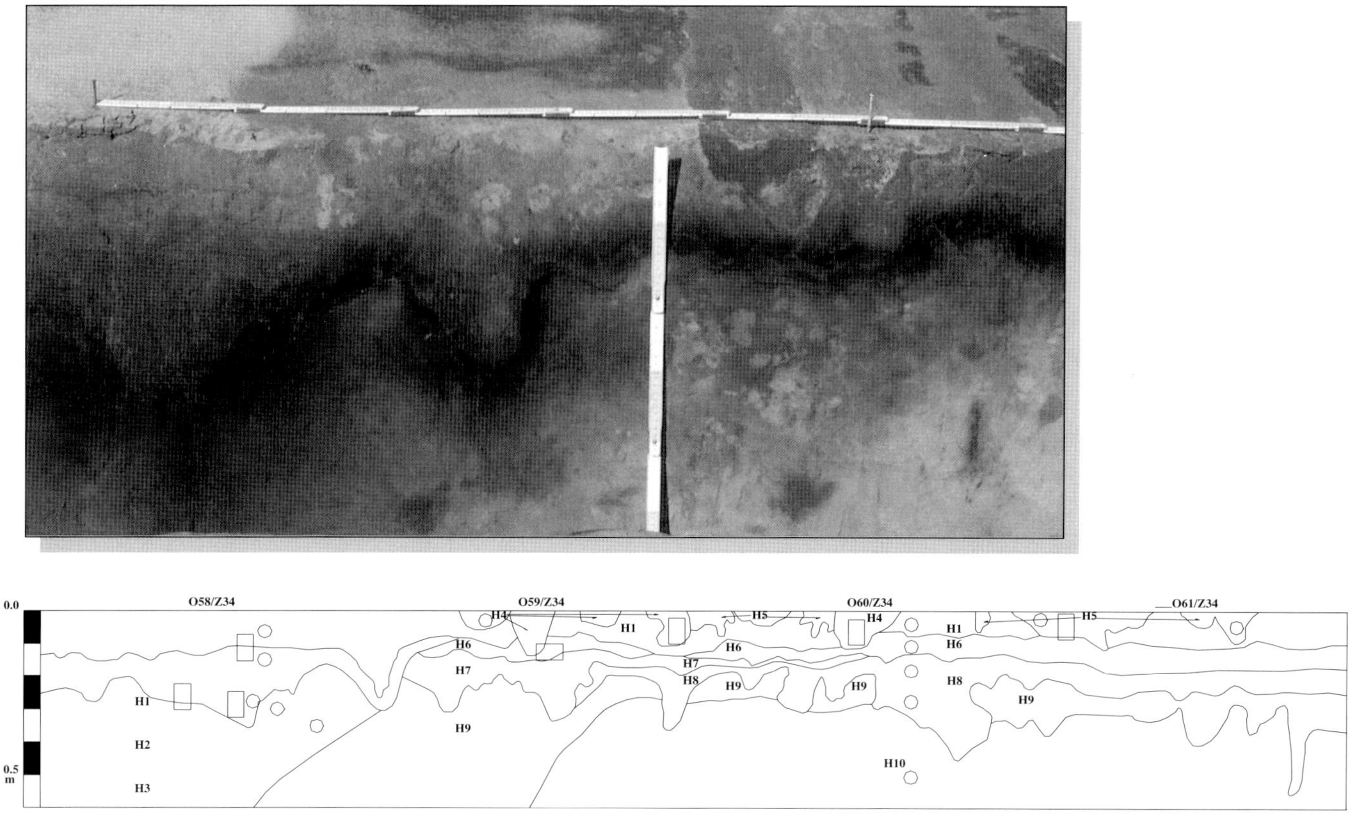

Figure 31: Verrebroek "Dok 1"; profile 19, photographic and schematic representation.

Horizon number	Symbol	Horizon depth (cm)	Profile description
H1	AE	(0)-7/28	Grey-light grey matrix; sand; distinct light grey mottles due to organic matter consumption by microbiological activity
H2	Bpp	7/28-9/37	Dark brown matrix; sand; Degraded Podzolised Soil (Postpodzol) development; not cemented anymore
H3	BC1?	18/36-30/(60)	Greyish brown matrix; sand
H4	Aa	-0/15	Dark - very dark grey matrix; humiferous sand; cart tracks; charcoal fragments; strong brown Fe concretions, root rust; locally bioturbated?
H5	E.Bbi	-0/11	White - light grey matrix; sand; bioturbated
H6	Bh?	7/15-11/18	Dark brown matrix; sand; cemented?
H7		11/18-13/32	Brown - strong brown matrix; sand; pale yellow mottles due to consumption of organic matter by microbiological activity
H8	Bh.s?	12/20-18/47	Strong brown matrix; sand
H9	BC2?	18/47-26/(60)	Brown - light brown matrix; sand; pale yellow mottles due to consumption of organic matter by microbiological activity; root rust
H10	C	25/57-(60)	Pale yellow matrix; sand

Table 13: Verrebroek "Dok 1"; profile 19, profile description.

Profile 19 (fig. 31, tab. 13, D.25, D.43)

As mentioned above, only H1, H4 and H5 are analysed. H1 corresponds to the AE-horizon of the Podzolised Soil and shows distinct light grey mottles. H4 has a recurrent appearance and abrupt smooth boundaries and corresponds to the surface horizon associated to the cart tracks of the Medieval field road. H5, the strongly bioturbated E.Bbi-horizon, has clear wavy boundaries and it is spatially associated to the field road cart tracks. The area is not covered with peat, which again calls for interpreting nutrient analysis results with care since they might be influenced by posterior nutrient leaching from the overlying flood sediments.

The pH of the three horizons is slightly alkaline. The Aa-horizon of the Medieval field road (H4) has a relatively high organic matter (OM) (OM_H4 = 3.38%) and nitrogen (N) (N_H4 = 0.07%) content when compared to H1 and H5, both eluvial horizons, and to the EB-horizon of reference profile P16; latter eluvial horizons have OM-contents roughly around 1% and N-contents around 0.02%. C/N-ratios are around 30 in the three horizons, comparable to or slightly lower than those commonly observed in AE- and EB-horizons of the reference Podzolised Soil profiles (P16 - P18). They thus indicate a comparable to slightly higher organic matter decomposition rate and concomitant microbiological activity. The highest total, organic and inorganic phosphate (P_2O_5) content appears in the Aa-horizon (H4) as well; phosphate content in the E.Bbi-horizon (H5) is also relatively high when compared to the surrounding E-horizon (H1). Phosphate is mainly inorganic in H4 (Pi/Po = 1.3) as opposed to H5 and H1 that have Pi/Po-ratios below 1. H4, the Aa-horizon, has an exceptionally high C/Po-ratio around 260 as compared to those of H5 and H1, the E-horizons that have C/Po-ratios around 100. H4 has a relatively high OM-content and concomitant C/Po-ratio, probably because the cart wheels left hollows within the road, easily capturing organic matter as e.g. fallen leaves or particulate organic matter. The relatively high phosphate contents with concomitant high C/Po- and Pi/Po-ratio in H4 once more suggest input of phosphorus as faunal excrements from draught animals, but indicate in the mean time a good mineralisation of that organic material. As mentioned earlier, high nutrient contents at the surface, especially phosphorus, coincide with high microbiological activity within the underlying horizons of the originally Podzolised Soil.

Tree-fall S90 (fig.145, tab.55, D.32, D.33, D.40, D.43)
At the level of this structure, the Bh-horizon surrounding the assumed tree-fall S90 (cf 3.3.1.2.) is particularly interesting. It is characterised by distinct pale yellow mottles (fig. 145), probably resulting from local organic matter consumption by microbiological activity. Micromorphological (tab. D.32) and chemical (tab. D.43) analysis have been performed on this horizon (H1).
It appears that the organic matter content in this horizon (OM_H1 = 1.37 - 1.13%) is, except in one case, lower than in the Bh-horizons of the reference Podzolised Soil profiles (OM_Bh_P16 - P18 = 3.41 - 1.24%). Parallel, the amount of (slightly) dusty organic fine mass coatings observed in thin section is lower than in the reference Bh-horizons: 2 to 5% in S90 as opposed to 5 to 10%, even 10 to 20% in the Bh-horizons of P16 to P18. Concluding, the lower organic matter content, reflected both in the chemical and micromorphological analysis, is in favour of the hypothesis that organic matter was decomposed posterior to podzolisation.

2.1.4.9. | Peat

Peat is mainly present in the depression positions (fig. 32). We recall that the site is located at the edge of the sand ridge of Maldegem - Stekene and hence is situated at an elevation where peat growth is at its ecological limit.

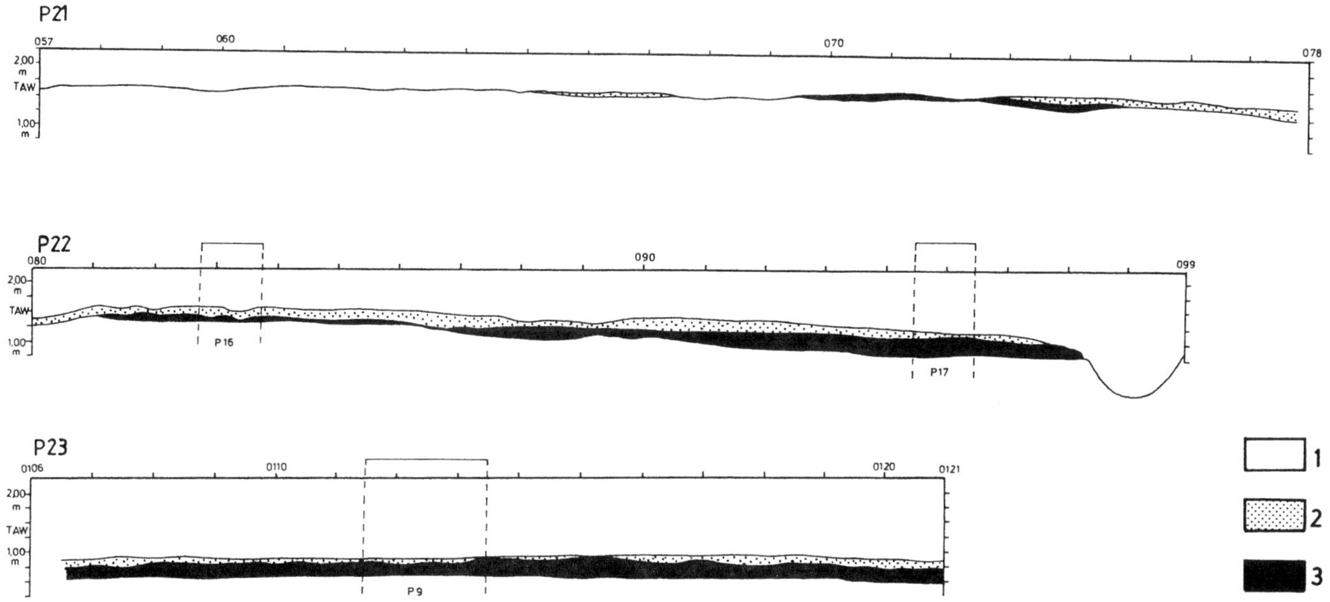

Figure 32: Verrebroek "Dok 1"; simplified stratigraphic section of the site, from top to slope. Key: (1) flood sediments, (2) Medieval cultivation horizon, (3) peat.

Around 6000 BP, sea level rise caused a strong lowering in river slope and led to concomitant soil waterlogging. The latter, together with low sediment supply, resulted in peat formation in the Lower Schelde alluvial plain (Kiden 1991). From 5000 BP onwards, at the end of the Atlantic, sea level rise slowed down and an almost closed coastal barrier system developed (Kiden, Verbruggen 1987; Louwye, Declercq 1998). Parallel, from about 4500 BP (Kiden, Verbruggen 1987; Kiden 1991), the rise of the coastal mean high water level (MHW) was faster than the local MHW in the Schelde-Rupel confluence area, controlled by the presence of an important floodbasin effect in the Lower Schelde (Kiden 1991). The latter is the result of a strong frictional dissipation of the tidal energy and/or the presence of large intertidal storage basins, leading to a considerable reduction of tidal amplitude as the tidal wave moves inland. Peat development in the alluvial plain occurred between MHW at sea and mean sea level (Kiden 1991). Summarising, peat growth in the Lower Schelde plain started roughly between 6000 and 4000 BP, depending on the elevation of the Pleistocene surface. The pollen spectra of the peat cover at Verrebroek "Dok 1" point towards Subboreal peat growth (cf 2.2.3.2.).

Peat formation in the Lower Schelde plain came to an end roughly between 2000 and 1500 BP (Minnaert, Verbruggen 1986; Verbruggen, Denys 1991). Downstream of Antwerp, peat formation was stopped by increased brackish water tidal sedimentation between 2500 and 1600 BP (Denys, Verbruggen 1989). Upstream of Antwerp by contrast, peat growth continued up to about 1500 BP and came to an end owing to increased clay deposition, presumably partly of tidal as well as fluvial origin (Kiden 1991). Verrebroek is located slightly downstream of Antwerp. Nevertheless, clear indications of brackish water are absent in the peat cover at Verrebroek.

The absence of *Fagus* and *Carpinus* in the palynological record of the peat horizon at the Verrebroek "Dok 1" site suggests that the top part is missing (cf 2.2.3.2.). Within this context, the question whether or not peat was exploited can be raised. Augustyn (1977; 1999) indeed reports on intensive peat exploitation in the Verrebroek area from the 12th till the 16th century. However Augustyn (1999) also mentions mixing of the peat with the underlying Pleistocene sand for agricultural purposes where peat exploitation itself was not profitable, i.e. where the peat horizon was too thin or of low quality. This was especially practiced in the vicinity of the Pleistocene sand ridges. The Medieval cultivation horizon indeed shows admixture of peat (cf 2.1.4.10.). Furthermore no clear pedomorphological evidences of past peat exploitation, such as remnants of peat cutting trenches, were found on the studied site. On the contrary, the peat surface was exceptionally flat and nearly perfectly horizontal. This was particularly evident in trench 11, where the peat thickness varied roughly from 0.20 to 0.75m (tab. 14) in relation with the palaeo-topography of the Upper Pleniglacial coversand ridge. Would there have been peat exploitation, it seems surprising that peat of a thickness up to 0.75m would not have been touched. The presence of an overlying surface horizon of peat mixed with sand and with nearly constant thickness of about 0.10m is indicative of a cultivation horizon (cf 2.1.4.10.). Agricultural land use is further evidenced by a deep ditch system that drained this land. Probably the upper part of the peat was removed by levelling operations and oxidised because of the lowering of the groundwater table for crop production. To conclude, the data of the Verrebroek "Dok" site point towards agricultural land use, rather than peat exploitation.

Profile	Horizontal distance on the vertical profile wall					
	0.0m	0.5m	1.0m	1.5m	2.0m	Mean
1	69	64	64	62	63	64
2	64	62	67	69	63	65
3	60	52	51	53	51	53
4	-	-	-	-	-	-
5	30	27	33	34	37	32
6	25	21	21	22	21	22
7	27	28	28	27	26	27
8	50	55	56	49	45	51
9	43	43	37	36	36	39
10	39	42	45	44	48	44
11	56	50	47	53	52	52
12	46	48	48	41	40	45
13	44	29	29	26	21	30
14	41	43	45	44	51	45
15	74	69	70	70	67	70

Table 14: Verrebroek "Dok 1"; variability of the peat thickness (cm) (location indicated in fig. 10, 24)

Profile 17, H5 (fig. 29, tab. 11, D.21, D.22, D.43)
Profile 18, H6 - H7 (fig. 30, tab. 12, D23, D.24, D.43)

Organic matter content in the peat horizons is 63% in H5 of P17 and 73% in H6 of P18. Average thickness for both horizons is 0.15 to 0.20m. Both contain frequent to common distinct root and leaf fragments that are (reddish to dark) brown (PPL) and only slightly birefringent in thin section. Amorphous brown organic fine mass is commonly (40 - 50%) observed (tab. D.21, D.23). Both peat horizons contain very few (< 2 - 5%) charcoal. Quartz content is low: very few (2 - 5%) in P17, H5 and few (5 - 10%) in P18, H6.

According to the diagnostic criteria of the World Reference Base for Soil Resources (WRB 1998, 35) the peat horizons can be described as organic histic O-horizons (Appendix B. Glossary).
In P18, the underlying H7 corresponds to the criteria of a histic horizon as well: it has an average thickness of more than 0.10m and its organic matter content is 56%. It has only few (5 - 15%) slightly birefringent organ residue and amorphous (dark) brown plant fragments. Despite of its higher quartz content (10 - 20% Q in H7 as opposed to 5 - 10% in H6), H7 probably belongs to the peat cover as well. The horizontal orientation is in favour of *in situ* peat formation under water, an interpretation also supported for peat covering the Mesolithic and Early Neolithic sites at the "Hoge Vaart" (Flevoland, The Netherlands) (Exaltus 2001, 18). The mineral material was possibly blown in during the intial phase of peat growth in a wetting environment due to ongoing rise of the groundwater table.

Few (5 - 10%) pyrite (FeS_2) framboids (cf tab. D.45) are observed in the discussed histic horizons. They confirm the wet conditions necessary for peat growth.

However, the presence of common amorphous dark organic fine mass forming the dominant part of the soil matrix points towards appreciable decomposition of the original peat cover. Physical disintegration of plant parts and their biochemical decomposition and biological granulation form an essential part of the pedological ripening process, initiated by peat drainage and aeration (Lee 1983). Open drainage ditches, cutting the peat horizon and the underlying Podzolised Soil, are indeed present here at regular distances.

Nutrient reserve (N, P, cations) is relatively high in the peat horizons. This can be partly due to the above-mentioned drainage and oxidation linked to agricultural land use, in the mean time providing continuous manuring. Stoklund (1999) indeed reports on intensive agriculture along the North Sea coast from Denmark to Flanders by means of peat manuring, providing nutrients and thus increasing the fertility, on wasteland. Pons (1987) more specifically refers to mesotrophic and eutrophic peat as highly productive fertiliser. Phosphate content is relatively high (P_2O_5t = 859 - 1157mg kg^{-1}) but is still mainly present as organic phosphate (Pi/Po_F+H = 0.3 - 0.5), which reveals a further nutrient mineralisation potential for this element. Particular attention should be paid to P18 where the sodium (Na) content in the peat horizon (H6) is exceptionally high in comparison to the other cations and the overlying horizon. The horizon underneath, considered as the organic surface horizon of the underlying Hydromorphic Podzolised

Soil (H7), has even still higher Na-content; its SAR-value of 12.2 even reaches the threshold value for soils with saline-sodic properties. When considering the profile of Na-content in depth reaching a peak in H7, it is probable that additional input from the surface caused Na-enrichment in this horizon. The colloidal character of both H6 and H7 accounts for high adsorption capacity, easily retaining exchangeable cations. As sodium moreover easily moves, it could have leached from the overlying flood sediments, as the reference profile for flood sediments (P3) indeed shows leaching tendency (cf 2.1.4.11.). Leenders *et alii* (1990) even mention historically documented salt production from peat deposits flooded with tidal sediments in the sandy area north of Antwerp. However, cation contents in general, especially sodium (Na^+) and magnesium (Mg^{2+}) are exceptionally high, both in H6 and H7, even when compared to other horizons with dominant (> 50%) organic matter content. As these horizons are affected by fire activity as shown by the presence of charcoal fragments, the high exchangeable Na- and Mg-content could reflect enhanced mineralisation too. With respect to leaching, we furthermore see that only the monovalent cation content of Na^+ and K^+ increases from H6 to H7, whereas the bivalent cation content of Mg^{2+} and Ca^{2+} decreases. This probably confirms preferential leaching of monovalent over bivalent exchangeable cations, as salts of the alkali metals sodium (Na) and potassium (K) are very soluble (Van Ranst 1991, 40).

2.1.4.10. | Medieval cultivation horizon

Spatially two distinct facies of the Medieval cultivation horizon have been observed on the site: (i) a humiferous sand facies on a sandy, partly to strongly eroded, Podzolised Soil lacking peat cover, and (ii) a facies consisting of the upper horizon of the peat horizon anthropogenically mixed with sand.

The soil surface of the peaty facies is exceptionally flat. The sand might have been imported in order to facilitate tillage practices on the peat soil. It is suggested that the terrain was levelled on purpose of land use for agriculture.

Stoklund (1999) parallel provides evidence for turf manuring as an element of intensive agriculture on the Danish island of Læsø and mentions use of latter method in extensive areas including The Netherlands, North-West-Germany, Jutland (Denmark), South-West-Norway and Scotland. In the borderland between The Netherlands and Germany, small part of the land was continuously used for rye cultivation ("ewiger Roggenbau"). Within this context, it is not excluded that peat admixture was partly aimed at manuring.

The humiferous sand cultivation horizon locally shows spade marks at its base. It can be referred to as a horizon tilled with a spade rather than a ploughed horizon. Other indications for agricultural land use within this tilled horizon are the limited presence of *Cerealia*, *Fagopyrum esculentum*, *Centaurea cyanus* and *Brassicaceae* pollen (cf 2.2.3.2.) and *Triticum sp.* phytoliths (cf 2.2.3.2.). However, the restricted presence suggests trial, rather than prolonged cereal cultivation at the spot. The presence of *Sphagnum*, *Menyanthes trifoliata* and *Typha latifolia* pollen, indicative of rather wet conditions (cf 2.2.3.2.), also points in this direction. Still, latter wet spectrum could reflect the admixture of peaty material or ditch-fill within the horizon. The recurrent presence of

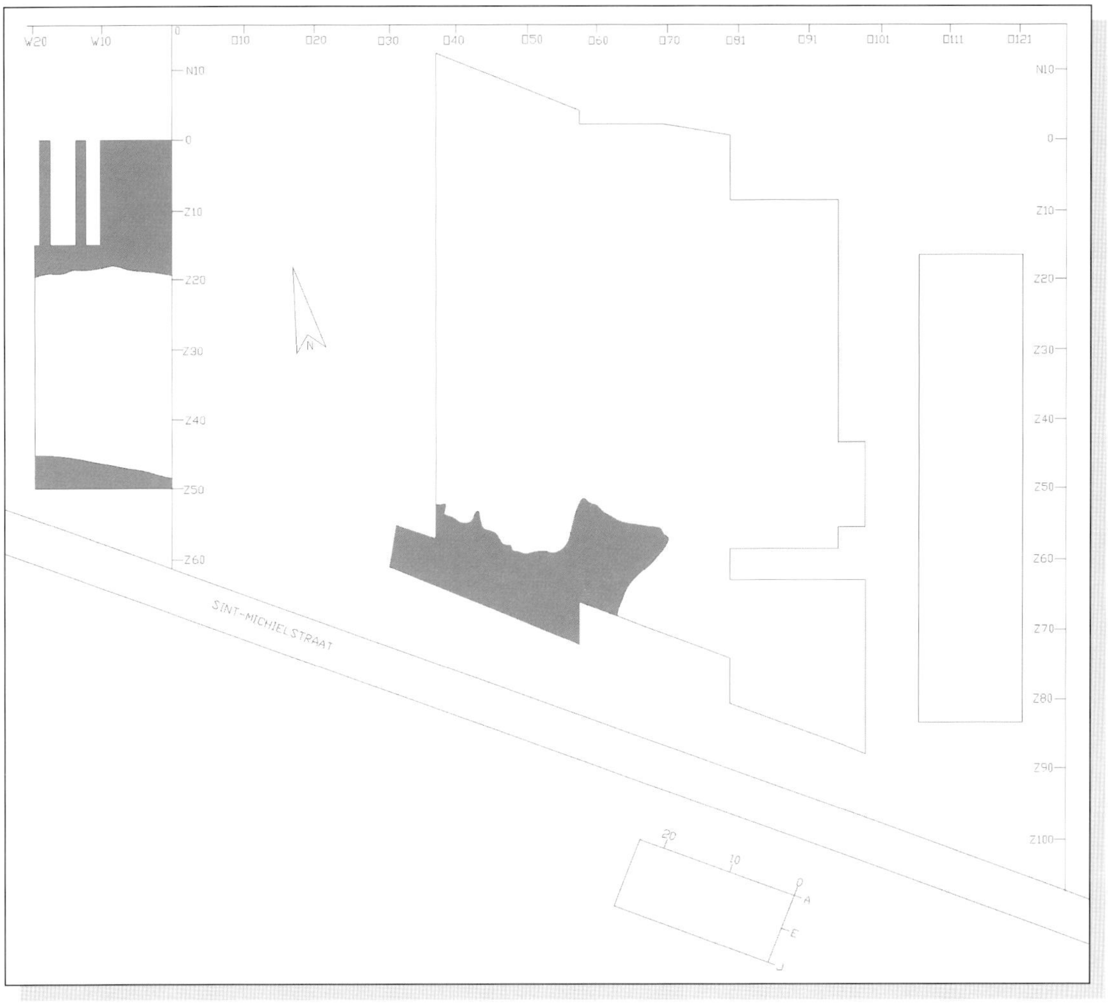

Figure 33: Verrebroek "Dok 1"; general excavation map marking the areas with truncated podzol.

open drainage ditches of about 3.0 to 3.5m width and 1m depth, corresponds to this agricultural land use. Artefacts found within these ditches reveal that tillage is linked to the 13-15th century Medieval period (Crombé *et alii* 1997).

On some spots, the Medieval cultivation horizon is completely lacking and flood sediments rest directly on a moderately to severely truncated Podzolised Soil (fig. 33). The boundary between these materials is very abrupt and wavy (up to 0.35m wide) to irregular (up to 0.20m deep). Most probably the soil erosion in these areas can be linked to the initial phase of flood on an originally vegetated surface soil from which plants and associated roots with adhering soil have been torn out. Erosion on the lower slopes of the coversand ridge was observed as well at the "Hoge Vaart" sections (Flevoland, The Netherlands) where the A- and in some cases the E-horizons (corresponding to the upper 10 - 15cm) were lacking (Spek *et alii* 2001a, b). Morphological evidences call for gradual erosion during recurrent flood events whereby water came down the coversand ridge in rills.

Figure 34: Verrebroek "Dok 1"; profile 3, photographic and schematic representation.

Horizon number	Symbol	Horizon depth (cm)	Profile description
H1	Ap	0-35/37	Dark greyish brown matrix; plough layer; calcareous shell fragments in the lowest 10cm
H2	B((g))1	35/37-61/64	Light brown - light yellowish brown matrix; 5-10% ligth grey reduced zones, mainly along bio-galleries; mixture of 20-30% complete subangular blockly and granular pedality; very porous; calcareous shell fragments; very few brick fragments; transition to H3: 5-10cm
H3	B(g)2	61/64-88/92	50% strong brown fine mottles; light grey reduced zones (<1cmø), mainly along bio-galleries; massive pedality, discontinous ped faces; very porous; calcareous shell fragments; mainly pseudogley
H4	Bg3	88/92-124/129	40% strong brown fine (few mm ø) scattered mottles; light grey reduced zones more extended than in H3; massive pedality; very porous; calcareous shell fragments
H5	2BCg	124/129-139/140	Alternating whitish grey sand and clay; stratified; clay with strong brown mottles; massive pedality; very porous; calcareous shell fragments
H6	2BCg.(fe)	139/140-141/144	70% strong brown mottles, mainly at the bottom; massive pedality; very porous; ceramic artefact
H7	3BCg.fe	141/144-144/150	85% strong brown mottles; massive pedality; very porous; calcareous shell fragments
H8	4Hb	144/150-156/160	Black matrix; humiferous; admixture of 5% sand; massive pedality, very few discontinuous ped faces

Table 15: Verrebroek "Dok 1"; profile 3, profile description.

Profile 16, H4 (fig. 28, tab. 10, D.19, D.20, D.40, D.43): Humiferous sand facies
Profile 17, H4 (fig. 29, tab. 11, D.21, D.22, D.40, D.43): Humiferous sand facies
Profile 18, H5 (fig. 30, tab. 12, D.23, D.24, D.40, D.43): Sandy peat facies
Profile 3, H8 (fig. 34, tab. 15, D.11, D.12, D.40, D.43): Sandy peat facies (cf 2.1.4.11.)

The similarity index diagram of P16 (tab. D.20) shows that similarity between the cultivation Ap-horizon (H4) and the underlying Podzolised Soil horizons (SI ≥ 80) is higher than the similarity between the transition horizon to the overlying flood sediments (H3) and the underlying soil (SI < 80). This characteristic confirms that cultivation started before the onset of flood events.

As mentioned above, two different facies are recognised: (i) a humiferous sandy facies, present in P16 and P17, and (ii) a facies consisting of peat horizon mixed with sand, present in P18 and P3. The coarse/fine related distribution and mineral and organic composition of the Ap-horizons confirm the existence of two different types. In the humiferous peat facies, a gefuric to enaulic related distribution was observed, corresponding to organic fine mass and particles embedded in sandy coarse material; in the sandy peat facies, a single to double spaced porphyric related distribution was observed, corresponding to coarse sand material embedded in organic material, mainly organic fine mass. Common (40 - 50% in P16, 30 - 40% in P17), but only frequent (10 - 20% in P18; 20 - 30% in P3) moderately sorted quartz is observed in thin section (tab. D.19, D.21, D.23, D.11) in the (i) dominantly mineral, respectively (ii) dominantly organic facies. Organic matter content evolves from 5% in P16, over 10% in P17, 17% in P18 to 50% in P3.

The pH of all Ap-horizons is neutral, varying from 6.5 to 7.3, and C/N-ratios range between 14 and 19. These chemical characteristics are in favour of a mull humus type that could develop under continued aerobic conditions, enabling soil organisms to coalesce organic aggregates into composites (Lee 1983). Latter composites are represented by the (i) frequently (10 - 20%) to (ii) commonly (30 - 40%) observed amorphous (dark) brown organic fine mass in thin sections of (i) P16 and P17 and (ii) P18 and P3 respectively.

As the amorphous organic fine mass partly has a stiple-speckled b-fabric (tab. D.19, D.21, D.23, D.11), it probably concerns an intimate mixture of organic matter and clay, the latter component however without distinct orientation pattern and thus showing only very slight birefringence. The Ap-horizons indeed contain (i) 4.5 to (ii) 6% clay (except for P3_H8 where particle size was not analysed). Presence of very few (< 2%) dusty, brown clay coatings to infilling fragments (cf tab. D.49) in both facies (except in P3, H8) points towards an anthropogenic reworking of this clay fraction. The clay could eventually correspond to decantation mud from the drainage ditches. Input of ditch decantation mud, mixed with the soil mineral material is indeed mentioned by Pons (1987) to increase fertility on poor land in The Netherlands and Flanders.

On H4 of P16, phytolith analysis was performed (cf 2.2.3.2.). In absolute quantities, very few (< 2%) phytoliths were observed. The phytolith identification gives a coherent assemblage of *Triticum sp.* and panicoid grasses. Weak presence of diatoms and sponge spicules were registered too.

2.1.4.11. | Flood sediments

Denys and Verbruggen (1989) described the particular conditions of non-sedimentation between the end of the peat growth (2000-1500 BP) and the onset of sedimentation in the last millennium. The gradual replacement of the Eastern Schelde by the Western Schelde as the estuary mouth of the river, is generally assumed to be responsible for an increase of tidal influence upstream. However, since the Great Clearances of the 11-12th century, peat digging followed by agricultural activities and dyke building resulted in a lowering of the surface and a reduction of the flood area of the Schelde. From that time onwards natural and strategic inundations regularly affected the polders. The occurrence of 13-15th century *archaeologica* finds in the Medieval cultivation horizon is evidence for a Late Medieval date for the overlying flood sediments. The strategic inundations of Farnèse (1583-1648) during the 80-years War indicate a historical rupture in the sedimentation and stabilisation sequence of the area. The dyke and polder plot pattern that originated at that time is nowadays still recognisable in the landscape.

Despite differences in place and duration of the floods, recurrent strata of calcareous clay and sand were deposited. This unit reaches more than 2m total thickness. Nowadays the polder is used as agricultural land. For this purpose, open ditches and tile tube drainage have been installed. The original sediment stratification is only preserved in the lower part of the alluvium. Two processes have largely erased the sediment stratification in the upper part of the soil: (i) bioturbation during the deposition and later associated with cropland under a lowered groundwater table, and (ii) swelling - shrinking processes. Better drainage conditions, due to tube drainage, have also enhanced the oxidation with the development of a brownish colour and structure Bw-horizon under the present-day plough horizon.

<u>Profile 3, H1 - H8</u> (fig. 34, tab. 15, D.11, D.12, D.40, D.43)

This profile is the reference profile for the Subatlantic (Late Medieval) flood sediments that buried the peat and/or coversands.

H8 shows the transition of the peat to the flood clay and sands. It has no visible distinct plant fragments at field observation, but has an organic matter content of about 50%. Most of the fine mass is dusty brown with an undifferentiated b-fabric and corresponds to amorphous organic colloidal fine mass. Only very few (2 - 5%) partly birefringent organ residue of roots and sclerotia were observed. Sclerotia are fungal hyphae of a fungus type living in symbiosis with nearly all tree roots in forests of boreal and temperate zones (Bal 1973, 53). Mycorrhizal associations between roots and fungi, such as e.g. *Cenococcum sp.*, thus produce sclerotia. Latter are quite frequent in some soils, usually in the lower part or the organic horizons in contact with the mineral material, or they may be preserved in peat (Fitzpatrick 1993, 219-220). Also very few (< 2%) carbonised organic matter fragments were observed. The horizon morphology doesn't correspond to that of *in situ* peat formation but rather points towards some reworked phase of the peat horizon underneath. In that sense, it could correspond to peat that was eroded during flooding, transported in suspension in small fragments and colloidal form and subsequently deposited elsewhere. However, as it has a smooth, nearly plane lower boundary, it is more likely that H8 represents a facies of the Late Medieval cultivation horizon wherein the peat is anthropogenically mixed with sand (cf 2.1.4.10.). H8 has a pH that is close to neutral.

Given the low ESP- (Exchangeable Sodium Percentage) (ESP_H8 - H6 = < 1%) and SAR- (Sodium Adsorption Ratio) (SAR_H8 - H6 = < 1) values in the lower horizons, the soil developed with the flood sediments doesn't show important impact of sodium and doesn't testify for influence of brackish water. However, according to Minnaert and Verbruggen (1986) and Verbruggen and Denys (1991), influences of the tidal system and hence of brackish water in the Subatlantic (2000 - 1500 BP) once stopped the Subboreal peat growth. As a consequence, we can suppose that, if this is correct, sodium is leached out of the system here, since the soil data don't reflect influence of brackish water.

The flood sediments in H7 and H6 contain in their coarse fraction, as in H8, frequent (20 - 30%) poorly sorted subrounded (polycrystalline) quartz and very few (< 2%) feldspars and mica, but have additional few (5 - 10%) calcite and very few (< 2%) carbonate bioclast fragments (tab. D.11). The frequently present (20 - 30%) fine mass shows two types: dominant (i) dusty brown mass with speckled b-fabric and few to frequent (ii) dusty brownish grey mass with Fe-depleted, speckled to crystallitic b-fabric (tab. D.11). These horizons show massive pedality, but are very porous at macroscopic level. In thin section (tab. D.11) they show an incomplete subangular blocky microstructure, eventually pointing towards beginning physical ripening. They have abundant strong brown iron mottles, corresponding to frequent (30 - 40%) amorphous iron (Fe) hypocoatings and nodules in thin section (tab. D.11). Both pedofeatures are developed within the matrix; hypocoatings directly adjoin pores or surfaces. The observed pedofeatures are in favour of gley *sensu stricto* along pores. Gley means that the matrix is reduced whereas the zone along the pores is oxidised. The top of H6 shows additional, though very few (< 2%), Fe-depletion hypocoatings in thin section (tab. D.11), indicating iron leaching. This type of gley *sensu largo* is called pseudogley and fits with reduction along decomposing roots and subsequent transport of reduced ferrous iron (Fe^{2+}) and final oxidation in the soil matrix as ferric iron (Fe^{3+}). It represents the onset of pseudogley, more pronounced in the overlying horizons. Pseudogley is related to dynamics of a perched water table. Organic carbon content in H7 and H6 (OC_H7 = 2.3%; OC_H6 = 1.7%) is about three to two times the amount of the overlying H5 to H2 (OC_H5 - H2 ≤ 1) and decreases from bottom to top, showing an oxidation gradient of the buried organic matter. Also free iron content in H7 and H6 is about two times higher than in the overlying horizons H4 to H2, which as opposed to H7 and H6 don't show iron mottling. Within this profile, higher iron content seems to be related to iron accumulation due to hydromorphism linked to a perched water table.

The flood sediments in H5 to H2 contain frequent (14 - 19%) clay, frequent (20 - 25%) fine silt, common (37 - 51%) medium to coarse silt, few to frequent (around 15%) very fine sand, and very few (< 5%) fine to medium sand (tab. D.12). A heavy sandy loam textural class can be attributed according to the Belgian textural triangle (Marechal, Tavernier 1974).

The groundmass in H5 to H2 consists of common (30 - 40%) well to moderately sorted quartz, few (5 - 10%) rounded glauconite, very few (2 - 5%) calcite fragments up to 0.1mm and carbonate bioclast fragments, frequent (15 - 20%) dusty reddish brown (PPL) mass with speckled b-fabric and frequent (15 - 20%) dusty brownish grey iron-depleted fine mass with speckled to crystallitic b-fabric.

These horizons have a channel to chamber microstructure, confirming the good porosity and aeration already observed at macrolevel and which is the result of faunal burrowing activities. The organic matter content varies gradually from 1.8% in H5 to 1.3 in H2 (tab. D.43), showing an oxidation gradient of the originally buried organic substances from bottom to top. The oxidation gradient, already starting in H7 (tab. D.43), provides evidence for physical ripening and concomitant chemical changes, such as acidification, decarbonatation or organic matter loss, in the flood sediments (de Bakker, Schelling 1989, 43), testifying for initial soil genesis. Very few (< 2%) root and sclerotium residue and very few (2 - 5%) amorphous plant fragments and organic fine mass are visible in thin section (tab. D.11). As for the plough horizon Ap (H1), pH in the underlying horizons H5 to H2 is moderately alkaline and the C/N-ratio is below 12, indicating good decomposition of organic matter. As mentioned by Brady and Weil (1999, 455), important calcium contents, freely available as cations on the exchange complex of these horizons, have a positive effect on lowering the C/N. The free iron content is around 1% and comparable to that of the plough horizon H1; the amorphous iron content is 0.7% and slightly higher than in H1. Micromorphological observations show frequent (10 - 20%) amorphous iron (Fe) hypo- and quasicoatings in H5 to H4 and few (5 - 10%) amorphous iron nodules and hypocoatings in H3 to H2. All of these pedofeatures are developed within the matrix; hypocoatings directly adjoin pores or surfaces, whereas quasicoatings are related to, but however don't directly touch pores or surfaces; nodules are not related to pores or surfaces (Bullock *et alii* 1985).

Along the horizon sequence H5 to H2, only very few (< 2%) manganese (Mn) nodules were observed (tab. D.11). This element only precipitates when oxidising conditions are high or in slightly to moderately hydromorphic conditions (Stoops, Eswaran 1985). Indeed, Mn^{2+} has a higher redox potential than Fe^{2+} which means that it has to move further than Fe^{2+} along the redox gradient before the redox is sufficient for its oxidation and precipitation. The minor presence of manganese nodules reveals that either (i) oxidation is not extreme within these horizons, or (ii) the flood sediments have low manganese content. All observed iron and manganese pedofeatures are rather faint due to ongoing bioturbation and homogenisation processes in the soil. Together with the above-mentioned arguments for good oxidation conditions within this soil, it seems more likely that the flood sediments originally contained little manganese. Manganese and manganese-containing sediments were probably transported further downstream with the Schelde River.

Presence of iron hypo- and quasicoatings, combined with low chroma in the matrix, are indicative of strong hydromorphic conditions, i.e. related to soil material remaining completely wet during several months (Stoops, Eswaran 1985). Latter conditions result in a strongly reduced soil matrix; macropores are the only sites that are still oxidised. Latter conditions are well developed in H5 and H4. In a first phase, reduction in the matrix as opposed to oxidation in pores induces iron transport from the matrix to the pores, giving rise to Fe-hypocoatings. Along decomposing roots, ferric iron (Fe^{3+}) might subsequently be reduced to ferrous iron (Fe^{2+}) and transported towards the matrix. Latter results in Fe-quasi-coatings, indicative of pseudogley. The tendency already observed in the top of H6 is thus extended towards the top. This type of gley is related to dynamics of a perched groundwater table. Strong hydromorphism is furthermore associated with long anaerobic conditions, which enables organic matter to be preserved in the soil material (Stoops, Eswaran 1985). H5 and H4 indeed have slightly higher organic matter contents than H3 and H2 (tab. D.43).

H3 and H2 mainly have nodules and furthermore also hypocoatings. Presence of nodules might indicate (palaeo-)surface gley, pointing towards an impermeable layer close to the surface or resulting from natural or artificial flooding during part of the year (Stoops, Eswaran 1985). Prior to dyke construction, flood dynamics were indeed formerly present at the site of Verrebroek.

Few (5 - 10%) clay-size calcite quasicoatings, coatings or infillings (cf tab. D.48) are present. They are secondary deposits; primary carbonate mass present in the groundmass as crystals or bioclasts weathered, dissolved, subsequently migrated and precipitated during alternating wet and dry soil conditions. The zones along former roots are likely environments for these precipitates, as roots take up nutrients and suck water from the water-filled pores.

H5 to H2 also contain very few (< 2% in H5 - H4, 2 - 5% in H3 - H2) dusty brown clay coatings reaching a thickness of 30µm in H2 and H3. The coatings, in fact consisting of clay and microscopically small fragments of organic matter, correspond to so-called "agricutans", as defined by Jongerius (1970). These features develop as a result of surface or internal slaking, i.e. the collapse of the structure of a cultivated horizon due to saturation with water. Parallel, but independent of agricultural land use, Stoops and Eswaran (1985) mention clay illuviation phenomena linked to hydromorphic conditions. With respect to this hypothesis, the higher mobility of clay is attributed to the absence of stabilising Fe oxyhydrates, provoking more easily dispersion of the fine material. An alternative hypothesis to clay dispersion enabling its transport, is that the water flow from the saturated surface horizons is at first "milky" with suspended soil particles of all fractions, but that it clears and only leaves the finest particles in suspension thereafter, as mentioned by Romans, Robertson (1983). This type of clay migration is different from clay migration driven by colloidal dispersion of the clay particles that only occurs within a pH interval of 5 to 7 (Schachtschabel *et alii* 1998, 388). The moderately alkaline soil pH in this soil indeed is not favourable for clay migration in colloidal dispersion.

All horizons contain very few (< 2%) carbonised particles up to 100µm; also very few brick fragments were observed in the field; water-transported alluvial and colluvial sediments commonly contain such material.

Total phosphate (P_2O_5t) contents are around 1400mg kg^{-1}, which is relatively high; the inorganic phosphate content is four to two times higher as the organic phosphate content (Pi/Po_H5 - H2 = 3.8 - 1.7). Nitrogen (N) contents are relatively high as well

(N_H5 - H2 = 0.10 - 0.17%). It is considered that the high productivity associated with the estuarine ecosystem is largely due to the detrital food chain (Boaden, Seed 1985, 81-82). The few quantities (2 - 5%) of amorphous (dark) brown organic fine mass observed in thin section (tab. D.11) provide indeed evidence for detritus, i.e. litter residue characterised by an advanced decomposition and colour change (dark brown), too.

H2 has a texture that is finer than sandy loam and higher clay content than the underlying horizons (< 3.9 μm_H2 = 18.7%, < 3.9 μm_H3 - H7 = 11.8 - 14. 3%). According to the similarity index diagram, textural difference is however most marked between H2 and H1 (SI_H2xH1 = 81, SI_H2xH3 - H7>85). Both H3 and H2 have a higher chroma (taken on crushed/rubbed air-dry sample) than the underlying horizons (Chroma_H7 - H4 = 2.4 - 3.0, Chroma_H3 - H2 = 3.1); latter chroma increase is gradual from bottom to top. The hue (taken on rubbed air-dry sample) is nevertheless slightly shifting towards yellower colours when moving from bottom to top (Hue_H7 = 9.4YR, Hue_H4 - H2 = 0.1Y), whereas redder hue is expected in a cambic Bw-horizon as opposed to the underlying horizons (WRB 1998, 28-29). It is however evident that H2 is completely mixed due to bioturbation as opposed to the underlying horizons that still show faint mottling and/or even the original stratification or lamination from the flood event. Furthermore H2 has a subangular blocky to granular pedality, whereas the underlying horizons are massive. Thickness of H2 is at least 0.15cm and its lower boundary is located deeper than 0.25m below the soil surface. Cation exchange capacity in H2 was not determined. As a consequence apparent exchange capacity (ACEC) could not be checked. It is however expected that H2 has comparable ACEC values as in H7, H6 and H1 that have values between 120 and 140cmol(+) kg^{-1}, well above the required 16cmol(+) kg^{-1}. Above characteristics confirm that H2 meets the requirements of a cambic Bw-horizon as defined by the World Reference Base for Soil Resources (WRB 1998, 28-29) (Appendix B. Glossary). Iron content in H2 is however not exceptionally high when compared to the other horizons; it is even lower than in the horizons with iron mottles related to pseudogley (H6 and H7). This indicates that the colour is not the result of a weathering process here but simply the result of homogenisation.

H1, the plough horizon or Ap-horizon, serves as a reference horizon with respect to nutrient availability. Its pH is moderately alkaline (pH_H1 = 8.1). The C/N-ratio is below the median value of 12 for arable horizons (Brady, Weil 1999) and indicates a good decomposition of the organic matter. The inorganic phosphate content is twice as high as the organic phosphate content, which again points towards a good decomposition and mineralisation of the organic matter and eventual direct input of inorganic phosphate to increase the soil fertility for agricultural purposes. The cation exchange capacity is rather low when compared to other arable soils in Belgium with comparable medium silt loam texture (Verloo 1992). The exchange complex is saturated with calcium, with an adverse effect on the other basic cations; contents of exchangeable calcium, sodium, magnesium and potassium are respectively very high, moderate, high and moderate, again when compared to values for other Belgian arable soils (Verloo 1992). The free iron (Fe_d) content is close to 1%; the amorphous iron (Fe_o) content is only about 0.6%.

Throughout the profile, we observe leaching of exchangeable cations (Na, Mg, K). Data on exchangeable calcium (Ca) are unreliable for leaching behaviour interpretation here, as free calcium carbonate is present throughout the flood sediment sequence. A differential leaching pattern is however not observed. The latter is possibly difficult to retrieve as the different flood sediment strata probably originally had different cation contents.

Differential leaching of elements occurs according to their differential mobility behaviour in soils. The simple salts in which the alkali metals sodium (Na) and potassium (K) are present, are very soluble (Van Ranst 1991, 40). As alkali metals, also alkaline metals as magnesium (Mg^{2+}) and calcium (Ca^{2+}) are only stable in water as cations, not in metallic form. The solubility of Mg^{2+}-salts is higher than that of Ca^{2+}-salts (Van Ranst 1991, 41).

2.1.5. | Doel "Deurganckdok-sector B"

2.1.5.1. | Pedo-lithostratigraphy

A schematic description of the reference profile (fig. 14) at Doel "Deurganckdok-sector B" from bottom to top is as follows.

1. Upper Pleniglacial coversands
2. Brunified Soil with a colour Bw.bi- (fig. 35) and in depth a clay accumulation Bt-horizon, with from top to bottom:
    - a 0.15 to 0.20m thick A-horizon;
    - no E-horizon;
    - a homogeneous brown (with a higher chroma than the underlying C-horizon), bioturbated Bw-horizon;
    - fragmented clay accumulation bands (Bt-horizon).
3. Final Mesolithic (Swifterbant) occupation, probably synchronous with the Brunified Soil development
4. Peat, containing distinct wood fragments

2.1.5.2. | Brunified Soil

Profile 1, H1 - H8 (fig. 35, tab. 16, D.26, D.27, D.41, D.44)

H1 corresponds to a peat horizon. H2 to H8 are the underlying mineral soil horizons.

The profile is situated within an area that shows evidence of Final Mesolithic human occupation. Traces of Final Palaeolithic occupation were discovered as well. Distribution of the Final Mesolithic (Swifterbant) artefacts is limited to the upper 0.15m, corresponding with the darker coloured surface A- horizons. The artefacts display no soil capping and lack any trace of posterior frost damage (Crombé *et alii* 2000). Final Palaeolithic (Federmesser) artefacts were mixed deeper down and were found within the upper 0.40m. They display soil capping, indicating alternating freeze-thaw cycles (Van Vliet-Lanoë 1985).

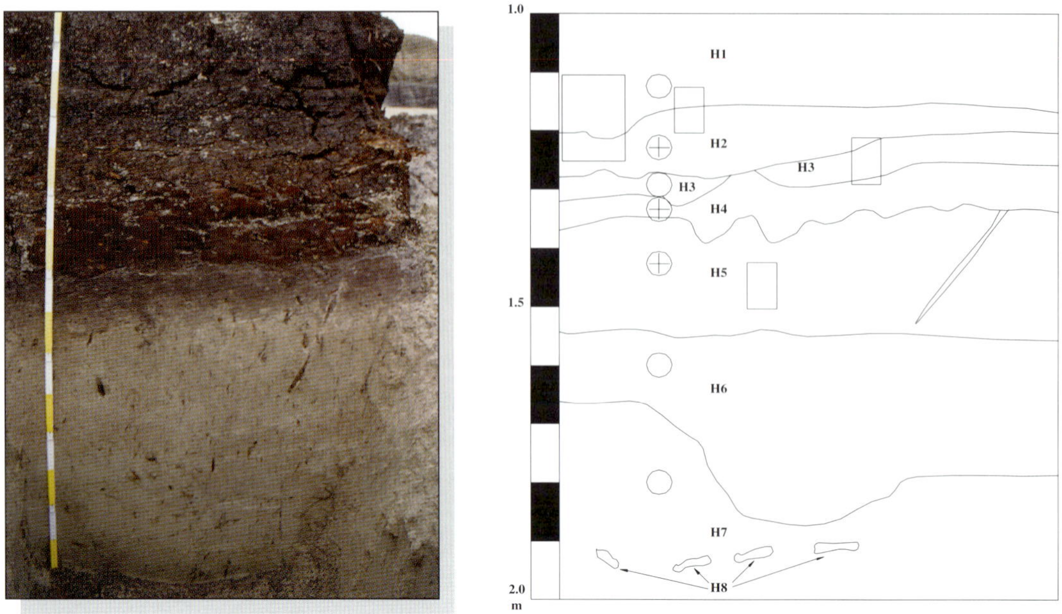

Figure 35: Doel "Deurganckdok - sector B" ; profile 1, schematic and photographic representation.

Horizon number	Symbol	Horizon depth (cm)	Profile description
H1	H	(100)-114/121	Very dark brown matrix; peat; distinct wood fragments
H2	A1	114/121-120/129	Dark grey - greyish brown matrix; humiferous sand; charcoal fragments; "salt and pepper"; Final Mesolithic (Swifterbant) ceramic artefacts
H3	A2	120/129-125/133	Light brownish grey matrix; sand; very homogeneous; Final Mesolithic (Swifterbant) ceramic artefacts
H4	ABbi	127/133-132/139	Transition; not homogeneous; Final Mesolithic (Swifterbant) ceramic artefacts
H5	Bbi1	132/139-153/155	Light grey matrix; sand; more humiferous than H6; common roots
H6	Bbi2	153/155-166/188	Light grey matrix; sand; common roots
H7	BC	166/188-(200)	Light grey matrix; sand; stratified; common roots
H8	Bt	190/194-191/195	Fragmented clay accumulation bands in H7

Table 16: Doel "Deurganckdok-sector B" ; profile 1, profile description.

The parent material (H7) shows a horizontal stratification to lamination. It consists of dominant (> 50%) poorly sorted, subrounded (polycrystalline) quartz, very few (2 - 5%) rounded glauconite that is partly, mainly chemically weathered and very few (< 2%) subrounded to rounded chert, feldspars and other mica minerals. The standard deviation of the particle size distribution (tab. D.41) (SD_H7 = 70.4µm) accordingly points toward a very poorly sorted sediment. The parent material has a median particle size of 158µm (tab. D.41), and is thus finer than the remobilised sands of Verrebroek "Dok 1" (cf 2.1.4.7.), but rather comparable in size to the Upper Pleniglacial coversands at Verrebroek "Dok 2" (cf 2.1.3.2, 2.1.3.4)

The mineral soil profile (H2 - H8) represents both a colour Bw-horizon (H5 - H6) and fragmented clay accumulation Bt-bands (H8). Morphological evidence of an eluviated E-horizon is however absent.

Texture throughout H3 to H6 shows strong similarity; similarity indices vary between 94 and 98 (tab. D.27). H2 and H7 have relatively lower similarity with the other horizons; their indices with the under-, respectively overlying horizons are below 95. H2 has higher clay and fine to medium silt content than the underlying horizons, whereas H7 has lower clay and silt content as compared to the overlying horizons. The higher clay content might be the result of physical weathering due to biological activity or frost activity. However, frost action is a less probable mechanism, since no other evidences were noticed. As H2 represents a human occupation surface, additional input of fine material linked to human activity is a further possibility.

Except for the covering peat horizon (H1) and the upper buried surface A-horizon (H2) that have a moderately acid pH (pH_H1 - H2 = 5.8 - 6.0), pH-values in P1 range between 4.2 and 5.4, fitting extremely to strongly acid soil conditions. Cation exchange capacity throughout H3 (A2-horizon) to H7 concomitantly is very low, ranging from 4cmol(+) kg^{-1} in H3 to 2cmol(+) kg^{-1} in H7 (BC-horizon). Exchangeable sodium percentages (ESP) in the mineral horizon are close to or above 15%, i.e. the threshold value for sodic soils, and don't show a gradient in either upward or downward direction. Sodium adsorption ratios (SAR) are however only around 6 to 7, being clearly below the threshold value of 13 for sodic soils. The extremely to strongly acid conditions moreover don't fit with sodic soil conditions. Exchangeable magnesium percentages (EMP) are around 50%, which is very high when compared to the reference Podzolised Soil profiles at Verrebroek that have values between 3 and 10.5%. Surprisingly, ESP and EMP are lowest in the horizon that corresponds to the Swifterbant living floor (H2); exchangeable sodium (Na) and magnesium (Mg) and calcium (Ca) contents are however highest here. Concomitantly, phosphate contents are highest as well; phosphate is mainly present as inorganic phosphate (Pi/Po_H2 = 1.3). Highest exchangeable sodium (Na), magnesium (Mg) and calcium (Ca) contents are also found in the Aa-horizons, the darkest part of the hearth-pit fillings that contain most charcoal fragments and powder (cf 3.3.2.). The relatively high sodium and magnesium percentages possibly result from input linked to human occupation, or more specifically from fire effect. Release of inorganic phosphate, magnesium and potassium indeed is commonly observed as a consequence of fire at the soil surface (Firmin 1986). Yet, sodium and magnesium contents are relatively high, whereas potassium content is relatively low. As a consequence, mere fire effect can hardly account for the relatively high cation contents.

Magnesium (Mg) release during glauconite weathering provides an alternative explanation. A theoretical composition of glauconite, $[Si_{7.5}\ Al_{0.5}]^{IV}\ [Fe^{3+}_{2.0}\ Al_{1.0}\ Fe^{2+}_{0.2}\ Mg_{1.0}]^{VI}\ O_{20}(OH)_4\ K_{1.1}$ (Van Ranst 1995, 115), reveals that Mg is present in octahedral position and becomes available on isomorphous substitution or chemical weathering. However, as Mg contents increase from bottom to top, natural chemical weathering of glauconite is excluded here. Furthermore, MgO is commonly only present in minor amounts of about 3 to 4% in glauconite (Van Ranst 1995, 118). Weathering induced by fire would also here be possible. Experiments (cf Appendix C. Methods) revealed that glauconite starts to loose its crystalline morphology at temperatures of about 500°C; temperatures of about 1100°C are needed in order to complete the transformation from crystalline to amorphous silicates in pure glauconite. During the latter transformation, cations incorporated in the lattice can be released. Latter could explain the higher contents of magnesium in the top as opposed to the bottom horizons. Indeed, the surface horizons parallel contain burnt bone, which was found clustered either in surface-hearths or hearth-dumps elsewhere on the site and with a processing temperature of 500 to 700°C (cf 3.1.3.3., 3.5.4.). However, the absence of carbonised organic fine mass in this profile suggests that burning didn't necessarily occur *in situ* here (cf tab. D.47). The carbonised ecofacts, such as very few charcoal (< 2%) and a fragment of burnt bone, observed in thin section (tab. D.26), could have been spread accidentally or intentionally from elsewhere on this particular spot. In latter case, the presumed high temperatures for bone burning are not guaranteed on site.

Decrease of exchangeable potassium (K) from bottom to top in the mineral soil horizons (tab. D.44) probably reflects mica chemical weathering. As a consequence, the clay coatings observed in H5 (tab. D.26) might be (partly) composed of mica clay.

Within the mineral horizons, $Al_o+{}^1/_2\,Fe_o$-contents are only around or slightly higher than 0.05%, confirming that podzolisation was not active here. As compared to the site of Verrebroek "Dok 1", peat growth started earlier here. Carbon dating at the base of the peat horizon gives 5050±55 BP at Doel, as opposed to 4690±30 BP at Verrebroek (cf 2.3.3.2., 2.3.4.). When comparing the texture of the parent material of both sites, it is evident that Doel has a finer texture than Verrebroek "Dok 1": the peak in the median particle size distribution lies at 158µm in Doel as opposed to 192µm in Verrebroek (P16); Doel has relatively more material in the clay, silt and very fine sand fraction too (tab. D.27, D.20, D.41). Parallel, soil genesis at the "Hoge Vaart" sections (Flevoland, The Netherlands) was mainly limited to brunification and showed only slight evidences of podzolisation; peat growth on the coversands started there already from 5500 BP onwards (Spek *et alii* 2001a).

Alternatively, H5, recognised as a Bw-horizon in the field, has very few (< 2%) thin (< 5µm) dusty brown-grey (PPL) clay coatings, reflecting *in situ* alteration typical for a cambic Bw-horizon (WRB 1998, 28-29). Both H6 and H5 have a stronger chroma (Chroma_H6 - H5 = 3.6 - 4.0), a redder hue (Hue_H6 - H5 = 8.6 - 8.2) and higher clay content (< 3.9µm_H6 - H5 = 2.3 - 2.5%) than the underlying H7 (Chroma_H7 = 3.5, Hue_H7 = 9.1YR, < 3.9 µm_H7 = 1.1%). Texture difference between the BC-horizon H7 and the overlying Bw-horizons H6 and H5 and the reasons behind were earlier commented when considering the similarity index diagram (tab. D.27). Though both H6 and H5 have definite characteristics of a Bw-horizon, their texture is too coarse to meet all the necessary requirements to be classified as a cambic Bw-horizon.

The buried surface A-horizons, H2 and H3, are considered to correspond to the Swifterbant living floor or human occupation surface. Characteristics of human occupation are most expressed in H2. It concomitantly has the highest phosphate content, mainly consisting of inorganic phosphate; it has a high Pi/Po-ratio (Pi/Po_H2 = 1.3) too. Both burnt bone and charcoal fragments were sporadically (< 2%) observed in thin section (tab. D.26), both possibly contributing to the dominantly inorganic phosphate content. Nitrogen (N) content is relatively high as well (N_H2 = 0.12%).

Organic matter is both in H2 and H3 frequently (5 - 20%) present as amorphous dark brown organic fine mass (cf tab. D.46), also here, as in P3 at Verrebroek "Dok 2" possibly bound to clay. C/N-ratios are concomitantly lowest (C/N_H2 - H3 $\leq$ 17). C/N-ratios under 17 are typical for well-aerated conditions, enabling soil organisms, especially earthworms, to mix and bind organic with mineral material (Lee 1983).

The combined presence of the altered Bw- and homogeneous A-horizons, the latter showing clear characteristics of mull humus type conditions, fit well with Duchaufour's concept of brunification and Brunified Soils (Duchaufour 1982, 270-305); P1 could according to his classification be designated as a Temperate Brown Soil.

Wet conditions are evidenced by presence of very few pyrite (FeS_2) framboids (< 2%) (cf tab. D.45) throughout the mineral soil profile and the relatively thick peat cover. Considering the presence of a Swifterbant living floor here, pyrite formation must be diagenetic and linked to subsequent wet conditions favouring peat growth. These continued wet conditions after peat burial guaranteed rather good protection of the archaeological site. As a consequence, even rather fragile artefacts and ecofacts, such as e.g. burnt bone, are well preserved (Appendix B. Glossary).

The peat horizon, H1, has an organic matter content of 72% and meets also the other requirements of a histic horizon (Appendix B. Glossary).

Pollen analysis was performed in H2, H4 and H5 of P1 (cf 2.2.4.). These horizons correspond with the A-, ABbi- and Bw.bi-horizons of a Brunified Soil. These horizons didn't reveal any pollen. The pollen-sterile context is in favour of well-aerated conditions with important biological activity (Havinga 1962) and fits with the bioturbated, homogeneous character of these Brunified Soil horizons.

2.1.6. | Doel "Deurganckdok - sector A"

Cyriel Verbruggen

During the whole period of the excavation of the "Deurganckdok", the geomorphologists of the ADW carefully followed up the stratigraphy of the exposed walls. A generalised profile was composed showing all the main stratigraphical units of the quaternary deposits (fig. 36, 37).

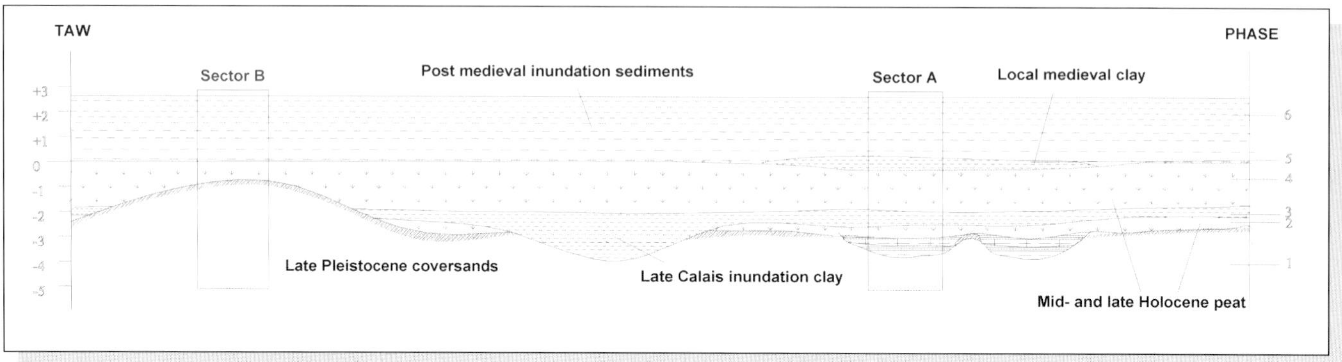

Figure 36: Doel "Deurganckdok - sector A"; generalised profile of the western foundation trench of the dock, marking the investigated sections within sector B and sector A.

Figure 37: Doel "Deurganckdok-sector A"; photographic and schematic representation.

Elaborate palaeobotanical research and ^{14}C-dating made it possible to reconstruct the palaeogeomorphological and –ecological conditions of the development of the different stratigraphical units. The following chronological and sedimentological phases could be distinguished.

1. Roughly until 6000 BP: a relative dry, sandy surface of Pleistocene coversands and restricted patches of Late Glacial dune sands in the neighbourhood of the River Schelde.
2. Between 6000 BP and 5700 BP: due to the rising sea level, the lower parts around the river change into a fen, causing peat accumulation.
3. Between 5700 BP and 4900 BP: increased marine influence induces a gradual inundation of the fen: deposition of organic clay to clayey peat. Only the highest landscape portions, such as the top of the sand ridge in sector B, remain dry.
4. Between 4900 BP and 1300 BP: continuous peat growth, evolving from an eutrophic fen into a mesotrophic one and finally ending as an oligotrophic bog.
5. Between 1300 BP and the 14th-15th century: the marine influence increases and peat growth has been stopped; the surface is covered by restricted silty sedimentation; beginning of dyke building in the 13th century and digging of peat.
6. From the 14th-15th century: onset of catastrophic inundations in the polders. Long lasting strategic inundation (1580-1640) and later inundations are responsible for the main deposition of clay and clay-sands, forming the present-day surface.

2.1.7. | Conclusions

2.1.7.1. | Verrebroek "Dok 2"

1. The parent material at Verrebroek "Dok 2" consists of Upper Pleniglacial coversands. They are dominantly horizontally laminated. This lamination points towards recurrent cycles of periodic sediment deposition. Any type of bioturbation processes hasn't visibly disturbed the sedimentary sequence afterwards. However, evidences of congeliturbation, such as the presence of few faint isolated ice vein casts with upward bended side laminae, were observed (P9). The coarse fraction of the parent material consists of dominant subrounded quartz and polycrystalline quartz, very few rounded, partly weathered glauconite and very few subrounded to rounded chert, feldspars, mica and eventually compound fragments of quartz, glauconite and dusty clay. The Upper Pleniglacial coversands have a median particle size between 125 and 155µm.

2. During the Late Glacial, a humiferous soil of 0.10-0.15m thickness developed within these coversands. Generally, the humiferous part of these Late Glacial soils showed (i) no, or very little, bioturbation, and (ii) an abrupt, somewhat wavy lower boundary. These Late Glacial soils developed according to a reverse graded bedding sedimentary structure (P11_H8-H4, P1_H6-H3), observed via the shift from the fine towards the coarse particle size classes. This phenomenon probably corresponds to ongoing suspension settling in a lake or dune slack environment. Additional wind ripple morphology is in favour of slow flow velocity conditions in shallow water in which the wind causes tiny waves. Particle size rupture within this

sedimentary structure indicates a shift in sediment transport from a small to a large scale, from subaqueous, wind-induced sediment transport in a dune slack towards direct aeolian sediment transport linked to dune dynamics. On the Final Palaeolithic site (P11), presence of fragmentary, mainly clustered charcoal was observed in the Late Glacial soil. These horizons also contain poorly oriented clay coatings or infillings that can be interpreted as fragments of former surface crusts. The Final Palaeolithic (Federmesser) flint artefacts indicating a former occupation surface occur in the same horizon. Based on a first analysis, around 60% of these artefacts display soil capping, indicating alternating freeze-thaw cycles. Heating traces occur on 15% of the artefacts. Human activity is also sustained by relatively high total phosphate contents, mainly present as inorganic phosphate. The parallel presence of relatively high exchangeable magnesium and potassium contents is in favour of nutrient release on organic matter carbonising. P1 and P3 showed presence of a peat horizon. The presence of horizontal elongated, yellow to reddish brown organ residues, partly showing birefringence, suggests *in situ* formation of this peat horizon. Pyrite framboids in and around the peat horizon are in favour of continued wet conditions. Nevertheless, biological activity evidenced by coprogenic organic fine mass particles or pellets, presumes that subaqueous conditions were occasionally interrupted by aerobic conditions, reflecting retardation or a break in the peat growth process. High total phosphate contents, cation exchange capacity and exchangeable cations indicate organic matter mineralisation and thus support the occurrence of oxidising conditions. The morphology of the organic matter present in overlying horizon(s) is in favour of reworked, rather than *in situ* deposited organic material. Based on pollen analysis and ^{14}C-dating, the Late Glacial soil at Verrebroek "Dok 2" can be attributed a Bølling/Earlier Dryas age.

At the bottom, spherical- to ellipsoidal-shaped structures composed of alternating laminae of white quartz sand-size grains and slightly humiferous sand were locally observed. As coarse-grained, heavier quartz sank into finer-grained lighter quartz, the sedimentary structures probably correspond to load structures, such as load casts or ball and pillow structures. These are known as post-depositional deformation structures linked to liquefied sediment and rapid sedimentation conditions. From the surrounding lamination following the sedimentary structure morphology, we conclude that pressure, either external or internal, played a crucial role in the development of these structures.

Involutions of the Late Glacial soil in the underlying sediment (P6) locally appeared. These involutions were recognised as congeliturbation structures and correspond to a particular type of flat-based involutions, called Drop Soil. The underlying sediment shows evidence of repeated freeze-thaw processes through (i) incomplete prismatic to subangular microstructure, probably the result of aggregate formation, (ii) whitish ice segregation eluvial fibres, witnesses of former ice lenses, and (iii) limpid, light green clay coatings, probably the result of physical glauconite weathering. Combined presence of (i) ice segregation eluvial fibres in the underlying sediment, (ii) relatively high organic matter content in the intruded Late Glacial soil A- and B-horizons, (iii) the reverse graded bedding sedimentary sequence within the Late Glacial soil, responsible for a differential particle size sequence, and (iv) the flattened bottom of the Late Glacial soil involutions, delivers sufficient arguments to interpret the observed morphology as a "Tropfenboden" or Drop Soil. Latter soil originated in a periglacial, even permafrost environment, more specifically within the water-saturated and seasonally freezing and thawing active layer. This environment represents conditions favourable for three important mass mouvement mechanisms: i) frost heave, ii) cryostatic pressure within the active layer, and iii) intergranular or pore water pressure within the water-saturated, hydrodynamically unstable, thawed active layer. However, occurrence of ice lenses parallel to the soil surface proves that liquefaction did not occur during thawing, thus counteracting the last mentioned mass movement mechanism typical for load casting. As a consequence, the observed involutions cannot be attributed to load casting.

Involutions of the parent material into the Late Glacial soil locally occurred (P11) as well. They show a linear, locally continuous pattern, whereby two involutions of about 0.10m wide occasionally appear parallel to one another, and whereby one crack follows a narrow frost crack or fissure. The frost fissure only witnesses frost action, no permafrost conditions. The parallel cracks are recognised as thermal contraction cracks, being the result of upward cracking with crack bifurcation within a permafrost environment. These permafrost conditions, either continuous or discontinuous, must have been active after the Late Glacial soil genesis. Additional mechanisms were responsible for the subsequent infilling of these cracks with concomitant upward mass flow. Interpreting the observed involutions as sand dykes, a type of water escape structure, implies accepting (additional) pressure at some time posterior to the Late Glacial (Bølling/Earlier Dryas) soil formation. This pressure could be either linked to (i) human impact, such as the Final Palaeolithic human occupation, or (ii) frost action, such as eventual freezing of the pond/lake water, or (iii) a combination of both. Diapiric upward movement, another type of water escape structure and driven by differential bulk density or differential frost susceptibility within the active permafrost layer during the thaw season, is an acceptable mechanism too.

3. Horizon(s) consisting of accumulation bands of very mobile organic matter and/or clay substances, from a few mm up to 0.30m thick, with abrupt upper and lower boundaries, were observed. Limpid strong brown organic fine mass coatings were visible in thin section. Precipitation probably occurred at the level of a texture jump or (temporary) water table. Most of this organic matter accumulated on top of the Late Glacial soil, forming in most profiles a continuous horizon of about 0.05-0.30m thickness. Several, often discontinuous organic matter accumulation bands were observed underneath the Late Glacial soil too. In both cases, their formation is diagenetic to Late Glacial soil formation. Morphologically, these accumulations resemble so-called "waterhard" or dopplerite. Chemical characterisation at the Verrebroek site was too limited to draw conclusions on this matter.

4. The basis of the redistributed aeolian sands, deposited on top of the Late Glacial soil, were locally recognised within the pedo-lithostratigraphic sequence. When deposited above the groundwater table, they are horizontally up to obliquely stratified; otherwise, when deposited in open water, they lack any stratification. The orientation of the oblique stratification indicates that the sediment was blown by western wind. These remobilised sands have a median particle size around 230µm and are thus coarser than the Upper Pleniglacial coversands.

5. Traces of a Brunified Soil with a colour Bw-horizon were only observed in particular positions. Brunified Soil genesis calls for climatic conditions where precipitation is lower than, or in balance with, the evapotranspiration. It probably corresponds to the soil evolution at the beginning of the Holocene, when Mesolithic hunter-gatherers settled periodically in the area. The horizon sequence typically consisted of (i) no, or only a slightly developed E-horizon, and (ii) a homogeneous brown, bioturbated Bw.bi-horizon. The A-, E- and ABbi-horizon(s) had frequent amorphous dark brown, probably humified, organic fine mass coatings and infillings; they parallel have higher clay to very fine sand contents. These horizons probably represent a mull humus type with clay-humic aggregates representing aged excrements. Latter complexes are typical for the brunification process and prevent the presence of true mobile complexes that are required for the podzolisation process. The underlying Bw.bi-horizons have higher chroma than the surrounding horizons, evidence of alteration. Their $Al_o+{}^1/_2\,Fe_o$-content is below 0.20%, as opposed to the Bh(.s)-horizons of the Podzolised Soils at Verrebroek "Dok 1" that have values above 0.20%. They showed weak evidence of clay illuviation, which concludes for clay migration as a secondary process superposed on the brunification process. They contain metal (aluminium)-humus compounds as well, probably indicating a weak, beginning tendency of podzolisation.

6. In the majority of the soil sections, an imperfectly to poorly drained Podzolised Soil developed.

7. Peat was locally present.

8. Calcareous flood clay and sand with variable thickness buried the coversands and peat, where present.
   The soil formation of the Podzolised Soil, the peat horizon and the calcareous flood sediments is discussed in detail at Verrebroek "Dok 1".

2.1.7.2. | Verrebroek "Dok 1"

1. Also at Verrebroek "Dok 1", the parent material consists of Upper Pleniglacial coversands. It shows a horizontal to oblique stratification or lamination. It consists of dominant poorly sorted, densely packed, subrounded quartz and polycrystalline quartz, very few rounded glauconite that is partly, mainly physically weathered, and very few subrounded to rounded chert, feldspars and other mica minerals; some laminae are however well sorted. The Upper Pleniglacial coversands have a median particle size between 100 and 115µm, and are thus slightly finer than at Verrebroek "Dok 2". Locally, wavy morphology (P14, P9) and presence of very dusty to impure dark brown clay coatings or infillings, amorphous very dusty organic fine mass and/or carbonised organic particles suggest fluviatile or fluvio-aeolian sediment input, rather than pure aeolian input. In these Upper Pleniglacial coversands, a number of soil forming processes occurred and gave rise to the development of soil horizons.

2. During the Late Glacial, a humiferous soil of 0.10 to 0.15m thickness developed within these coversands. This soil locally contains charcoal, mostly in powdery form and shows no or very little bioturbation, except for the presence of roots up to several centimetres diameter, linked to an overlying, younger Podzolised Soil. It has an abrupt, somewhat irregular lower boundary. Locally it is composed of two units, separated by aeolian sediment input. In P9, characteristics of the older unit are in favour of a Gyttja soil type. It has a high nutrient status and nutrient ratios reflect good mineralisation. Amorphous dusty organic fine mass coatings, infillings or particles, possibly of coprogenic origin, clay-size carbonate coatings and infillings and spherulitic carbonate crystals and pyrite framboids are present. The carbonate can be calcite, $CaCO_3$, or siderite, $FeCO_3$. The observed characteristics call primarily for reducing, water-saturated conditions, followed by aerobic conditions within this unit of the Late Glacial soil. The younger unit consists of peat with distinct plant fragments and common amorphous dusty brown organic fine mass, calling for strongly humified conditions. As the Gyttja Soil, it contains pyrite and carbonate coatings, fillings and crystals. These characteristics fit within Kubiena's definitions of Peat Gyttja. The Late Glacial soil sequence in P9 calls for continuous wet and anaerobic conditions over a litho-stratigraphic sequence of almost half a metre. The presence of gyttja and peat could eventually correspond to the infilling of former river gullies; latter would fit with the fluviatile sediment character of the parent material. Based on pollen analysis and ^{14}C-dating, both units of the Late Glacial soil at Verrebroek "Dok 1" could be attributed an Allerød age.

3. Particular light olive brown mottles and dark grey mottles, synchronous to the Late Glacial soil, were locally observed (P11) in the upper unit of the Late Glacial soil. Charcoal fragments, carbonised organic fine mass coatings and infillings and carbonate coatings, infillings and crystals call for presence of ash and are in favour of an *in situ* fire event. The C/N ratio fits with values around 60 commonly reported for wood ash. The olive brown mottles have high inorganic phosphate, iron and aluminium contents and show parallel dusty fine mass with undifferentiated b-fabric and amorphous and poorly oriented calcium iron phosphate infillings. As the analysed phosphate contents are quite high, phosphate input might eventually be associated with an occupation event reflecting deposition of animal refuse and/or ashes; the latter were attested in this particular feature. Eventually, remnants of game trapped in an occasional forest fire might be a possible explanation for the high phosphate content too. No archaeological evidences of human occupation were observed at first sight here. Seen the particular, albeit very local interest of these phosphate and ash accumulations, we suggest investigating comparable features from an archaeological point of view during future excavations, possibly clarifying eventual human impact.

4. A particular feature posterior to the Late Glacial soil but prior to the poorly developed Podzolised Soil extending from about 0.50 to 0.95m below the Mesolithic occupation surface, was locally observed (P14). Presence of amorphous limpid cracked to slightly dusty organic matter coatings and relatively high inorganic phosphate and free iron content suggests amorphous iron phosphate accumulations. They possibly developed prior to podzolisation and correspond then to local phosphate discharge. It should be questioned whether we could expect a burial place

here, eventually linked to Early Mesolithic occupations. Alternatively, high quantities of bones and possible other phosphate-rich material, as e.g. flesh, must have been dumped on the Early Mesolithic occupation surface. Posterior migration and accumulation in depth following the groundwater table of these phosphates, eventually in association with iron, might have resulted in the observed olive yellow zones.

5. Horizon(s), consisting of accumulation bands of very mobile organic matter and/or clay substances, with abrupt upper and lower boundaries, were observed. Precipitation probably occurred at the level of a texture jump or (temporary) water table. Most of this organic matter accumulated on top of the Late Glacial soil, forming in most profiles a continuous horizon of about 0.05m thickness. They are described in detail at Verrebroek "Dok 2".

6. The basis of the redistributed aeolian sands, deposited on top of the Late Glacial soil, observed at Verrebroek "Dok 1", lacks any visible lamination or stratification. These remobilised sands have a median particle size between 180 and 200µm, and are thus coarser than the Upper Pleniglacial coversands in which the Late Glacial soil units at Verrebroek "Dok 1" developed.

7. Traces of a Brunified Soil with a colour Bw-horizon, with (i) no E-horizon, and (ii) a homogeneous brown, bioturbated Bw.bi-horizon were only observed in particular positions, such as in tree-fall structures, at Verrebroek "Dok 1". The Brunified Soil genesis and appearance is discussed in detail at Verrebroek "Dok 2" and further commented under the natural disturbance features (cf 3.3.1.2.).

8. Podzolisation occurs in regions where precipitation exceeds evapotranspiration, at least seasonally. At Verrebroek, it is not excluded that the effect of fire, either natural or anthropogenic, on these already poor sandy soils led to strong leaching conditions and further acidification, thus inducing podzolisation. The intra-site soil variability shows topo-hydrosequences with Podzolised Soils of up to 0.5m thickness. Well-developed, imperfectly drained Podzolised Soils with a distinct, very light-coloured E-horizon appear in the slightly higher palaeo-landscape positions; weakly developed, poorly drained Podzolised Soils with nearly no E-horizon occur in the lower relief positions. In latter areas the later development of the overlying peat horizon is thickest. In such positions, the podzolisation was controlled by a permanent, or almost permanent, acid water table causing by reduction a mobilisation and subsequent lateral and/or vertical removal of iron by groundwater flow. The resulting Hydromorphic Podzolised Soil can be called Groundwater Podzolised Soil. The surface of the Podzolised Soil usually has (i) a black H-horizon, with a gradual boundary to the overlying peat and (ii) an A- or H+E-horizon, that is weakly developed and composed of black organic pellets (H) and bleached quartz sand grains (E). The dark brown Bh(.s)-horizons at Verrebroek "Dok 1" typically have a C/N-ratio around 25 and higher aluminium (Al) than iron (Fe) contents. $Al_o + {}^1/_2 Fe_o$-contents in the Bh(.s)-horizons vary between 0.13% and 0.37% and are thus lower than the required 0.50% for a spodic horizon; they are however above 0.20% in the non-Hydromorphic Podzolised Soils, as opposed to those of the Bw.bi-horizons in the Brunified Soils. Most probably, the parent material contained only few aluminium and/or iron. In the imperfectly drained Podzolised soils (P16, P17), the

Bh(.s)-horizons have the highest Al- and Fe-contents of the profile, whereas in the poorly drained Podzolised Soil (P18), Al only shows a relative peak at this level. Under the best drainage conditions (P16), Al accumulated deeper than Fe, whereas in the next profile along the topo-hydrosequence (P17), the reverse pattern was observed, confirming the already poorer drainage conditions. In case the Bh.s-horizon locally shows patchy, strong decomposition of the organic matter due to diagenetic microbiological activity, a Degraded Podzolised Soil originated (P19). High nutrient contents at the surface, especially phosphorus, coincide with high microbiological activity within the underlying horizons of the originally Podzolised Soil. Based on pollen analysis, the E-horizon of the Podzolised Soil can be attributed a Late Boreal age.

9. Peat, up to 0.75m thick, and containing non-transported wood fragments was only observed in palaeo-depression positions. Clear pedo-morphological evidences of past peat exploitation, such as remnants of peat cutting trenches, were absent. The peat surface was exceptionally flat and nearly horizontal. The overlying surface horizon consisting of a mixture of peat and sand with nearly constant thickness of about 0.10m is indicative of a cultivation horizon. The data suggest agricultural land use, rather than peat exploitation. The pollen spectra of the peat cover at Verrebroek "Dok 1" point towards Subboreal peat growth.

10. Spatially two distinct facies of the cultivation horizon, of 0.10 to 0.20m thickness, were present. On the ridges lacking peat cover, the surface horizon consisted of humiferous sand, showing locally spade marks at the base. In depression positions, peat was anthropogenically mixed with sand. The recurrent presence of open drainage ditches of about 3.0 to 3.5m width and 1m depth, fits with this agricultural land use too. Presence in both facies of dusty brown clay coatings to infilling fragments might reflect input of ditch decantation mud mixed with the soil mineral material, to increase fertility on poor land. Also the proper peat admixture could reflect some manuring practice. Based on artefact retrieval, the cultivation horizon could be attributed a Late Medieval age.

11. Calcareous flood clay and sand with variable thickness (0.4 - 1.5m) covered the coversands and peat, where present. They have been observed in detail in P3_H1 - H8 and show (i) a weak stratification at the base, (ii) a strong bioturbation in the upper 1.0 to 1.3m, (iii) presence of vertically oriented *in situ* shells and shell fragments, (iv) presence of few brick fragments, (v) a high porosity and in general a massive pedality, except for the local development of a colour and structure Bw-horizon with subangular blocky structure in the upper part, and (vi) a present-day plough horizon, which is homogeneous and has a sharp lower boundary, having already some decarbonatation. Sodium apparently leached out of the system, as the soil data don't reflect influence of brackish water. Different types of hydromorphism have been observed: (i) gley *sensu stricto* along pores in the lower 0.10m, and (ii) pseudogley in the overlying horizons. Pseudogley is related to dynamics of a perched water table. An oxidation gradient, evidenced by a decrease in organic matter content from bottom to top, is present over the different horizons. It indicates physical ripening and concomitant chemical changes, such as acidification, decarbonatation or organic matter loss. Primary carbonate mass is present in

the groundmass as crystals or bioclasts. On weathering, dissolving and subsequent migration and precipitation during alternating wet and dry soil conditions, secondary carbonates are deposited as calcite quasicoatings, coatings or infillings. Dusty brown clay coatings are especially present in the upper 1.0m. The type of clay migration is different from clay migration driven by colloidal dispersion of the clay particles. These dusty clay coatings develop as a result of surface or internal slaking due to saturation with water. As the iron content in the cambic Bw.bi-horizon is not exceptionally high when compared to the other horizons, the colour is not the result of a weathering process, but of homogenisation.

2.1.7.3. | Doel "Deurganckdok-sector B"

1. The parent material consists of Upper Pleniglacial coversands. It shows a horizontal stratification to lamination and consists of dominant poorly sorted, subrounded quartz and polycrystalline quartz, very few rounded glauconite that is partly, mainly chemically weathered and very few subrounded to rounded chert, feldspars and other mica minerals. The parent material has a median particle size of 158µm, and is thus finer than the remobilised sands of Verrebroek "Dok 1", but rather comparable in size to the Upper Pleniglacial coversands at Verrebroek "Dok 2".

2. Within these coversands, a Brunified Soil with a colour Bw-, and in depth, a clay accumulation Bt-horizon, were observed. The pedostratigraphy shows (i) a 0.15 to 0.20m thick A-horizon, (ii) no E-horizon, (iii) a homogeneous brown, bioturbated Bw.bi-horizon, (iv) fragmented clay accumulation Bt-bands. Within the mineral horizons, $Al_o + \frac{1}{2} Fe_o$-contents are only around are or slightly higher than 0.05%, confirming that podzolisation was not active here. As compared to the site of Verrebroek "Dok 1", peat growth started earlier here. Carbon dating at the base of the peat horizon gives 5050±55 BP at Doel, as opposed to 4690±30 BP at Verrebroek. When comparing the texture of the parent material of both sites, it is evident that Doel has a finer texture than Verrebroek "Dok 1": the peak in the particle size distribution lies at 158µm in Doel as opposed to 192µm in Verrebroek (P16); Doel has relatively more material in the clay, silt and very fine sand fraction too. Parallel, soil genesis at the "Hoge Vaart" sections (Flevoland, The Netherlands) was mainly limited to brunification and showed only slight evidences of podzolisation; peat growth on the coversands started there already from 5500 BP onwards (Spek *et alii* 2001a).

3. On this soil, a Final Mesolithic (Swifterbant) occupation settled, probably synchronous with the Brunified Soil development, leaving a living floor or human occupation surface.

4. The Brunified Soil is covered with peat, containing distinct wood fragments and indicating sustained wet conditions. Wet conditions are also evidenced by presence of very few pyrite framboids in the peat, but also throughout the underlying mineral soil. These wet conditions after peat burial guaranteed rather good protection of the archaeological site. As a consequence, even rather fragile artefacts and ecofacts, such as e.g. burnt bone, are well preserved.

2.1.7.4. | Inter-site comparison

With regard to the different units of Late Glacial soil observed at Verrebroek "Dok 1" and "Dok 2", it is worthwhile to relate them to type sections from Flanders and The Netherlands. At Verrebroek "Dok 2", the Late Glacial soil was attributed a Bølling/Earlier Dryas age; at Verrebroek "Dok 1", the two units observed were attributed an Allerød age. Humiferous sandy soils of both Bølling and Allerød age within the same vertical profile have been reported in Stabroek, situated 15km northeast from the site (De Coninck *et alii* 1966). Only the Bølling stage has here been referred to as the Stabroek soil (Paepe, Vanhoorne 1967; Paepe 1968). The occurrence of both facies in the same profile also appears at Opgrimbie (Meuse valley, Limburg, Eastern Belgium) (Paulissen, Munaut 1969) and in the Lutterzand section (Dinkel valley, Twente, Eastern Netherlands) (Van Der Hammen, Wijmstra 1971; Stapert, Veenstra 1988). In the Lutterzand section, the palaeo-soil of Bølling age is referred to as the Wierden Member, and the Usselo Bed or Soil is of Allerød age; both form a loamy light-coloured band with fragipan characteristics (Van Der Hammen, Wijmstra 1971). This sequence has however not been observed at the studied sites in Verrebroek. The reported configuration can be linked with a palaeo-inversion of the relief in between the Bølling and Allerød soil formation periods. During the relatively cold and dry conditions of the Earlier Dryas (Verbruggen *et alii* 1996), all the soils developed on the dune positions outside the dune slacks have been eroded and the deflated sediment buried the depression soils. At the end of the Earlier Dryas period, the dune slacks of the new landscape were rather situated in the position of the former dune ridges. Under the deteriorating climate of the Younger Dryas (Bohnke *et alii* 1987; Verbruggen *et alii* 1996), again the soils occurring at the level of the mires were buried under drifting sand originating from eroded dune ridges, but compared to the Netherlands and other northern regions the aeolian activity was restricted, probably due to a more resistant vegetation cover. Elaborate research of Late Glacial palaeoecological conditions around the whole coversand ridge in Northern Flanders is reported in Vanhoorne and Verbruggen (1975) and Verbruggen (1979).

With respect to the detailed pedological study performed on the Late Glacial soil units and laterally variable facies, it is worthwhile mentioning that Late Glacial soils often have been found in auger observations or on soil sections during archaeological excavations, but never have been investigated in that much detail with respect to their soil morphological, physical and chemical characteristics.

Throughout the text, regular reference is made to the pedo-lithostratigraphical sequence at the Mesolithic to Early Neolithic site of the "Hoge Vaart" (Flevoland, The Netherlands) (Hogestijn, Peeters 2001), as it has a comparable soil and environment genesis as the Verrebroek "Dok 1" site.

2.2. | Pollen and phytolith analyses

Koen Deforce, Vanessa Gelorini, Cyriel Verbruggen & Luc Vrydaghs

2.2.1. | Moerbeke "Moervaart"

Cyriel Verbruggen

Taking into account its geomorphological structure, the deepest zones of the former Late Glacial mire of the "Moervaart" depression are to be found close to the northern border along the edge of the coversand ridge Verrebroek (Stekene) - (Maldgem) Gistel (cf 1.2.2.). The maximum thickness of the gyttja is here about 1.90m. It is assumed that this part contains the most complete sequence and that its base represents the very beginning of the sedimentation. It has been palynologically investigated twice.

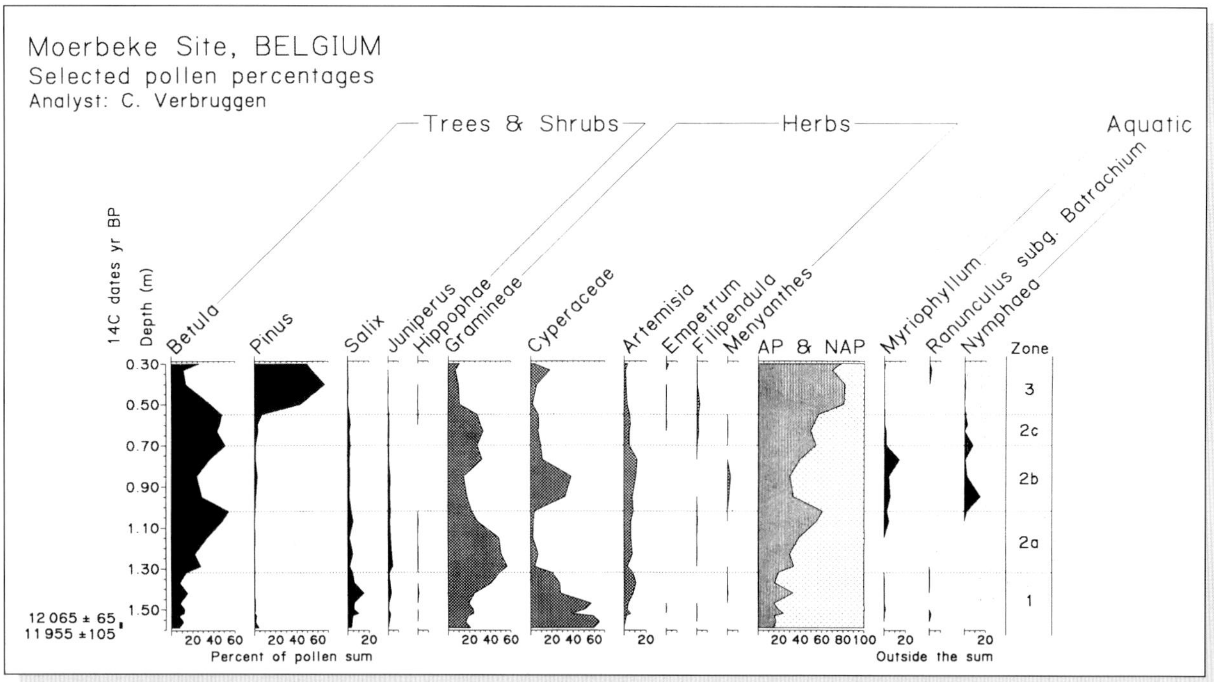

Figure 38: Moerbeke "Moervaart"; pollen diagram of the sequence (Verbruggen *et alii* 1996). The pollen sum does not include aquatic taxa and spores.

The first study, published in 1971 (Verbruggen 1971), shows a pollen sequence from the early Oldest Dryas-Ia- to the second part of the Allerød-IIb- (fig. 38). In the seventies the base of this sequence, corresponding to biozone 1 (Cyperaceae paz), was radiocarbon dated by means of two peat samples. The obtained results – 12065 ±65 BP (GrN-6376) and 11955±105 BP (GrN-6032(3)) – were very similar and situated the start of the sequence around 12000 uncal. BP. Unfortunately no further dating was done on the above-lying, younger layers of the "Moervaart" sequence at that time.

A second analysis of 27 levels was conducted on the sequence used for the ^{14}C-research (cf 2.3.1.). These pollen data allowed identification of the pollen assemblage zones and the biozonation of the sequence necessary for the ^{14}C research. These zonations are in agreement with those of the sequence analysed in 1971.

2.2.2. | Verrebroek "Dok 2"

Koen Deforce

2.2.2.1. | Sampling and methods

The palaeo-soil of Verrebroek "Dok 2" was sampled for pollen analysis from two different locations. The samples were taken with metal boxes, open at one side, which were pushed into the exposed profile. Sub-samples from these boxes were taken in the laboratory and treated with HCl, KOH and subsequently acetolysed (Faegri, Iversen 1975) and mounted in glycerine jelly.

The first sampling location – profile 11 – was just next to a Final Palaeolithic knapping area (cf 2.1.3.). The samples from this location did not contain any pollen, however. The second – profile 1 – was about 140m to the west of the knapping site, in a lower-lying part of the palaeo-soil. Here, the palaeo-soil was richer in organic material and contained a layer of sandy peat. In this part of the profile, some of the samples did contain pollen, although in a very low density and rather poorly preserved. From the samples for pollen analysis, two further sub-samples were taken for radiocarbon dating (bulk samples of sandy peat). For a detailed description of the sampled profile and the stratigraphical context, see 2.1.3.

2.2.2.2. | Results[1]

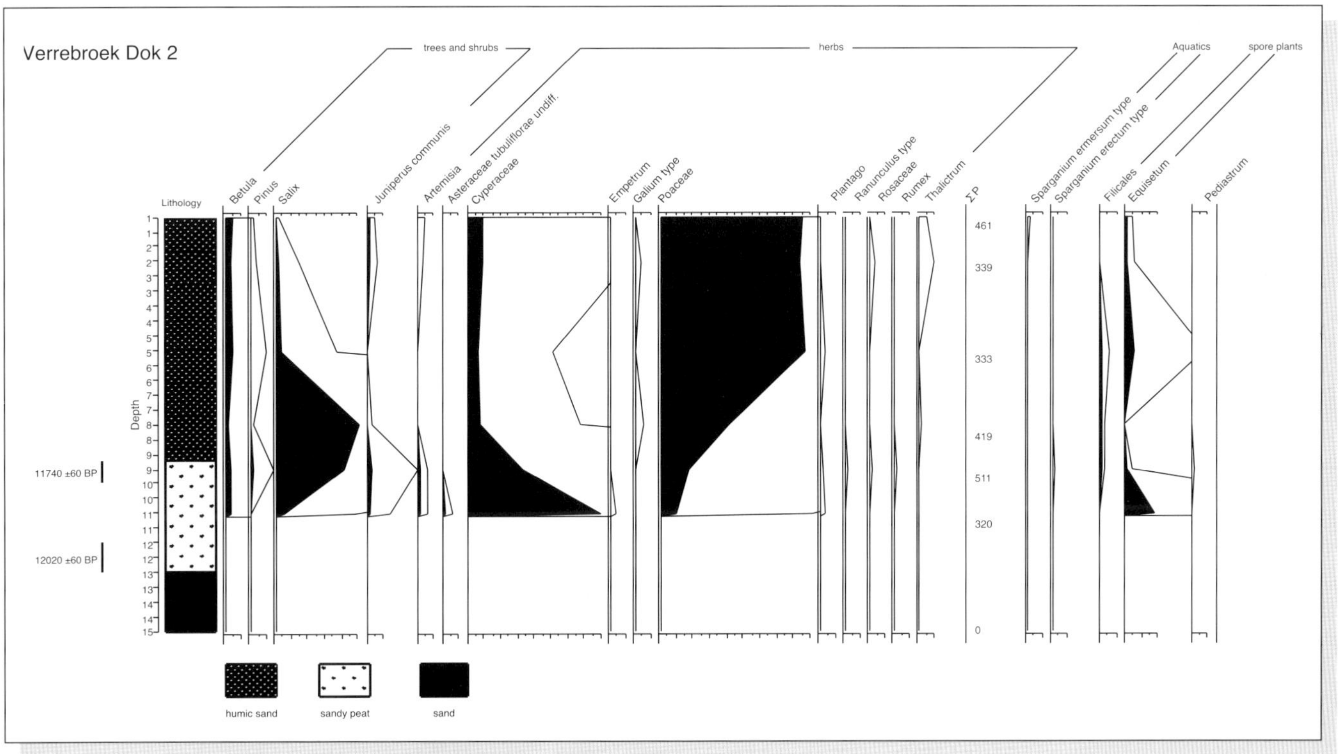

Figure 39: Verrebroek "Dok 2"; pollen diagram of the Late Glacial soil.

1 The scientific and English botanical nomenclature is based on respectively Lambinon *et alii* (1998) and Stace (2001).

The results are presented in a pollen diagram (fig. 39). As a basis for the calculation of the percentages of the individual taxa, a pollen sum (ΣP) based on the total of pollen from trees, shrubs and herbs was used, from which aquatics and spores were excluded. The base and the top of the sandy peat layer were radiocarbon dated to 12020±60 BP (NZA-11013) and 11740±60 BP (NZA-11021) respectively (cf 2.3.2.).

The Pleniglacial coversand in the lowermost part of the sampled profile did not contain any pollen. In the lower part of the sandy peat layer, pollen were preserved but the concentration was far too low for analysis. Therefore, the pollen diagram starts in the second half of the sandy peat layer.

In the lower part of the diagram, Cyperaceae (sedge family) are dominant, with values up to ±80%. *Equisetum* (horsetail) (± 15%) has high percentages as well. *Betula* (birch), *Salix* (willow) and *Juniperus* (juniper) are the only trees and shrubs, all showing very low percentages. *Pinus* (pine) is absent at this level.

In the second level, *Salix* becomes the dominant species (± 45%). *Juniperus* reaches its highest value (only ± 3%). *Pinus* appears in the diagram but with very low percentages (± 1.5%). *Pediastrum* occurs at this level in very small amounts.

In levels 10cm to 8cm, Cyperaceae values decrease dramatically and Poaceae (grass family) become dominant. *Salix* shows a sharp decrease to ± 4%, *Betula* and *Pinus* decrease from 3% and 1.5% to 1.5% and 0.2% respectively.

In the uppermost part of the diagram, there is very little evolution. The percentages of *Betula* remain constant, *Pinus* and *Salix* show a light decrease and *Juniperus* reappears in the diagram. Within the herbs, Poaceae dominate with ± 80%.

2.2.2.3. | Interpretation

The sampled site must have been a wet depression during the period that the sandy peat was formed, with local vegetation consisting of a sedge marsh. This is illustrated by the high Cyperaceae percentages in the corresponding part of the pollen diagram and the occurrence of *Sparganium erectum* (branched bur-reed type). The occurrence of very small amounts of *Pediastrum* in this part of the diagram indicates the presence of relatively little open water (Komárek, Jankovská 2001). The surrounding vegetation must have been sparse and almost treeless. *Salix* has high percentages in the first half of the diagram, but this might indicate (one of) the dwarf shrubs *Salix herbacea* (dwarf willow), *Salix reticulata* (net-leaved willow) or *Salix polaris* (polar willow), which have a maximum height of only 20cm (Stace 1997), as well as taller growing species. *Betula* percentages are rather low. As for *Salix*, its presence might indicate the shrub, *Betula nana* (dwarf birch), of which macro remains have frequently been found in Late Glacial deposits (Hoek 1997), as well as *Betula* trees (*Betula pendula*/*pubescens* or silver/downy birch).

Seen its low percentages, we can conclude that *Pinus* did not grow in the region during the period covered by the pollen diagram of Verrebroek 'Dok 2'. *Pinus* is a very high pollen producer with a good dispersion capacity. According to Lang (1994), pollen percentages of *Pinus* have to reach a minimum of 20% to indicate that *Pinus* was actually growing locally.

The openness of the vegetation cover is also indicated by the high mineral content of the peat. There must have been at least patches without vegetation, allowing the wind to transport the sand.

After 11740±60 BP, the peat formation stopped, probably due to a prohibitively high sedimentation rate of sand. As a consequence a layer of humic sand accumulated on top of the sandy peat in the depression. This must have been caused by further opening up of the vegetation cover which means that the climate must have become drier and/or colder at that time. This evolution towards a drier climate is also illustrated by a strong decrease of the Cyperaceae percentages and the increase of Poaceae, which indicate a lowering of the water level in the depression.

2.2.2.4. | Discussion

The peaty palaeo-soil was analysed for its pollen content in order to make a vegetation and landscape reconstruction for the period corresponding with the excavated Final Palaeolithic archaeological site, situated only about 140m from the sampling site (Crombé *et alii* 1999c). Based on the stratigraphy, it was presumed that the investigated peaty soil profile corresponds chronologically with the more sandy soil profile containing the artefacts. This is confirmed by the radiocarbon datings. The knapping site was dated 11900±90 BP while the base and top of the sandy peat layer were dated 12020±60 BP and 11740±60 BP respectively. A third sample from the sandy peat layer was dated 11930±90 BP (cf 2.3.2).

As there is little variation in the percentages of *Betula* and *Pinus* in the diagram of Verrebroek "Dok 2", it is difficult to correlate this diagram with the existing regional biozonations for the Late Glacial period. This is because these biozonations are mainly based on the variation of the percentages of *Betula* and *Pinus* (Van Geel *et alii* 1989; Hoek 1997; Verbruggen 1977, 1979, 1999).

When we compare the diagram of Verrebroek "Dok 2" with other Late Glacial pollen diagrams from northern Belgium, there is a very good correlation with the diagram from Jabbeke (Jabbeke 1) (Vanhoorne, Verbruggen 1975) and also with the lowermost part of that from Snellegem "Molenbroek" (Verbruggen 1979). Apart from the slightly higher percentages for *Betula* in these two diagrams, they show a similar vegetation evolution. In both diagrams, a *Salix* maximum of ± 40% occurs, Cyperaceae are dominant at the start of the *Salix* rise and they are replaced by Poaceae by the time of the *Salix* maximum. *Betula* has low values (± 10%) and *Pinus* has very low values (± 2-3%).

Like Verrebroek "Dok 2", both sites formed shallow depressions during the Late Glacial period (Vanhoorne, Verbruggen 1975; Verbruggen 1979). This might explain why their pollen diagrams are fairly easy to correlate with that of Verrebroek "Dok 2", in contrast with the Late Glacial pollen diagrams from large depressions like Moerbeke "Moervaart" (cf 2.2.1.) and deep abandoned riverbeds like Vinderhoute "Kale" (Verbruggen *et alii* 1996). The pollen spectra observed in different Late Glacial sites seems to be strongly dependent on the type and size of the deposition site. Bottema and Mook-Kamps (2000) come to the same conclusion for several Late Glacial diagrams from different deposition sites in the northern Netherlands. Another parallel is that Jabbeke and Snellegem are, just like Verrebroek, situated on the long coversand ridge stretching out from Verrebroek (Stekene) to (Maldegem) Gistel (cf 1.2.2;1.2.4.). This means that the three sites had a similar abiothic landscape during the Late Glacial with seemingly a similar vegetation cover as a consequence.

Verbruggen (1979) places the corresponding part of the pollen diagram for Snellegem-"Molenbroek" in the second part of the Oldest Dryas (Ia2), the Bølling *sensu stricto* (Ib) and the Older Dryas (Ic).

If we compare this diagram with the regional biostratigraphical zonation for the Late Glacial period of Hoek (1997), which is based on pollen diagrams from the Netherlands, northern Belgium and western Germany, the diagram of Verrebroek "Dok 2" correlates best with the end of sub-zone 1b and the first part of sub-zone 1c. These sub-zones can be considered equivalent to the end of the Bølling *sensu stricto* and the Earlier Dryas zone as defined by van Geel *et alii* (1989) (Hoek 1997). The difference with the schematic pollen diagram of Hoek (1997) is the absence of Heliophillous taxa such as *Hippophaë* (sea buckthorn) and *Helianthemum* (rock-rose). This might be because the depression at the Verrebroek site was rather small, resulting in an overrepresentation of the local vegetation (Jacobson, Bradshaw 1981). As a consequence, small pollen producers growing in the drier parts of the landscape such as these will have little chance to occur in the pollen diagram.

However we cannot exclude the possibility that the Verrebroek "Dok 2" diagram correlates with the sub-zone 2a2 (Hoek 1997), which can be considered equivalent to a part of the Allerød zone as defined by van Geel *et alii* (1989), because both sub-zones are known to have a parallel vegetation development. Although one would expect higher *Pinus* and *Betula* percentages in the Verrebroek "Dok 2" diagram in that case. The end of the organic deposition, due to a covering by aeolian sand is assumed to coincide with a climatic deterioration. Taking into account the stratigraphical position, the vegetation composition and the radiocarbon dating, these phenomena have to be situated in the Older Dryas period.

2.2.3. | Verrebroek 'Dok 1'

2.2.3.1. | Late Glacial soil

Koen Deforce

Sampling and methods

As at the site of Verrebroek 'Dok 2', a palaeo-soil occurred in the coversand on which the Early Mesolithic site of Verrebroek Dok 1 was situated. In view of its stratigraphical position, it was likely that this palaeo-soil was of Late Glacial origin.

As this palaeo-soil showed much variation in thickness and organic content, it was sampled for pollen analysis where it was most pronounced, i.e. at profile 9 (cf 2.1.4.). On this spot, the palaeo-soil consisted of two thin peat layers, separated by a layer of humic sand. The lowermost peat layer was ± 5.5cm thick and had a high minerogenic content (sand). At the base of this sandy peat was a layer of humic sand (± 6cm). The uppermost peat layer was ± 11cm thick and consisted of almost pure organic matter. The intervening layer of humic sand was ± 7.5cm.

The samples were taken with metal boxes, open at one side, which were pushed into the exposed profile. Subsamples from these boxes were taken in the laboratory and treated with HCl, KOH and subsequently acetolysed (Faegri, Iversen 1975) and mounted in glycerine jelly. From the samples for the pollen analysis, four subsamples were taken for radiocarbon dating (bulk samples of peat and sandy peat; cf 2.3.3.1.). For a detailed description of the sampled profile and the stratigraphical context, see 2.1.4.

Results

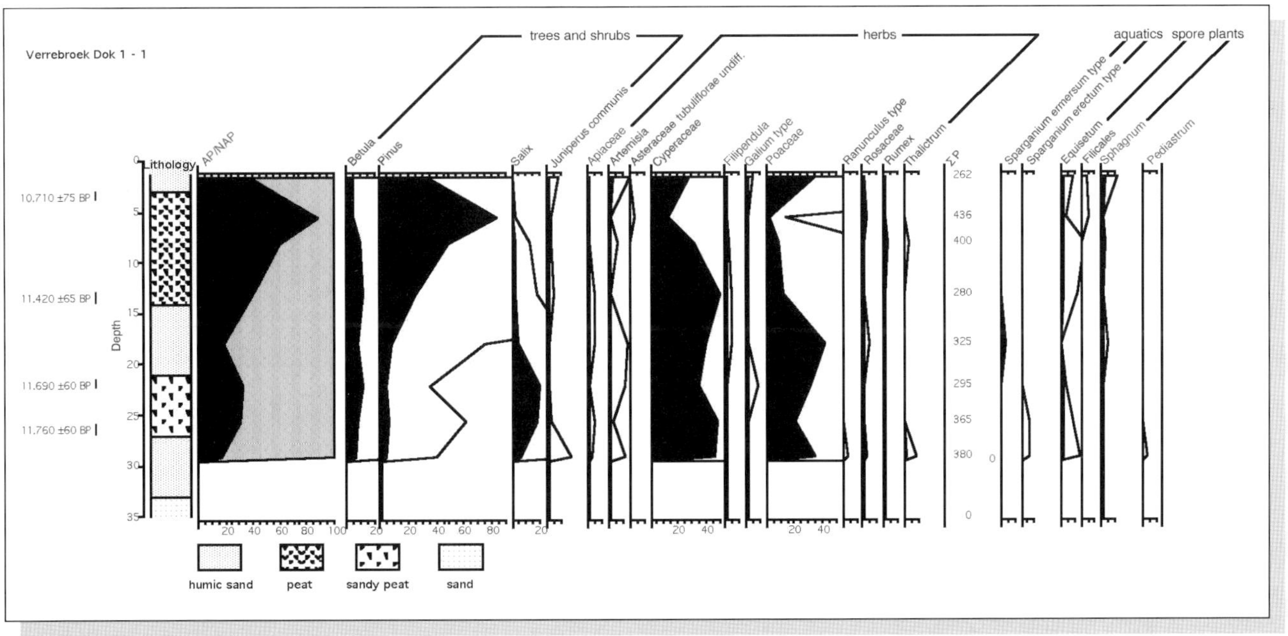

Figure 40: Verrebroek "Dok 1"; pollen diagram of two Late Glacial soils.

The results are presented in a pollen diagram (fig. 40). As a basis for the calculation of the percentages of the individual taxa, a pollen sum (ΣP) based on the total of pollen from trees, shrubs and herbs was used, from which aquatics and spores were excluded.

The lowermost samples (35cm - 30cm, sand and humic sand) did not contain any pollen. Therefore the diagram starts in the upper part of the humic sand layer. The lowermost part of the diagram is characterised by low percentages of arboreal pollen (pollen from trees and shrubs) (=AP). Only four taxa of trees and shrubs occur at this level: *Betula* (birch) (5.5%), *Pinus* (pine) (4%), *Salix* (willow) (5%) and *Juniperus* (juniper) (1.5%). Cyperaceae (sedge family) (47%) and Poaceae (grass family) (34%) are dominant. *Pediastrum* occurs only at this level and in very small amounts.

In the levels corresponding with the sandy peat layer (27cm – 21.5cm), *Salix* percentages rise to ± 18% while *Betula* rises to 10.5% at the top of the peat layer. *Pinus* first rises to 6% and then falls to 3.5%.

In the levels corresponding with the humic sand between the two peat layers (21.5cm – 14cm), the arboreal pollen percentages decrease as a consequence of the decrease of *Salix*, *Betula* and *Juniperus*, the last of these being entirely absent in this part of the diagram. *Filipendula* (meadowsweet) appears and Poaceae percentages show a strong increase. *Artemisia* (mugwort) increases as well.

At the base of the upper peat layer, *Pinus* values have increased to ± 20%. *Salix* continues to decrease and *Juniperus* reappears in the diagram.

In the second half of the upper peat layer, arboreal pollen percentages continue to rise mainly due to the rise of *Pinus*, which reaches a maximum of ± 80% in this part of the diagram. *Betula* percentages decrease to ± 5%. Cyperaceae and Poaceae show a strong decrease.

In the uppermost level in the diagram, corresponding with the humic sand overlying the upper peat layer, the AP curve shows a sharp decline. This is caused by the decrease of *Pinus* and the small decrease of *Salix*, which nevertheless actually disappears on this level. *Juniperus* shows a small increase. Within the herbs, *Artemisia*, Cyperaceae and Poaceae increase.

Interpretation
Before 11760±60 BP, the surrounding vegetation was very sparse, *Salix* and *Betula* being the only occurring trees. Part of the pollen of these two taxa might originate from dwarf shrubs however (cf 2.2.2.). *Pinus* percentages are too low to conclude that it was part of the vegetation. According to Lang (1994), pollen percentages of *Pinus* have to reach a minimum of 20% to conclude that *Pinus* was actually growing locally. *Juniperus* is a coniferous shrub that must have grown on the drier parts in the landscape, but, in view of its rather low percentages, it cannot have been abundant.

From 11760±60 BP, a sedge marsh developed in the depression and the accumulation of (sandy) peat started. The high percentages of *Salix* indicate that it must have been abundant in or at the border of the depression. The rise of arboreal pollen, especially those of *Betula*, together with the decrease of *Juniperus* and *Artemisia* indicate a slightly denser vegetation cover on the higher grounds. The presence of *Pinus* in the diagram must be seen as long-distance transport. As the peat contains a lot of sand, the vegetation cover must still have been open enough to allow sand to be blown into the depression.

Shortly after 11690±60 BP, the arboreal vegetation becomes sparser. The peat accumulation stopped due to a prohibitively high sedimentation rate of sand, indicating a more open vegetation, and only humic sand was deposited in the depression. The decrease of Cyperaceae percentages and the rise of Poaceae might indicate a lowering of the water level in the depression.

Around 11420±65 BP, peat formation restarted and the arboreal pollen show a strong increase. *Pinus* exceeds 20% and even reaches a maximum of 80% in the upper part of the upper peat layer, indicating that it was now part of the vegetation.

After 10710±75 BP, peat accumulation stopped again. The arboreal pollen percentages show a strong decrease and those of *Juniperus, Artemisia*, Cypeaceae and Poaceae increase indicating an opening up of the vegetation. Humic sand was deposited on top of the peat.

As pollen from submerged aquatics such as *Potamogeton* (pondweed) and *Myriophyllum* (water-milfoil) and aquatics of deeper water such as *Nuphar* (yellow water-lily) and *Nymphaea* (white water-lily), which are frequently found in Late Glacial lacusterine sediments, are absent throughout the pollen diagram, no deep water but only a shallow marsh existed at Verrebroek "Dok 1" between 11760±60 BP and 10710±75 BP.

Discussion

Based on the vegetation evolution, there are two different possible correlations for the lower half of the diagram, dated between 11760±60 BP and 11690±60 BP:

1° It can be correlated with the diagram of Verrebroek "Dok 2" (cf 2.2.2.). The curves of the various taxa show a similar evolution in both diagrams, though the percentages of *Betula* and *Pinus* are higher at Verrebroek "Dok 1" than at Verrebroek "Dok 2", while *Salix*, Poaceae, Cyperaceae and *Equisetum* have lower percentages at "Dok 1" than at "Dok 2". This is probably because the peat at "Dok 2" was formed in a smaller depression than "Dok 1" (or on the edge of a bigger depression with "Dok 1" being closer to the centre of that same depression). In a small basin (and on the edge of a basin), the local vegetation is over represented in the pollen rain while the regional vegetation will be underestimated (Jacobson, Bradshaw 1981), which explains the differences in the percentages of "Dok 1" and "Dok 2". The base of the peat layer at "Dok 2" is dated 12020±60 BP and at VD1 at 11760±60 BP (base of the lowermost peat layer). This seems to contradict the above made correlation but seeing that the diagram of "Dok 2" only starts in the middle of the peat layer while that of "Dok 1" starts below the base of the peat, it is possible that the diagram of "Dok 2" and the lower part of "Dok 1" pertain to the same period. In this case, they could both be correlated either with the end of sub-zone Ib and sub-zone Ic of the regional biostratigraphical zonation for the Late Glacial period of Verbruggen (1979) or with subzone 1b and 1c of the regional biostratigraphical zonation for the Late Glacial period of Hoek (1997) (cf 2.2.2). These sub-zones can be considered equivalent to the end of the Bølling *sensu stricto* and the Earlier Dryas zone.

2° The second possibility is a correlation with the IIa period (Verbruggen 1979); this is the first part of the Allerød.

The difference between these two possible correlations is the stratigraphic position of the humic sand layer between the lower and the upper peat layer: the first interpretation points to an Older Dryas age, while the second one presumes the sedimentation of (aeolian) sand in the pool during the IIa phase of the Allerod.

The upper half of the diagram, dated between 11.420±65 BP and 10.710±75 BP and corresponding with the upper peat layer, can be correlated with the zone II(b) of the zonation for the Late Glacial period of Verbruggen (1979) which is attributed to the Allerød. This zone is characterised by a rise of the arboreal pollen percentages, especially those of *Pinus*.

2.2.3.2. | Holocene peat

Pollen analysis

Koen Deforce

Sampling and methods

The profile investigated - profile 20 (cf 2.1.4.) - was 65cm long and consisted of peat (lower 35cm) and humic sand (upper 30cm). The peat was deposited on top of the coversand in which a podzol soil profile had developed. The upper half of the peat layer contained a lot of wood fragments. On top of the humic sand, 125cm of clay was deposited, in which a present-day plough horizon was present.

Results

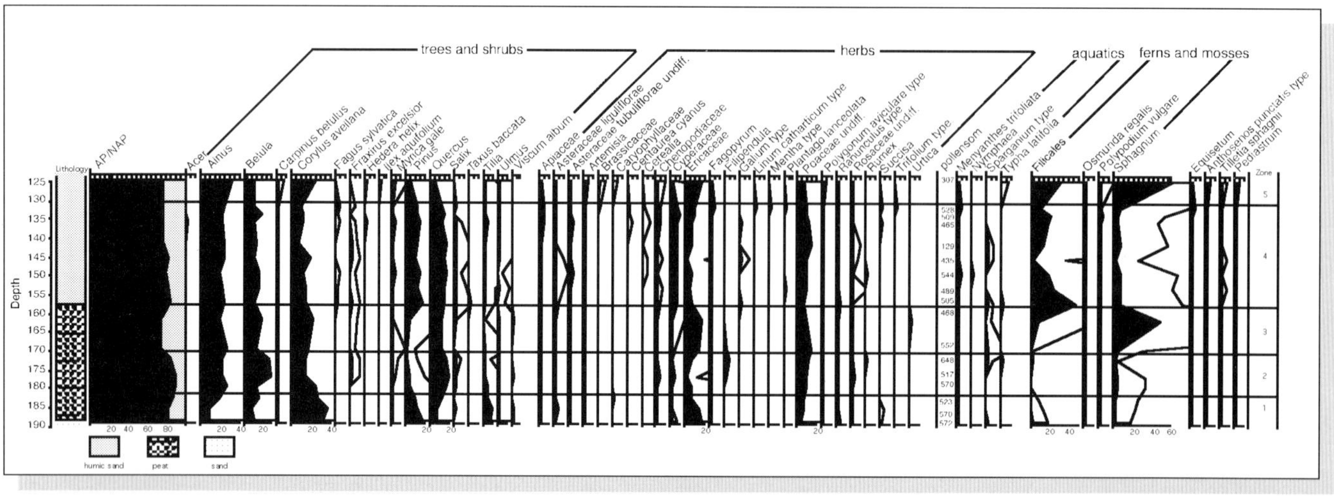

Figure 41: Verrebroek "Dok 1"; pollen diagram of the overlying Holocene peat.

The results are presented in a pollen diagram (fig. 41). As a basis for the calculation of the percentages of the individual taxa, a pollen sum (ΣP) based on the total of pollen from trees, shrubs and herbs was used, from which aquatics and spores were excluded. For the presentation and discussion of the results, the pollen diagram is divided into five zones based on major changes in the curves of the various taxa. The base of this peat layer was radiocarbon dated to 4690±30 BP (KIA-17992) (cf 2.3.3.2.).

Zone 1: 188cm – 181cm:
The arboreal pollen percentages are high in this zone (80-87%). *Corylus* (hazel) is dominant with percentages up to 37%. *Pinus* (pine) and *Tilia* (lime) have high values as well. *Alnus* (alder), *Betula* (birch) and *Quercus* (oak) have relatively low values. Ericaceae (heather family) have the highest percentages within the herbs. Poaceae (grass family) have relatively low percentages in this part of the diagram.

Zone 2: 181cm – 170cm:
Corylus, Pinus, Tilia and Ericaceae decrease while the percentages of *Alnus, Betula* and *Quercus* rise. *Fraxinus* (ash) and *Myrica* (bog-myrtle) appear in the diagram. Poaceae show a slight increase and *Filipendula* (meadowsweet) appears in the diagram.

Zone 3: 170cm – 158cm:
The percentages of *Betula* and *Quercus* are markedly lower in this zone than in the previous one. The percentages of Cyperaceae (sedge family), Ericaceae and Filicales (true ferns), on the other hand, increase. *Sphagnum* (peat moss) shows a very strong increase, up to ca. 50%. *Menyanthes* (bogbean) appears in this zone.

Zone 4: 158cm – 130cm:
The percentages of most taxa show little variation within this zone except for the uppermost part of the diagram. *Alnus* is dominant. *Pinus*, Cyperaceae and Filicales have high values as well. A lot of taxa appear for the first time in the diagram in this zone, e.g. *Fagus* (beech), *Carpinus* (hornbeam), *Taxus* (yew) and also Cerealia (cereals), *Fagopyrum* (buckwheat) and *Centaurea cyanus* (cornflower).

Zone 5: 130cm – 125cm:
Alnus en *Corylus* rise while *Pinus* en *Quercus* decrease. Cyperaceae decrease as well and the percentages of Ericaceae, Filicales, *Polypodium vulgare* (polypody) and *Sphagnum* increase.

Interpretation
In zone 1 of the diagram, the high percentages of trees which grow in relatively dry habitats, such as *Corylus* and *Tilia,* indicate that the vegetation on at least a part of the Pleistocene sand ridge (on which the site of Verrebroek "Dok 1" is situated) was at this point not much affected by the rising groundwater level. The high Ericaceae values and the presence of *Sphagnum* indicate a wet and relatively nutrient-poor (mesotrophic) local environment. The sampling spot was situated on one of the higher parts of the landscape. Peat growth must have started earlier in the adjacent lower lying parts and later overgrown the sampling point with the rising of the groundwater. This could explain the rather mesotrophic character of the peat in zone 1. This nutrient-poor character may also be a consequence of the substrate being a podzol soil.

At the start of zone 2, *Corylus*, *Pinus*, *Tilia* and Ericaceae decrease while the percentages of *Alnus*, *Betula*, *Quercus* and *Fraxinus* rise. *Tilia* and *Corylus* are replaced by *Quercus* and *Fraxinus* as a consequence of the rising groundwater. The increase of first *Alnus* and later of *Betula* indicates the development of alder and birch carr vegetation respectively in the lower lying parts while the rise of Cyperaceae, *Sparganium* (bur-reed) and *Typha latifolia* (bulrush) reflect wetter local conditions.

The rise of *Myrica*, *Sphagnum*, Ericaceae, *Menyanthes* and Filicales in zone 3 indicates the evolution towards a more nutrient-poor environment (Lambion *et alii* 1998; Weeda *et alii* 1985).

In view of the sharp sedimentological boundary, the presence of spade marks between the peat and the humic sand (cf 2.1.4.10.) and the abrupt change in the curves of a lot of the taxa between zone 3 and zone 4, there must be a hiatus between these two zones. The presence of pollen from Cerealia (cereals), *Fagopyrum* and *Centaurea cyanus* in zone 4 and 5 indicates agriculture. The last two taxa are typical for medieval and post-medieval times athough both have been found in older deposits as well (Leenders 1987; Van Haaster 1997; Pals, Van Geel 1976). The high percentages of Cyperaceae, Ericaceae, Filicales and *Sphagnum*, especially in zone 5, must come from peat that has been mixed with the sand.

The presence of taxa such as Chenopodiaceae (goosefoot family), Asteraceae (daisy family) and *Artemisia* (mugwort), normally considered to be halophytes, in this part of the diagram might indicate a marine influence. However, as they occur together with Cerealia and other anthropogenic indicators such as *Plantago lanceolata* (ribwort plantain) and Brassicaceae (cabbage family) (Behre 1981), it is far more likely that in this case the pollen from them derives from ruderal plants associated with agricultural practises or other anthropogenic activities.

Discussion

The base of the peat layer is dated 4690±30 BP in profile 1 at ca. 0.20/0.25m TAW (cf 2.3.3.2.). This date corresponds with the pollen spectra in zone 1 of the pollen diagram. Based on the low *Ulmus* percentages and the absence of *Fagus*, the start of the peat growth in Verrebroek "Dok 1" can be situated in the beginning of the Sub-boreal period (Verbruggen 1977; Verbruggen *et alii* 1996). The start of the peat growth in the nearby Doel is radiocarbon dated a lot further back in time (cf 2.3.6.2.). Munaut (1967) and Mys *et alii* (1983) even place the start of the peat growth in the lower lying parts in the north of the province of Antwerp at the start of the Atlantic period. But as the site of Verrebroek "Dok 1" is situated on a Pleistocene sand ridge, it is logical to assume that peat formation started a lot later here.

Another difference from other pollen diagrams from nearby localities is the absence of a 'long' oligotrophic phase in the uppermost part of the peat. Zone 3 of the pollen diagram is characterised by relatively high percentages of *Sphagnum* and Ericaceae but only in the upper 13cm of the peat. The diagrams from Doel (Minnaert, Verbruggen 1986; Denys, Verbruggen 1989), Zandvliet (Munaut 1967) and Oorderen (Munaut 1967) all contain an oligotrophic phase represented by 40 to 100cm of peat. The same diagrams have a continuous curve of *Fagus* and *Carpinus* in the upper part of the investigated peat sequence, indicating that the peat growth continued well into the Sub-Atlantic period. This is not the case in Verrebroek "Dok 1". Fagus only occurs in the diagram in the levels corresponding with the humic sand, *Carpinus* only in the uppermost level of the humic sand. This means that the top of the peat at Verrebroek "Dok 1" must be placed in the Sub-boreal period. There is no reason to assume, however, that peat growth stopped a lot earlier at Verrebroek than at the other above-mentioned localities. The conclusion that must be drawn is that the original top part of the peat at Verrebroek "Dok 1" is missing. However, the results of the palynological analysis do not allow us to determine whether this is the consequence of peat extraction or simply of drainage, which would have caused the oxidation of the upper part of the peat.

No features typical of past peat extraction were observed in the exposed profiles and the presence of the humic sand on top of the peat indicates that at least part of the peat has been mixed with the Pleistocene substrate, probably for agricultural purposes (cf 2.1.4.10.). The high values of Ericaceae and *Sphagnum* observed in the levels corresponding with the humic sand also points towards a mixing with peat. In particular, the high percentages of Ericaceae and *Sphagnum* in zone 5 suggest that the original oligotrophic phase of the peat has been mixed with the sand. The presence of drainage ditches, dated to the 13-15th centuries, also points towards agriculture (cf 3.1.2.2.). However, given the low percentages of Cerealia, *Fagopyrum* and other indicators of arable farming, it is not likely that cultivation took place on the investigation site itself.

Phytolith analysis

Luc Vrydaghs

Sampling and methods

A small quantity of sediment was put in suspension in demineralised water with dispersing agent (calgon). For each sample, two sections were observed at amplification x40. Three horizontal and three vertical lines were scanned.

Absolute quantification is based on visual frequency charts as given in Bullock *et alii* (1985). In a next step relative presence of the different morphotypes observed is quantified.

Based on the conducted analysis, there is no possibility to give grass identification. Taxa identification below the subfamily level requires at least reference material, extensive morphological description and statistical analysis (cf Fredlund, Tieszen 1994).

Results

On H4 of profile 16 (cf 2.1.4.10.), phytolith analysis was performed. In absolute quantity, very few (< 2%) phytoliths were observed; 31 individuals, distributed over 7 morphotypes, were counted. The elongate morphotype is dominant. The phytolith identification gives a coherent assemblage of *Triticum* sp. (wheat) and panicoid grasses. Weak presence of diatoms and sponge spicules were registered too. The absolute phytolith presence is evaluated as very weak. When taking the phytolith identification of Triticum sp. and panicoid grasses into account, an absolute presence of 5 to 10% phytoliths within the suspension preparation can be expected.

2.2.3.3. | Podzol horizons

Pollen analysis

Cyriel Verbruggen

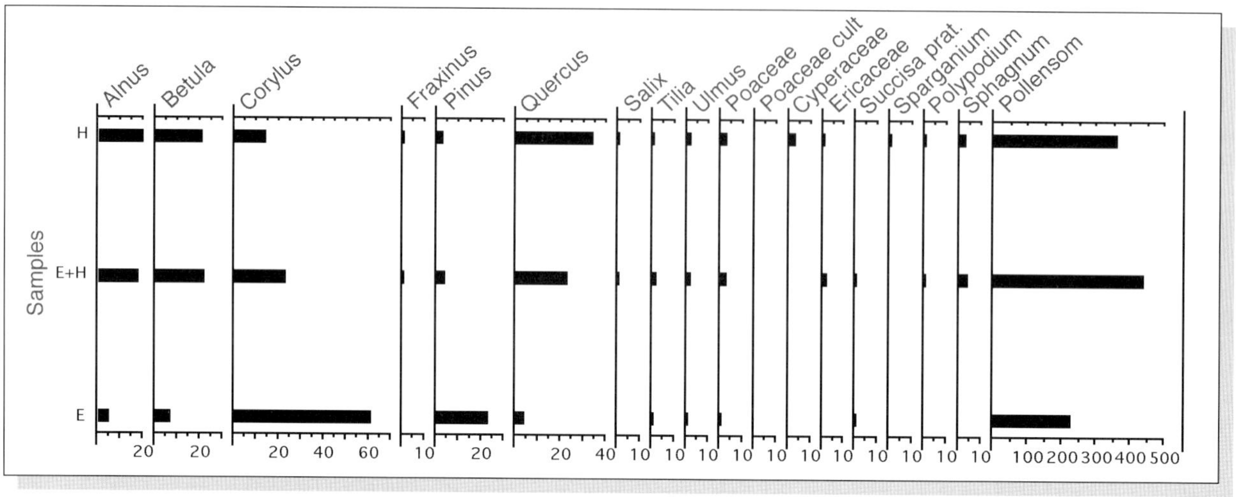

Figure 42: Verrebroek "Dok 1"; pollen diagram of the podzol horizons.

Three pollenspectra (fig. 42) were counted from the horizons H2 and H3a and H3b, corresponding with the H(+E)-, H+E- and E(+H)-horizons of the Podzolised Soil within the soil profile 9 (cf 2.1.4.7.). It is obvious that the pollen spectrum of E differs significantly from those of E+H and H. It can be interpreted as of late Boreal age, while the latter two are of early Subboreal age. In horizon E, *Corylus* (hazel) and *Pinus* (Pine) dominate, pointing to dense forest and dry soil conditions. The horizons E+H and H have *Quercus* (oak), *Corylus*, *Betula* (birch) and *Alnus* (alder) as main species. They also show the presence of Ericaceae (heather family) and *Sphagnum* (peat moss), which are absent in E, indicating that now we are dealing with a more open forest and a decrease of the previous eutrophic conditions.

Phytolith analysis

Luc Vrydaghs

On the A-horizon of the podzolised soil of profile 16 (H6), phytolith analysis was performed (cf 2.1.4.7.). In absolute quantity, very few (< 2%) phytoliths were observed; 21 individuals, distributed over 4 morphotypes, were counted; the dumbbell and elongated morphotypes are dominant. The phytolith identification gives a coherent assemblage of panicoid grasses. Some pollen grains were registered too. The absolute phytolith presence is evaluated as very weak. However, based on the presence of the identified panicoid grasses, an absolute presence of 5 to 10% phytoliths within the suspension preparation is expected as grass is an important phytolith accumulator. Besides morphology, the major characteristic of a grass phytolith spectrum is the presence of hundreds of phytoliths, i.e. much higher than the observed 2%. This is attributed to the "decay-in-place tendency" (Rovner 2001)[2].

2 The "decay-in-place tendency" of opal phytolith release into soil contexts provides many special opportunities in archaeological research. Comparisons of phytolith assemblages from small-scale contexts can identify specific cultural behaviours and ethno botanic activities in the past. In addition, phytoliths can help identify very subtle micro ecological conditions within and between sites that are absent or difficult to detect in other data systems, e.g. pollen. The "decay-in-place tendency" is defined as the tendency of phytoliths to remain in plant tissue until the plant is deposited in the soil and decays to release its phytoliths (Rovner 2001, 119).

2.2.4. | Doel "Deurganckdok-sector B"

Cyriel Verbruggen

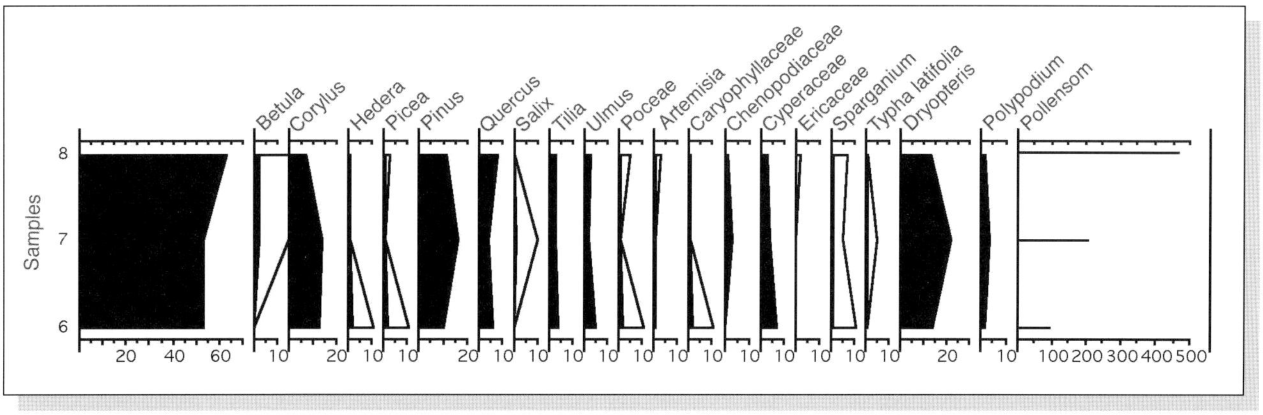

Figure 43: Doel "Deurganckdok-sector B"; pollen diagram of the top of the coversand and the base of the overlying peat.

A soil sample of 35cm was taken from profile 1 at the point of contact of the coversand deposits and the above-lying peat (cf 2.1.5.1.). The sample was palynologically investigated by means of 8 levels. The lower 5 levels are completely pollenless (fig. 43). The upper three contain pollen with an increasing preservation towards the top.

In view of the increasing dominance of *Alnus* (alder) and Cyperaceae (sedge family), the pollenspectra point to increasing wetness. The high values of *Pinus* (pine) are exceptional in this late Atlantic forest. The presence of Chenopodiaceae (goosefoot family) can be explained either by the neighbourhood of human activity or by marine influence.

The lack of pollen in the lower levels is in contradiction with the abundance and good preservation of organic material there. A rapid accumulation seems to be the only explanation for this particular situation.

Separate from this first investigation, three levels (Doel H5, PxH2, PxH4 - profile 1) from the same layer were submitted by pedologists. These levels also are completely pollenless despite the abundance of organic material. These results definitely confirm those of the first investigation.

2.2.5 | Doel "Deurganckdok-sector A"

Vanessa Gelorini & Cyriel Verbruggen

2.2.5.1. | Sampling and methods

In the vicinity of the prehistoric sites of Doel "Deurganckdok-sector B/C" a Holocene peat profile was sampled for palynological investigation (cf 1.3.6; 2.1.6.). The pollen data from the peat section and the intervening clayey Calais deposit indicate vegetational changes in the lower Schelde area from the Late Atlantic (KIA-17640: 5885±35 BP) to the Subatlantic (KIA-17637: 1340±25 BP)(cf 2.3.5.). As the radiocarbon dating of the lowest level of the sequence coincides with the Final Mesolithic/Neolithic phase of human occupation at Doel "sector B/C", only this part of the section i.e. the Late Atlantic, will be discussed in this contribution. A final report on the palaeoecological study of the entire profile is in progress.

Forty-five samples of the entire profile were prepared for pollen analysis. Seventeen of these were concentrated in the section under discussion. For preparing both peat and clayey deposits the standard extraction method of Faegri *et alii* (1989) was applied, i.e. HCl- and KOH-treatments, sieving, acetolysis, glycerol and mounting in glycerine jelly. The removal of the clay gradient had been obtained through additional density gradient separation. Pollen counting was done at a magnification of x400 and x1000 using an Olympus microscope with phase contrast facilities. If the pollen concentration was strong enough, counting continued until a minimum of 500 pollen grains had been achieved. The pollen grains and spores were identified to the lowest possible taxonomic level with the aid of standard pollen and spore keys (e.g. Punt *et alii* 1984-1995; Faegri *et alii* 1989; Moore *et alii* 1991) and the reference collection. For the microfossil types (non-pollen palynomorphs) reference was made to descriptions and illustrations defined by van Geel and others (Bakker, van Smeerdijk 1980, 95-163; Pals *et alii* 1980, 371-418; van Geel 1976; van Geel *et alii* 1981, 367-448; van Geel *et alii* 1986, 31-61; van Geel 2001, 99-119; van Geel *et alii* 2003, 797-940). In the pollen diagram, pollen and spore counts were expressed in the form of percentages of the pollen sum i.e. the total sum of trees and shrubs (arboreal pollen, AP) and herbs (non-arboreal pollen, NAP). Due to their local presence in mire environments (overrepresentation) aquatics, ferns and mosses and non-pollen palynomorphs were excluded from the pollen sum.

2.2.5.2. | Results

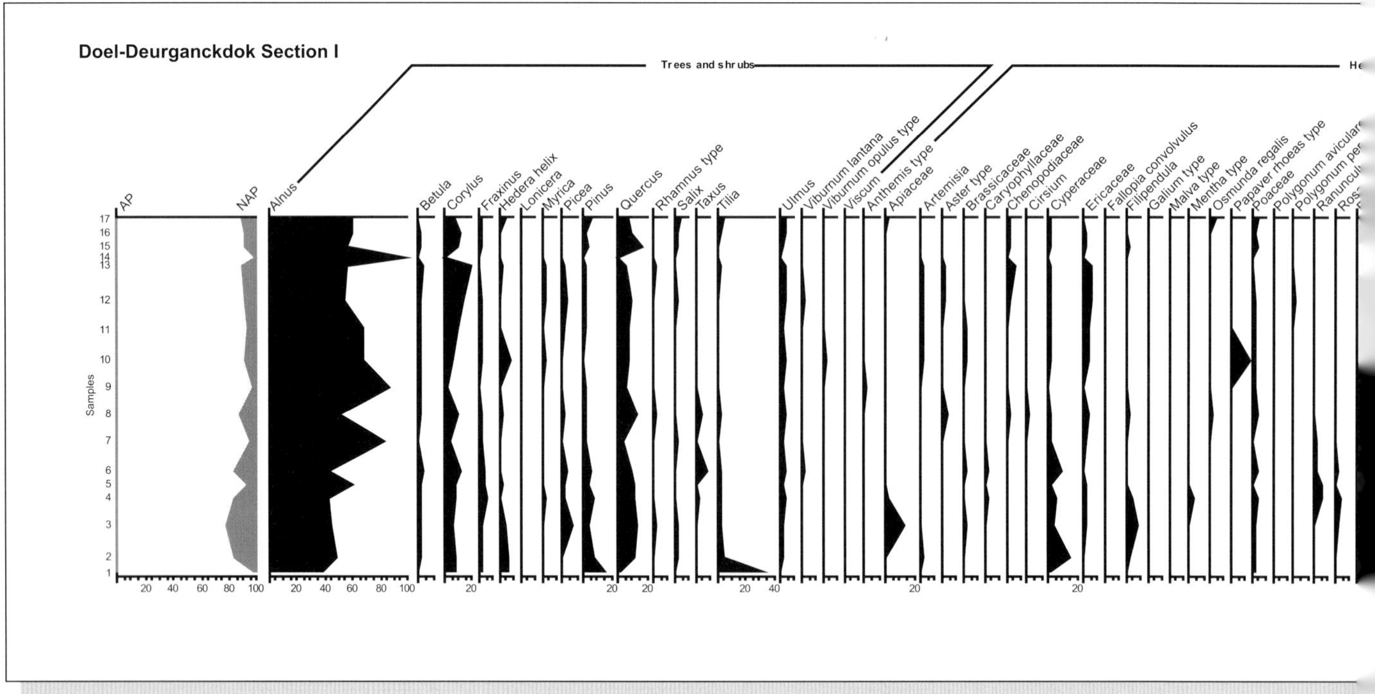

Figure 44: Doel "Deurganckdok-sector A"; pollen diagram of the lower peat levels and Calais-deposits in section 1.

Pollen data (fig. 44) are described following the chronological and sedimentological phases (fig. 37).

- 3,2m TAW to -2,8m TAW: Phase 2: Sample 1 to 3

Arboreal pollen is dominant. *Alnus* (alder, ca. 45%) is the most prominent taxus, with *Quercus* (oak) at ca. 12% and *Corylus* (hazel) at ca. 7%. *Tilia* (lime) pollen is one of the major types at the beginning of the zone (about 30%), but sharply declines to ca. 4% towards the top. The same evolution, but less pronounced, is apparent in the pollen values of *Pinus* (pine). *Betula* (birch), *Fraxinus* (ash), *Hedera helix* (ivy), *Lonicera* (honeysuckle), *Salix* (willow) and *Ulmus* (elm) occur consistently, though at low values. *Rhamnus* type (buckthorn type) and *Picea* (spruce) are also recorded. Herbaceous pollen, with Cyperaceae (sedge family) and Poaceae (grass family) as the most abundant taxa, are represented in small numbers. Other herbs are scarce, except for Apiaceae (carrot family), which has a high value in sample 3. Aquatics, such as *Typha latifolia* (bulrush), *Lythrum salicaria* (purple loosestrife), *Sparganium emersum* type (unbranched bur-reed type) and *Sparganium erectum* type (branched bur-reed type), are in a minority. At the base of the peat accumulation Filicales (undifferentiated fern spores cf *Dryopteris*), *Polypodium vulgare* (polypody) and *Sphagnum* (peat moss) are well represented. Also *Diporotheca rhizophila* and *Glomus* type attain high values. *Ustulina deusta* occur in sample 2, while *Bactrodesmium* type, type 18 and type 547 are typical of sample 3.

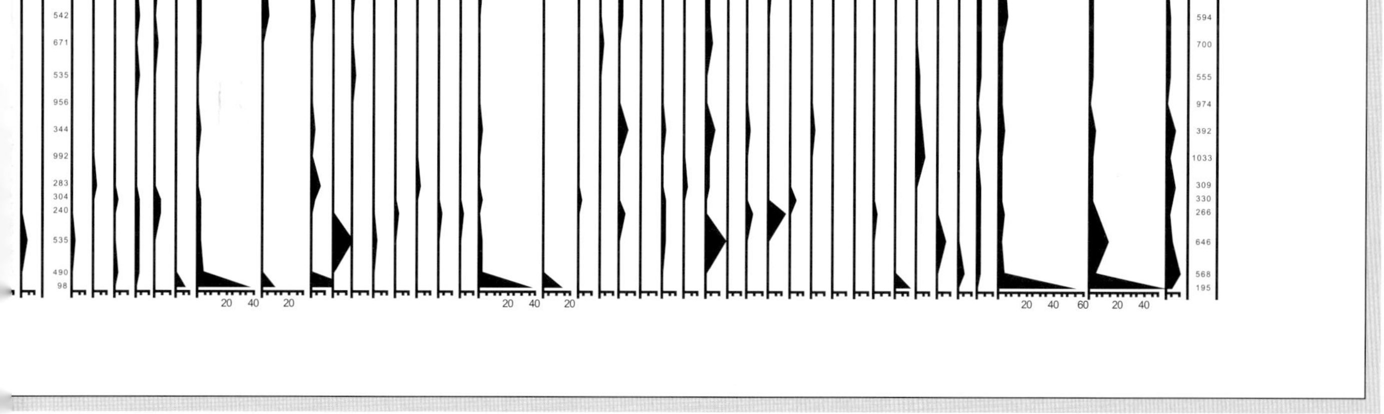

- 2,8m TAW to -1,9m TAW: Phase 3: Sample 4 to 13

This zone is still dominated by arboreal pollen (maximum 95%), but some fluctuations occur in the pollen value of *Alnus*, *Corylus* (hazel) and *Quercus* (oak); and also *Tilia*, *Ulmus* and *Fraxinus* to a lesser degree. In comparison with the basement peat zone *Alnus* shows an extreme increase (with peaks of up to 85%). *Quercus* and *Corylus* are of secondary importance. *Myrica* (bog-myrtle) is present and reaches up to 5%. *Taxus* (yew), *Viburnum lantana* (wayfaringtree) and *Viburnum opulus* type (Guelder-rose type) are scarcely represented. Herbaceous pollen, mainly Poaceae, Cyperaceae and Chenopodiaceae (goosefoot family), accounts for only 10%. Cyperaceae falls gradually to 2% while Chenopodiaceae increases to 7% towards the end of the zone. Ericaceae (heather family) forms less than 2%. Percentages of aquatics, ferns and mosses are low. Type 18 and type 201 are regularly present, though also at very low values.

- 1,9m TAW to -1,7m TAW: Phase 4: Sample 14 to 17

Arboreal pollen makes up more than 90%. *Alnus* has a very high frequency (98%) in the first sample (S14), but is stagnant at 60% in the following samples. *Quercus* and *Corylus* are still well represented. *Pinus*, *Tilia* and *Ulmus* attain percentages to 5%. *Betula* and *Fraxinus* have very low values. Poaceae and Chenopodiaceae (goosefoot family) are the major herbaceous types. Ericaceae and other herbs are virtually absent. Among the aquatics only *Sparganium erectum* type and *Sparganium emersum* type are attested. High Filicales and *Polypodium vulgare* percentages (maximum 25%) occur from sample 17 *et seq* (cf not discussed).

2.2.5.3. | Interpretation

Phase 2

The basal zone reflects the local presence of an *Alnus* dominated fen carr, producing eutrophic peat (also indicated by *Diporotheca rhizophila*; see van Geel *et alii* 1986; van Geel *et alii* 2003, 881; Pons 1992, 9). Taxa commonly associated with *Alnus* in fen carr are consistently recorded e.g. *Salix*, *Rhamnus* type, *Lonicera*, Cyperaceae, *Polypodium vulgare*, *Typha latifolia* (Waller 1994, 98). *Bactrodesmium* type, a fungal spore, is particularly found on wood and bark of *Alnus* or other deciduous trees (van Geel *et alii* 1986). Also *Ustulina deusta*, a mild parasite, occurs and causes soft-rot of wood (van Geel 2002, 106). *Glomus* type is related to soil erosion (van Geel 2002, 107), probably before the start of peat accumulation.

On the drier Pleistocene, sandy soils, outside the alluvium, a dense mixed oak forest (*Quercetum mixtum*) is present. Deciduous trees, such as *Quercus*, *Tilia*, *Ulmus*, *Pinus* and *Fraxinus*, are regularly attested. The high values of *Tilia*, *Pinus*, ferns and mosses in the humic soil at the base of the peat are most likely due to the poor pollen preservation (cf low pollen sum) as these pollen and spores (particularly *Polypodium vulgare*) are very resistant to decay and easy to detect (Janssen 1974, 18-20; Havinga 1984b, 541-558; Waller 1994, 25). In contrast to wind-pollinated trees (e.g. *Alnus* and *Quercus*), *Tilia* and *Fraxinus* are both insect-pollinated species and are therefore generally assumed to be underrepresented (Janssen 1974, 21). The influence of dispersal bias on their percentages, through a marked decline with increasing distance from sources (regional components), is decisive (Janssen 1974, 78; Waller 1994, 95). The shade-tolerant tree *Corylus* is attested on the deciduous woodland edges. The occurrence of *Picea* in this part of the section is probably related to atmospheric long-distance transport, while in clay deposits it rather originates from alluvial and (peri)marine influences.

Despite the presence of light-demanding species, such as Ericaceae and Poaceae, patterns of forest clearances are hard to distinguish. Evidence for cultivation and/or cattle breeding has not been found. The Neolithic *landnam* (by Iversen and Troels-Smith, see Janssen 1974, 89-90), famous for his *Ulmus*-fall, is even not applicable to this region. Perhaps the role of Final Mesolithic/Neolithic man in the landscape of the Schelde basin was passive (transition zone, restricted by the natural wet conditions) rather than the more active one which may be presumed on drier sandy soils. The visibility of these human activities has possibly been disguised by the pollen-influx of local wind-pollinated trees (e.g. *Alnus*) in the fen environment.

The environmental reconstruction of this zone is strictly corresponding to the vegetation pattern of the Atlantic, in which *Quercus*, *Alnus*, *Corylus*, *Tilia* and *Ulmus* occur as the most prominent taxa (Janssen 1974, 55-57; Verbruggen *et alii* 1996, 565, 573; Verbruggen 1999, 108). Similar palynological results from this period are also recorded on other sampling sites at Doel (Minnaert, Verbruggen 1986, 203) and in the surrounding areas (e.g. Melsele and Oorderen) (Van Roeyen *et alii* 1991, 45-46; Munaut 1967, 39).

Phase 3

While pollen assemblages from stratified peat sections are reliable for interpretation, those of (peri)marine deposits or alluvial clay sediments are rather problematic. As a result of mobilisation and long-distance transport by water pollen and spores from clay transitions originate from various regional sources. The secondary or inwashed component (see Moore *et alii* 1991, 13) is almost superior to the local component. Consequently, a local or even regional environmental reconstruction of these pollen assemblages is more questionable. For example the pollen curve of some tree species (e.g. *Alnus*, *Corylus*, *Quercus* and *Ulmus*) in the discussed pollen diagram shows marked fluctuated patterns throughout the entire clay deposit. *Myrica* is allochthonous as it cannot be associated with fen vegetation or deciduous woodland on sandy soils; it is generally established in ombrothrophic bogs (Barkman 1992, 175; Waller 1994, 100; Verbruggen *et alii* 1996, 572; Lambinon *et alii* 1998, 104).

On the other hand, important information can be retrieved concerning the palaeohydrology of a region. The early tidal influence in the lower Schelde basin during the Late Atlantic/Early Subboreal has already been noted by Verbruggen and others (Minnaert, Verbruggen 1986, 203-204; Van Roeyen *et alii* 1991, 45; Verbruggen, Denys 1995). Evidence of clastic brackish deposits was found from pollen and diatoms data in clayey deposits at Doel and Melsele. In particular high values (5% and more) of Chenopodiaceae and the presence of *Armeria* (thrift) are related to the influence of brackish water (Minnaert, Verbruggen 1986, 203-204; Van Roeyen *et alii* 1991, 45; Verbruggen, Denys 1995). Similar values of Chenopodiaceae, which can be attributed either to a brackish environment or human activity, are also attested in the peaty clay deposits of phase 3 discussed above. Taking the lack of explicit anthropogenic indicators into account, the first hypothesis remains the most conceivable. Referring to the chronosequence of the Belgian and Dutch coastal plain, this inundation period has to be situated in the late Calais period. It is obviously the most transgressive and the only one that penetrated so deep into the lower Schelde valley. According to the ^{14}C-datings (cf 2.3.5.) it has to be dated between 5700 and 5000 BP, but the most intensive phase appears to be in the second half of this period. The duration of the inundation and the thickness of the clay on this spot point to a closer vicinity of the river than in Doeldok (Minnaert, Verbruggen,1986) and in Melsele (Van Roeyen *et alii* 1991; Verbruggen, Denys 1995) where the inundation is only registered between 5400 and 5000 BP. The end of the marine influence is synchronous on the three sites and probably rather abrupt (cf 2.3.6.2.).

Phase 4

At the beginning of this zone the re-establishment of *Alnus* dominated fen carr has been completed. Despite the slight increase in the values of *Corylus*, *Quercus* and Poaceae the vegetational conditions are similar to these of phase 2. An *Alnus* fen woodland dominates the local environment. On the drier sandy soils a deciduous forest is attested. Anthropogenic influences have once more not been detected.

To the end of the part of the phase 4 peat, discussed in this chapter, high Filicales (undifferentiated) and *Polypodium vulgare* values are recorded. Despite the indications for poor pollen preservation, they herald the mesotrophic peat growth in the following section (not discussed).

2.2.5.4. | Conclusions

The palynological investigation of the lowest part of a peat sequence in the vicinity of the excavation site at Doel strongly suggests that the natural landscape of the Late Atlantic and Early Subboreal in the lower Schelde alluvial plain was only very slightly determined by Final Mesolithic/Neolithic man, if at all. Perhaps human activity was restricted to the drier Pleistocene, sandy soils.

2.3. | Radiocarbon dating

Mark Van Strydonck
(With contributions from Cyriel Verbruggen & Nathalie Van der Putten)

2.3.1. | Moerbeke "Moervaart"

Mark Van Strydonck, Cyriel Verbruggen & Nathalie Van der Putten

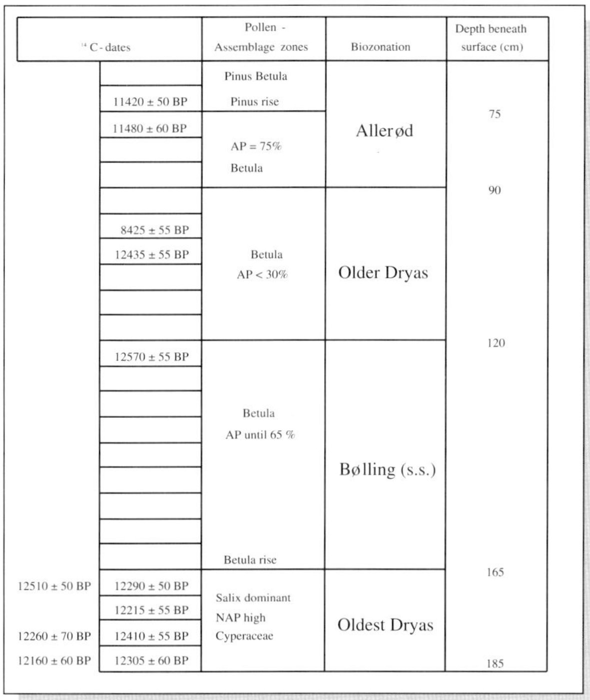

Figure 45: Moerbeke "Moervaart"; position of the radiocarbon dated samples within the biozonation obtained by pollen analysis.

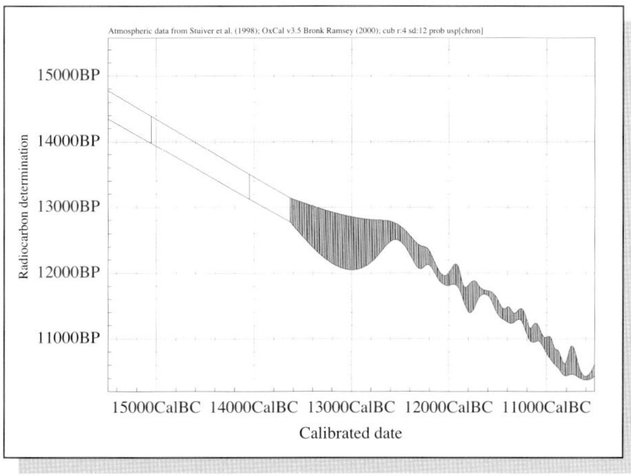

Figure 46: calibration curve INTCAL98.

Previous palynological research (Verbruggen 1971, 1979) on sediments from the "Moervaart" depression proved the value of the sequence for the reconstruction and dating of the palaeo-environmental conditions around the coversand ridge Verrebroek (Stekene) - (Maldegem) Gistel. In the frame of the research conducted in the *Waasland Scheldepolders*, situated only 25km to the northeast of the "Moervaart", it was decided in 2001 to reinvestigate this specific palaeo-lake sequence, with special reference to absolute chronology.

A drilled sequence was sampled by means of an 11cm diameter PVC-tube. After opening the tube, 5cm thick slices were sieved and analysed for macro-remains (tab. 17). Not all levels of the chalk-rich gyttja produced enough organic material for dating. Care was taken to avoid as much as possible submerged water plants, which incorporate dissolved CO_2 from the water in their photosynthesis. This is necessary because the reservoir age of lakes can be quite different from the atmosphere. On three levels it was possible to date two samples of different botanical origin. Figure 45 presents the results of 12 datings and their 9 levels situated in the biozonation obtained by the pollen analysis (cf 2.2.1.). It is obvious that the ages do not increase with depth at all in the time span between the Older Dryas(Ic) and the onset of the infilling in the Oldest Dryas(Ia). Compared to the sedimentation rate of the upper part of the sequence, i.e the Allerød part (IIa and IIb), the Ia - Ic span under consideration must have lasted between 1000 and 2000 years. However the ^{14}C-datings do not show any time evolution. This "impossibility" for radiocarbon dating within the Ia-Ic period is a generally known phenomenon.

Indeed, according to the currently accepted curve (INTCAL98) (Stuiver et alii 1998) there is a large "bulb" around 13000 cal BC (fig. 46). Possibly this problem can be resolved when the new calibration curve (INTCAL04)

(Reimer *et alii* 2002), announced two years ago and presented at the radiocarbon conference in Wellington (September 2003), becomes available. This new curve, made up with many more data points, is much smoother in that period. Unfortunately the curve will not be available before mid 2004. This means that the "Moervaart" dates must be revised in the near future.

^{14}C-dates in BP	macroscopic remains seeds and fruits	other	depth beneath surface (cm)
11420 ± 50	*Betula alba/pendula*: 1 *Scirpus lac./mar./tab.**+++	*Alnus* sp. (root)	75-70
11480 ± 60	*Myriophyllum spicatum* +++ *Scirpus lac./mar./tab.** + *Potamogeton* sp. +	*Alnus* sp. (root)	80-75
	*Scirpus lac./mar./tab.**: 2	*Alnus* sp. (root)	90-85
8425 ± 55	*Carex sp. (triconvex)*: 1 *Scirpus lac./mar./tab.**: 1 *Myriophyllum spicatum:* 8 *Betula alba/pendula:* 1	*Alnus* sp. (root)	100-95
12435 ± 55	*Carex* sp.: 1 *Carex* sp. *(triconvex)* *Scirpus lac./mar./tab.** + *Betula alba/pendula* + *Myriophyllum* sp. +++	*Alnus* sp. (root)	105-100
12570 ± 55	*Betula alba/pendula:* 5 *Carex* sp.: 2 *Potamogeton* sp.: 2 *Scirpus lac./mar./tab.**: 1	**Bryophytes**	125-120
	Carex sp.: 2		135-130
	*Scirpus lac./mar./tab.**: 3 *Potamogeton* sp.: 1 *Carex* sp.: 1		160-155
12290 ± 50 **12510 ± 50**	*Potamogeton* sp. +++ *Carex sp. (triconvex)* ++ *Scirpus lac./mar./tab**. +	**Bryophytes**	170-165
12215 ± 55	*Potamogeton* sp. +++ *Scirpus lac./mar./tab.** +++ *Carex* sp. *(triconvex)* ++ *Betula alba/pendula* +		175-170
12260 ± 70 **12410 ± 55**	*Eleocharis palustris/uniglumis* *Scirpus lac./mar./tab.** *Carex* sp. *(triconvex)* *Potamogeton* sp. *Betula alba/pendula* *Potentilla anserina* *Chenopodiaceae* sp.	***Salix* sp. (twig)**	180-175
12305 ± 60 **12160 ± 60**	*Scirpus lac./mar./tab.** +++ *Carex* sp. *(triconvex)* + *Eleochris palustris/uniglumis* ++ *Potamogeton* sp. +	**monocotyledon**	185-180

Table 17: Moerbeke "Moervaart"; list of radiocarbon dates and macroscopic remains per level; Key: Abundance scale seeds & fruits:
0 - 5: exact number
5 - 20: +
20-50: ++
> 50 +++
* *Scirpus lacustris/maritimus/tabernaemontanii*
Underlined and **bold** refer to the dated taxa.

The dates in table 18 (except the one in bold) must only be considered as estimates based on the new calibration data. To make it clear that these are no definite results, no standard deviation is added and the results are expressed in cal BP.

lab. reference	BP date	estimated calibrated date or 68.2% confidence	95.4% confidence
KIA-18757	11420±50	13100 cal BP	
KIA-18756	11480±60	13150 cal BP	
KIA-18759	8425±55	**7580BC (0.93) 7450BC 7390BC (0.07) 7380BC**	**7590BC (1.00) 7330BC**
KIA-18758	12435±53	14150 cal BP	
KIA-18750	12570±55	14200-14500 cal BP	
KIA-18747	12290±50	14100 cal BP	
KIA-18752	12510±50	14200 cal BP	
KIA-18748	12215±55	14000 cal BP	
KIA-18755	12410±55	14200 cal BP	
KIA-18751	12260±70	14000 cal BP	
KIA-18754	12160±60	13950 cal BP	
KIA-18753	12305±58	14100 cal BP	

Table 18: Moerbeke "Moervaart"; list of radiocarbon dates.

2.3.2. | Verrebroek "Dok 2"

From profile 1 of this site (fig. 8) the base and top of a peaty Late Glacial soil (cf 2.1.3.4.) were dated (NZA-11013) 12020±60 BP and (NZA-11021) 11740±60 BP respectively. A third sample of this soil within the same profile was dated (UtC-9435) 11930±90 BP. Unfortunately the exact stratigraphic position of this sample within the peaty layer is no longer known.
The calibrated results of these three dates are represented in figure 47 and table 19.

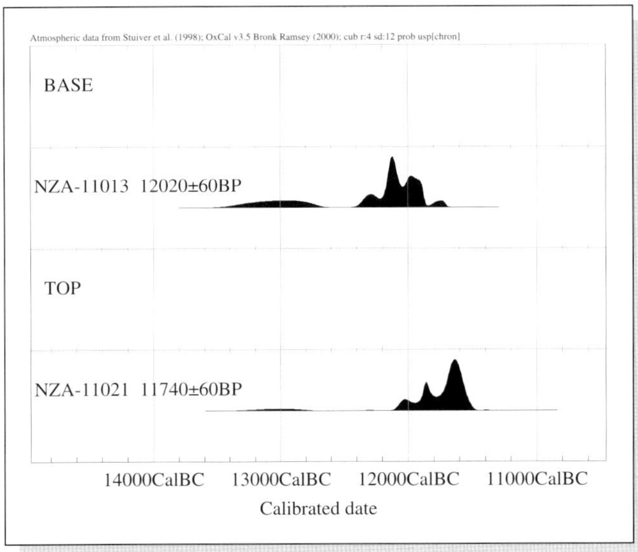

Figure 47: Verrebroek "Dok 2"; calibrated radiocarbon dates from the top and the base of the Late Glacial soil.

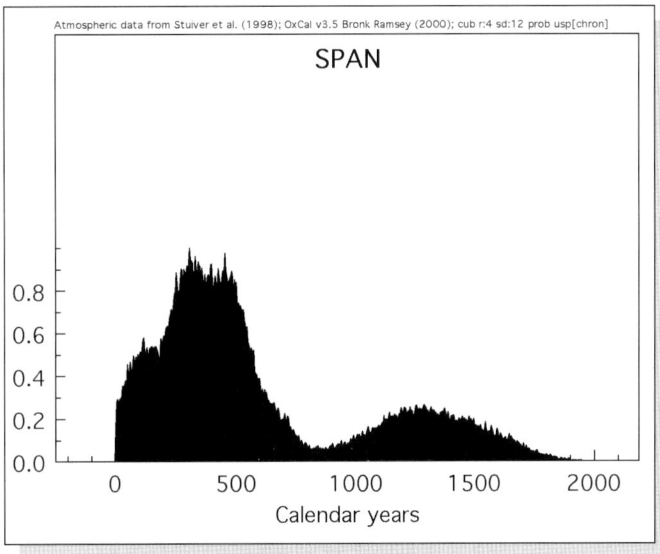

Figure 48: Verrebroek "Dok 2"; probability distribution of the duration of the Late Glacial soil.

As in the "Moervaart" sequence (cf 2.3.1.), the dates suffer from the irregularity in the calibration curve. The segment of the probability distribution around 13000 cal. BC should be considered as an artefact. As a consequence the SPAN can only be ≤ 800 years (fig. 48). Once again it has to be noted that this conclusion is based on the general shape of the new curve and not on the data.

lab. reference	dating material	BP date	68.2% confidence	95.4% confidence
BASE				
NZA-11013	Organic fraction of peat	12020±60	12340BC (8.2%) 12220BC 12210BC (60.0%) 11870BC	13400BC (20.9%) 12700BC 12400BC (74.5%) 11700BC
TOP				
NZA-11021	Organic fraction of peat	11740±60	11900BC (13.8%) 11800BC 11750BC (54.4%) 11540BC	13100BC (2.7%) 12900BC 12200BC (92.7%) 11400BC
SPAN				
			0 (68.2%) 650	-100 (74.6%) 800 900 (20.8%) 1700
NON-LOCALISED				
UtC-9435	Organic fraction of peat	11930±90	12170BC (62.1%) 11840BC 11780BC (6.1%) 11690BC	13300BC (14.2%) 12700BC 12400BC (81.2%) 11600BC

Table 19: Verrebroek "Dok 2"; list of radiocarbon dates from the top and the base of the Late Glacial soil.

2.3.3. | Verrebroek "Dok 1"

2.3.3.1. | Late Glacial soils

From profile 9 (fig. 24) the base and top of two successive humiferous to peaty layers (cf 2.1.4.2.) were dated (NZA-11018) 11760±60 BP, (NZA-11019) 11690±60 BP and (NZA-11521) 11420±65 BP, (NZA-11520) 10710±75 BP respectively. The calibrated results are represented in figure 49 and table 20.

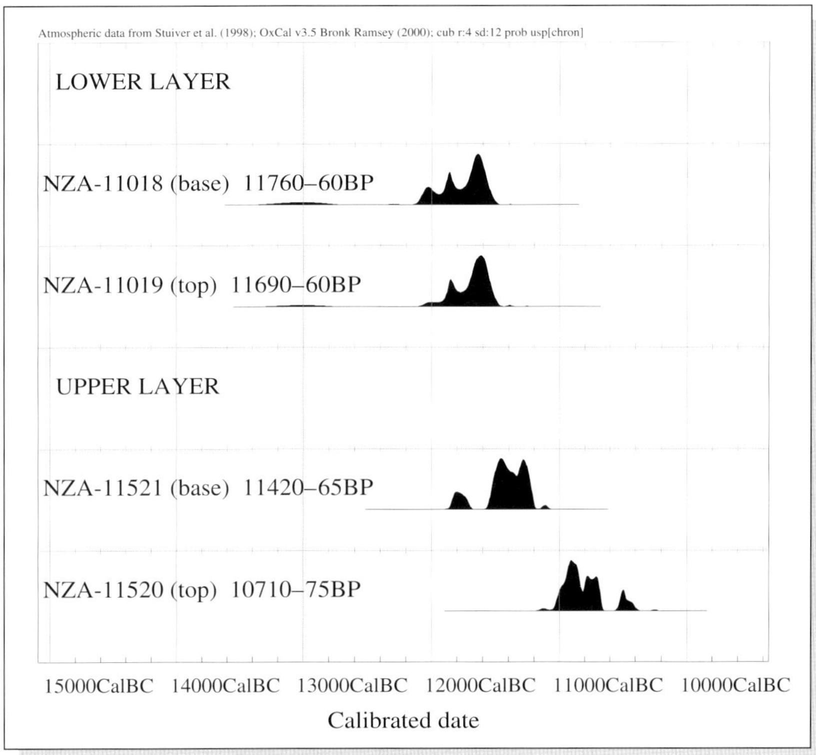

Figure 49: Verrebroek "Dok 1"; calibrated radiocarbon dates from the top and the base of both Late Glacial soils.

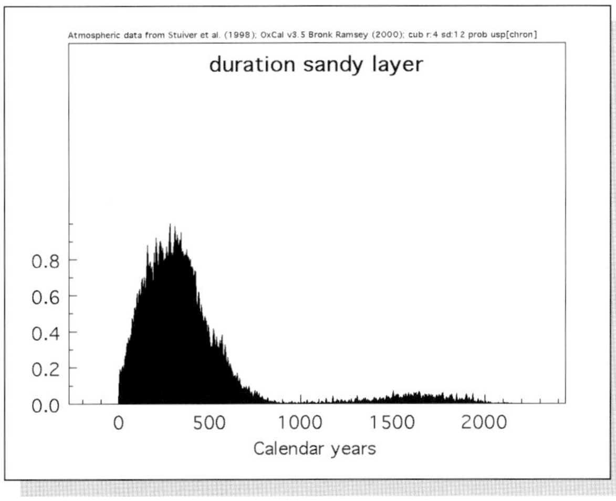

Figure 50: Verrebroek "Dok 1"; duration of the sandy layer situated between the two Late Glacial soils.

Figure 50 represents the limits of duration for the deposition of the intercalated sandy layer. As in the "Moervaart" sequence (cf 2.3.1.), it needs to be remembered that the older part of NZA-11019 is an artificial effect caused by the shape of the calibration curve. The difference in age between the end of the lower and the start of the upper peat layer can vary between 0 and 800 years.

lab. reference	dating material	BP date	68.2% confidence	95.4% confidence
LOWER LAYER (base)				
NZA-11018	Organic fraction of peat	11760±60	12050BC (4.1%) 12010BC 11900BC (15.1%) 11800BC	13200BC (3.8%) 12800BC 12200BC (91.6%) 11400BC 11760BC (49.0%) 11550BC
LOWER LAYER (top)				
NZA-11019	Organic fraction of peat	11690±60	11880BC (9.7%) 11820BC 11730BC (58.5%) 11530BC	13100BC (1.4%) 12900BC 12100BC (94.0%) 11300BC
UPPER LAYER (base)				
NZA-11521	Organic fraction of peat	11420±65	11510BC (43.9%) 11340BC 11330BC (24.3%) 11230BC	11850BC (11.6%) 11700BC 11600BC (83.8%) 11200BC
UPPER LAYER (top)				
NZA-11520	Organic fraction of peat	10710±75	10980BC (41.2%) 10830BC 10810BC (27.0%) 10680BC	11050BC (84.4%) 10650BC 10550BC (11.0%) 10400BC
SPAN				
			80 (68.2%) 460	-100 (93.8%) 800 1300 (1.6%) 2000

Table 20: Verrebroek "Dok 1"; list of radiocarbon dates from the top and the base of the Late Glacial soils.

2.3.3.2. | Holocene peat

From profile 1 at Verreboek "Dok 1" (fig. 24) the bottom of the peat layer (cf 2.1.4.9.), overlying the coversand deposits which yielded the Early Mesolithic industry, was dated at (KIA-17992) 4690±30 BP. The sample was taken at ca. 0.20/0.25m above sea level. The result is represented in table 21 and figure 51.

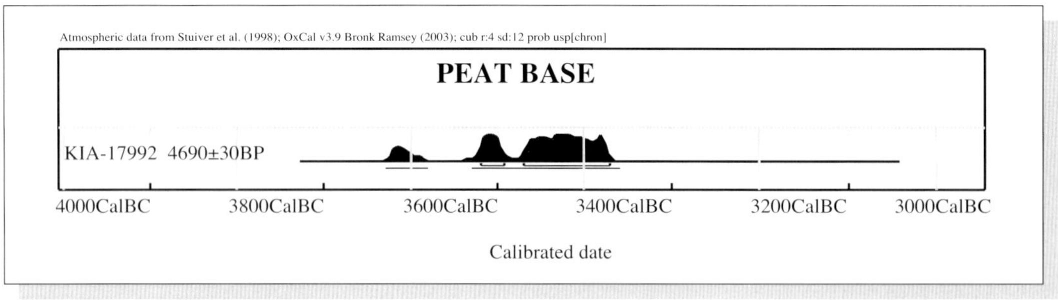

Figure 51: Verrebroek "Dok 1"; calibrated radiocarbon date from the base of the overlying Holocene peat.

lab. reference	dating material	BP date	68.2% confidence	95.4% confidence
KIA-17992	Organic fraction of peat	4690±30	3520BC (14.4%) 3490BC 3470BC (53.8%) 3370BC	3630BC (11.1%) 3580BC 3530BC (84.3%) 3360BC

Table 21: Verrebroek "Dok 1"; radiocarbon date from the base of the Holocene peat layer.

2.3.4. | Doel "Deurganckdok-sector B"

From profile 1 (fig. 14) the base of the peat, overlying the Swifterbant find-layer, was dated (cf 2.1.5.). The sample was taken near the top of the sand ridge at ca. - 1.00m TAW. The sample was wet sieved and the fraction between 0.5 and 0.25mm was dated (fig. 52 and tab. 22).

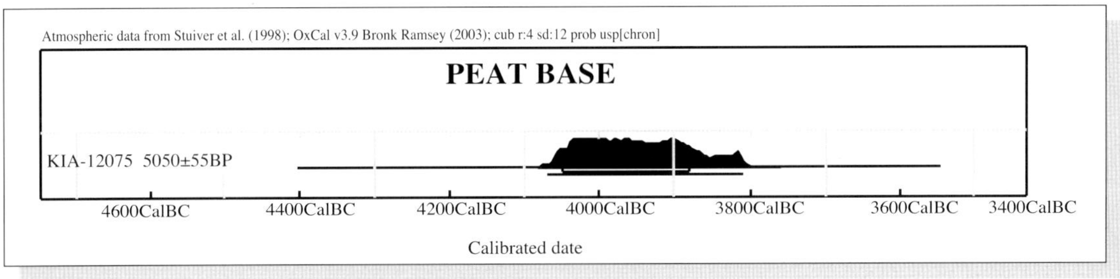

Figure 52: Doel "Deurganckdok-sector B"; calibrated radiocarbon date from the base of the Holocene peat.

lab. reference	dating material	BP date	68.2% confidence	95.4% confidence
KIA-12075	Organic fraction of peat	5050±55	3950BC (68.2%) 3780BC	3970BC (95.4%) 3710BC

Table 22: Doel "Deurganckdok-sector B"; radiocarbon date from the base of the Holocene peat.

2.3.5. | Doel "Deurganckdok-sector A"

In sector A four samples from a deep peat section, which was pollen analysed in detail (cf 2.1.6.; 2.2.5.), were radiocarbon dated (fig. 53 and tab. 23). Sample KIA-17640 represents the beginning of the lowermost peat growth. KIA-17639 dates the end of this peat growth and the beginning of the Calais deposits. Sample KIA-17638 comes from the lowest level of the covering peat and KIA-17637 dates the top of the peat.

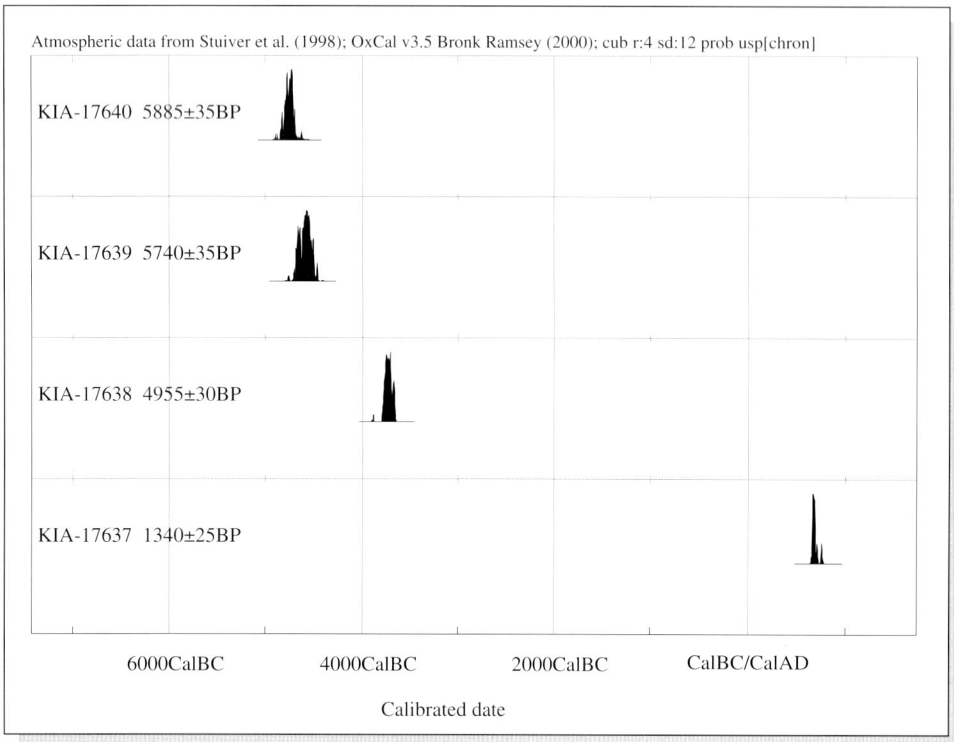

Figure 53: Doel "Deurganckdok-sector A"; calibrated radiocarbon dates from the Holocene peat layers.

lab. reference	dating material	TAW (m)	BP date	68.2% confidence	95.4% confidence
KIA-17640	Organic fraction of peat	-3.21	5885±35	4795BC (68.2%) 4710BC	4850BC (94.4%) 4680BC 4640BC (1.0%) 4620BC
KIA-17639	Organic fraction of peat	-2.84	5740±35	4680BC (12.9%) 4630BC 4620BC (53.2%) 4530BC 4510BC (2.1%) 4500BC	4690BC (94.4%) 4490BC 4470BC (1.0%) 4460BC
KIA-17638	Organic fraction of peat	-1.89	4955±30	3775BC (68.2%) 3695BC	3790BC (95.4%) 3650BC
KIA-17637	Organic fraction of peat	-1.00	1340±25	658AD (68.2%) 689AD	640AD (86.4%) 720AD 740AD (9.0%) 770AD

Table 23: Doel "Deurganckdok-sector A"; list of radiocarbon dates from the Holocene peat layers.

2.3.6. | Conclusions

2.3.6.1. | Late Glacial soils

The radiocarbon dates obtained at Moerbeke, Verrebroek "Dok 1" and Verrebroek "Dok 2" seem to correspond very well with other Late Glacial dates realised during previous research in Sandy Flanders (Verbruggen 1971, 1979; Heyse 1979). Except for a few "older" dates (tab. 24), the start of most Late Glacial sequences in this study-area is situated between 12400/12500 and 11800/11750 uncal. BP years. Due to problems related to the calibration of radiocarbon dates it is currently not possible to work out a finer absolute chronology for the Late Glacial in Sandy Flanders. It is hoped that the publication of the new calibration curve INTCAL04 will offer better prospects.

site/location	lab. reference	BP date
Vinderhoute "Kale I"	GrN-6062	12655±70
Stabroek	GrN-2458	12460±140
Stabroek	GrN-3052	12340±120
Stabroek	GrN-3049	12330±120
Zelzate	GrN-4782	12300±100
Verrebroek "Aven Ackers"	IRPA-577	12120±140
De Klinge II	GrN-2993	12000±110
Oedelem "Geite"	GrN-6073	12010±65
Moerkerke "Maleveld"	GrN-6045	11950±65
De Klinge	GrN-2891	11920±80
Jabbeke	GrN-6654	11900±90
Snellegem "Molenbroek"	GrN-6061	11780±70
Adegem "Balgerhoeke I"	GrN-6935	11780±100
Oedelem "Geite"	GrN-6714	11770±65
Roksem	GrN-129-67	11740±130
Sijsele "Floré"	Antw. 127	11490±180
Oedelem "Vliegend Paard"	GrN-6715	11395±65

Table 24: list of radiocarbon dates related to the initial stages of the Late Glacial (Earliest Dryas/Bølling) in Sandy Flanders.

2.3.6.2. | Early Holocene deposits

In the frame of the recent research in the *Waasland Scheldepolders* three radiocarbon dates have been obtained from the base of the Holocene peat covering the coversand deposits. Taking into account previous radiocarbon dating from the study-area (tab. 25-27), the start of the peat growth can be situated between ca. 5900/5800 BP or ca. 4750/4650 cal. BC in the lowermost parts (ca. 2 to 3m below actual sea level) and ca. 5000/4500 BP or 3780/3100 cal. BC on the higher parts of the coversand landscape (ca. 1 to 2m above actual sea level).

Apart from the difference in stratigraphic position, both dates for the deposition of the Calais-deposits at Doel "Deurganckdok-sector A" fit fairly well with previous dates obtained from the nearby locations of Doel "Doeldok" (tab. 25) and Melsele "Hof ten Damme" (tab. 26). All these dates point to increased marine influence, resulting in the deposition of organic clay to clayey peat, between ca. 5700 (ca. 4550 cal. BC) and 5000/4900 BP (ca. 3680 cal. BC), i.e. contemporaneous with the Swifterbant and Michelsberg occupations in sector B (cf 3.2.4.) and C (cf 3.2.5.). However, it should be emphasised here that both the starting and ending date should be considered as extremes, since none of the dated sampled directly dates the Calais-deposits. All so-called Calais-dates were indeed obtained on peat samples taken immediately under or above the clayey deposits. Hence, these dates should be interpreted as *terminus post quem* or *terminus ante quem* respectively.

lab. reference	dating material	TAW (m)	BP date	68.2% confidence	95.4% confidence
TOP OF UPPER PEAT LAYER					
IRPA-455[1]	Organic fraction of peat	ca. 0.00	2050±70	170BC (68.2%) 30AD	350BC (2.7%) 310BC 210BC (91.4%) 90AD 100AD (1.3%) 130AD
IRPA-456	wood	ca. 0.00	3000±70	1380BC (9.5%) 1330BC 1320BC (58.7%) 1120BC	1410BC (95.4%) 1010BC
TOP OF CALAIS DEPOSIT[2]					
IRPA-457	Organic fraction	ca. -1.38	5350±70	4320BC (7.1%) 4290BC 4260BC (61.1%) 4040BC	4340BC (92.4%) 4030BC 4020BC (3.0%) 4000BC
BASE OF CALAIS DEPOSIT					
IRPA-454	Organic fraction	ca.-1.50	4900±60	3760BC (68.2%) 3630BC	3910BC (1.8%) 3870BC 3800BC (87.0%) 3620BC 3590BC (6.6%) 3520BC
BASE OF LOWER PEAT LAYER					
IRPA-458	Organic fraction of peat	ca. -2.12	5490±80	4450BC (68.2%) 4240BC	4500BC (86.1%) 4210BC 4200BC (4.7%) 4160BC 4130BC (4.5%) 4040BC

1 IRPA: samples measured by ß-decay counting at the Royal Institute for Cultural Heritage, Brussels
2 The "inversion" of these dates obtained for the top and the base of the Calais deposits at Doel "Doeldok" so-far remains unsolved.

Table 25: Doel "Doeldok"; list of radiocarbon dates (Dauchot-Dehon *et alii* 1984; Minnaert, Verbruggen 1986).

lab. reference	dating material	TAW (m)	BP date	68.2% confidence	95.4% confidence
BASE OF PEAT LAYER (from different soil profiles)					
IRPA-951	Organic fraction of peat	1.65/ 1.66	4080±60	2860BC (13.6%) 2810BC 2740BC (1.2%) 2720BC 2700BC (44.9%) 2560BC 2530BC (8.5%) 2490BC	2870BC (16.8%) 2800BC 2790BC (78.6%) 2460BC
IRPA-947	Organic fraction of peat	0.50/ 0.52	4300±60	3020BC (68.2%) 2870BC	3100BC (83.3%) 2850BC 2820BC (12.1%) 2690BC
IRPA-944	Organic fraction of peat	-0.15/ -0.19	5350±50	4250BC (10.5%) 4220BC 4200BC (38.7%) 4160BC 4130BC (12.5%) 4110BC 4070BC (6.5%) 4050BC	4250BC (18.0%) 4210BC 4200BC (77.4%) 4040BC
TOP OF CALAIS DEPOSIT					
IRPA-946	Organic fraction	0.20/ 0.24	5000±40	3910BC (11.6%) 3870BC 3800BC (56.6%) 3700BC	3950BC (28.9%) 3830BC 3820BC (65.1%) 3690BC 3680BC (1.4%) 3660BC
BASE OF CALAIS DEPOSIT					
IRPA-942	Organic fraction	-0.48/ -0.56	5300±70	4230BC (15%) 4180BC 4170BC (48.3%) 4040BC 4020BC (4.9%) 4000BC	4330BC (5.5%) 4280BC 4260BC (89.9%) 3970BC

Table 26: Melsele "Hof ten Damme"; list of radiocarbon dates from the (Final) Mesolithic site (Van Strydonck *et alii* 1995).

lab. reference	dating material	TAW (m)	BP date	68.2% confidence	95.4% confidence
VERREBROEK "Aven Ackers"					
IRPA-607	Organic fraction of peat		4930±70	3780BC (68.2%) 3640BC	3950BC (93.4%) 3630BC 3560BC (2.0%) 3530BC
VERREBROEK "Paardenkerkhof"					
KIA-14320	Organic fraction of peat		4375±30	3020BC (68.2%) 2920BC	3090BC (9.8%) 3050BC 3040BC (85.6%) 2900BC

Table 27: radiocarbon dates from the peat base at two sites in Verrebroek (resp. Dauchot-Dehon *et alii* 1986; Van Strydonck *et alii* 2002).

2.4 | General conclusions

2.4.1. | Late Glacial landscape

Pedological, palynological and palaeobotanical investigations at the sites of Moerbeke "Moervaart", Verrebroek "Dok 1" and Verrebroek "Dok 2" have yielded important information concerning the Late Glacial environment in the eastern district of Sandy Flanders. The data obtained confirm the conclusions of earlier studies that the Late Glacial in this area was a very dynamic period in which the landscape was dominated by shallow, at least periodically inundated, depressions (mires/lakes/dune slacks) alternating with dry coversand ridges and dunes.

At two locations at Verrebroek, remnants of Subaqueous Soils were found, one dating back to the Bølling/Earlier Dryas, another to the Allerød. They indicate the presence of lakes or dune slacks in palaeo-depression positions. Both palaeo-lakes existed only briefly, as they were filled in with aeolian sand during the Earlier and the Younger Dryas respectively. As a consequence, the pollen diagrams from these soils represent only a very short period, which makes it difficult to correlate them with the existing regional pollen assemblage zones. However, based on common trends in the pollen curves, it seems likely that the diagram of Verrebroek "Dok 2" can be correlated either with zone Ib and the first part of zone Ic of the regional biostratigraphical zonation for the Late Glacial period in Sandy Flanders proposed by Verbruggen (1979), or with subzone 1b and subzone 1c of the regional biostratigraphical zonation for the Late Glacial in The Netherlands published by Hoek (1997). Taking into account the relatively high percentages of *Pinus* (pine) in the upper part of diagram 1 of Verrebroek "Dok 1", this can be correlated with zone IIb of Verbruggen (1979), which is attributed to the Allerød. For the lower part of the diagram, however, the results of pollen analysis do not allow a decision as whether it represents the same period as the diagram of Verrebroek "Dok 2", which would imply that it can be correlated with zone Ib and the first part of zone Ic of Verbruggen (1979), or whether it should be correlated with zone IIa, which is the first part of the Allerød. Stratigraphy and pedological data point towards a discontinuous sedimentation but continuous wet, subaqueous conditions. In this respect the situation at Verrebroek "Dok 1" differs from that at Verrebroek "Dok 2". Its aeolian sediment input (e.g. H13 in P9) occurred subaqueously and is only a few centimetres thick, whereas at "Dok 2" (e.g. H2 to H1 in P1, H5 to H2 in P11) the input consists of several decimetres of both subaqueously and non-subaqueously deposited sand. The latter data rather support an Allerød age for both parts of the Late Glacial soil at "Dok 1". Unfortunately the available radiocarbon dates cannot definitively elucidate this problem due to calibration problems.

At Verrebroek "Dok 2" congeliturbation structures (Drop Soil) and other particular involutions where probably colder conditions are involved (sand dykes or diapirs, load cast structures) affected the Late Glacial Bølling soil. They indicate periods with periglacial conditions.

A third investigated palaeo-lake, situated at Moerbeke "Moervaart", clearly underwent another evolution. This much larger palaeo-lake apparently existed almost throughout the whole Late Glacial, probably until the end of the Allerød, and was never sealed by wind-blown sand. Unlike the two dune slacks at Verrebroek, this palaeo-lake is still visible in the present-day landscape.

Spatial analysis of contemporaneous prehistoric sites, mainly belonging to the *Federmesser* and related cultures (Crombé, Verbruggen 2002), indicates a dense concentration of sites along the edges of these former lakes and dune slacks. At Moerbeke "Moervaart", for example, at least 12 Final Palaeolithic surface-sites are known along the northern bank of the lake (fig. 7). At Verrebroek "Dok 2", a small knapping site was excavated at the eastern limit of the Bølling lake. This concentration of campsites near lakes is most likely linked to the high resource potential of these zones. Indeed, most sites are situated in transitional areas with a great diversity of habitats, including both dry and wooded hunting zones, lacustrine fishing, and presumably also fowling zones. In terms of productivity and diversity, these lake areas were most likely the richest in Sandy Flanders, especially in spring and autumn/winter when fish and birds respectively would have been particularly abundant.

2.4.2. | Early Holocene landscape

At the start of the Holocene, during the Preboreal/Boreal, when the main occupation of Verrebroek "Dok 1" took place, the lake landscape must have already changed into a dry forested landscape. Absence of sediments containing contemporaneous pollen at the excavated site, except for the E-horizon of the Podzolised Soil, hampers a detailed reconstruction of the former vegetation.

Absence of pollen might, however, indicate aerated, bioturbated conditions as well, which fits with the hypothesis of Brunified Soils for the Preboreal to Boreal. Remains of Brunified Soils were only observed in particular positions, such as in tree-fall structures at Verrebroek "Dok 1" or in relatively finer sediments at Verrebroek "Dok 2" and Doel "Deurganckdok". Brunified Soils have an organic-rich surface A-horizon, lack an eluvial E-horizon and have a homogeneous brown, bioturbated B-horizon. They are characterised by high biological activity and distinct weathering, and thus reflect particular environmental conditions. The climate must have favoured limited leaching conditions, such that precipitation and evapotranspiration were roughly in balance. Brunified Soils must once have been present all over the area. However, subsequent podzolisation processes erased most of the soil characteristics resulting from brunification.

Palynological analysis of the eluvial E-horizon of the Podzolised Soil at Verrebroek "Dok 1" indicates that, possibly from the Late Boreal onwards, climatic conditions promoted strong leaching and were thus favourable for podzolisation. At Verrebroek, it is not impossible that fire, either natural or anthropogenic, by opening the canopy, enhanced the leaching conditions and accelerated acidification, thus inducing podzolisation on these already poor sandy soils. Whether there is a relation between the numerous charcoal-features, i.e. hearth-pits and charcoal-pits, excavated at Verrebroek "Dok 1", and such fires remains to be further investigated (cf 3.6.). However, one may note that these features do not appear earlier than the second half of the Boreal.

The earliest traces of a gradual drowning of the landscape are found in the northeastern part of the study area situated closest to the River Schelde, in particular at Doel "Deurganckdok". Based on the available radiocarbon dates, we can deduce that the wettening of the landscape, as a result of a gradual rise of the sea level, started around ca. 4750/4650 cal. BC (5900/5800 BP). At that time (Late Atlantic) fenlands came into existence for the first time, changing the landscape in a radical way. Based on pollen analysis, mixed deciduous forests (oak, lime, elm, hazel, birch) can be reconstructed on the elevated zones, while intermediate wetland forests and alder forests can be assumed in the lower parts. From ca. 4550 cal. BC (5700 BP) to ca. 3680 cal. BC (4900 BP) peat growth was locally interrupted by increasing marine influence, which resulted in the deposition of clayey sediments (so-called Calais-deposits). During this period hunter-gatherers belonging first to the Swifterbant culture (sector B) and later to the Michelsberg culture (sector C) settled on the top of a narrow coversand ridge, which probably formed one of the few dry places within the inundated area that was still appropriate for settling down and building a campsite. A radiocarbon date from the base of the peat covering the sand ridge in "sector B", however, indicates that ultimately, around ca. 3770 cal. BC (5000 BP), this (lower) part of the ridge also became too wet for human occupation. This probably explains why the later, Michelsberg occupants settled higher-up on this sand ridge.

In the southern and topographically higher area of the *Waasland Scheldepolders*, i.e. in the surroundings of Verrebroek, fenlands clearly penetrated much later. According to the available radiocarbon dates, peat growth didn't start here until 3780/3100 cal. BC (5000/4500 BP) (Sub-boreal). Because of their high elevation some landscape portions, such as the top of the large sand ridge at Verrebroek "Dok 1", were probably never or hardly covered by peat at all. Here peat, up to 0.75m thick, was only observed in palaeo-depression positions and on the dune slopes.

Despite the fact that the top of the sand ridge at Verrebroek remained "dry", no traces of human activity later than the Late Mesolithic (Atlantic) were found during the excavations. According to the radiocarbon dates on hazelnut shells (cf 3.6.), the latest activity in prehistoric times took place around ca. 5260 cal. BC (6300 BP). After this date the area apparently remained beyond human influence until Medieval times. Clear pedomorphological evidence of past peat exploitation, such as remnants of peat cutting trenches as found at other places in the *Waasland Scheldepolders,* is absent at Verrebroek. According to B. Augustijn (1999), however, peat extraction must have been important and extended allover Sandy Flanders. At Verrebroek "Dok 1" the peat surface is exceptionally flat and nearly horizontal. The overlying surface horizon, consisting of a mixture of peat and sand with a nearly constant thickness of about 0.10m, is indicative of a cultivation horizon. The data suggest agricultural land use, rather than (or after) peat exploitation, during the Late Medieval period (13th-15th century).

These activities probably came to an end as a result of flooding, which led to the deposition of calcareous clay and sand with variable thickness (0.4-1.5m) on top of the coversands and peat, where present. Once protected by dikes and drained, these flood sediments underwent several soil processes from the stabilised surface on. Among these, decarbonatation, structure formation and strong bioturbation resulted in the development of a soil with a colour- and structure-B-horizon.

III.

FEATURES

3.1. | Archaeological description

Philippe Crombé, Yves Perdaen & Joris Sergant

3.1.1. | Verrebroek "Dok 2"

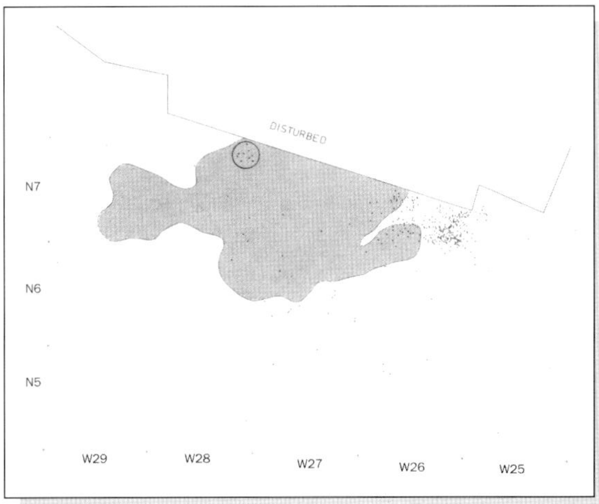

Figure 54: Verrebroek "Dok 2"; distribution plan of the artefacts (small dots) and the charcoal patch (grey shaded). Encircled area: small cluster of burnt bones.

Figure 55: Verrebroek "Dok 2"; view of the soil section at profile 11.

At this location an irregular shaped charcoal patch was found in the same stratigraphic position as a small cluster of Final Palaeolithic flint artefacts, i.e. on the top of a Late Glacial (Bølling/Earlier Dryas) humiferous soil (Crombé *et alii* 1999c). This charcoal feature (fig. 54, 55), which was partly destroyed by the harbour works, measures about 3 x 1.5m and is situated adjacent to the artefact cluster. It consists mainly of ash and dispersed charcoal fragments; only locally *in situ* burnt branches could be identified. A discrete cluster of burnt bone fragments is located at the NW limit of the feature.

Whether this charcoal feature is directly related to the human occupation at this spot (humanly induced fire) or whether it is the result of a natural forest fire remains unclear.

3.1.2. | Verrebroek "Dok 1"

At the Verrebroek "Dok 1" site, structural features of different ages have been uncovered. They can be classified into two main groups: 1° features dating back to the period prior to the covering of the coversand landscape (= prehistoric features), and 2° features dating back to the period after the formation of peat and/or the deposition of clay.

3.1.2.1. | Prehistoric features

The features of prehistoric age can be further subdivided into two main groups: 1° features containing a lot of charcoal in their filling, and 2° other features. The charcoal features include two subtypes: 1° typical hearth-pits (N=20), and 2° charcoal-pits (min N=40). The category of "other prehistoric features" is represented by surface-hearths (min N=50), hazelnut-pits (N=2), tree-falls and tree-sway structures.

Charcoal features

Typical hearth-pits

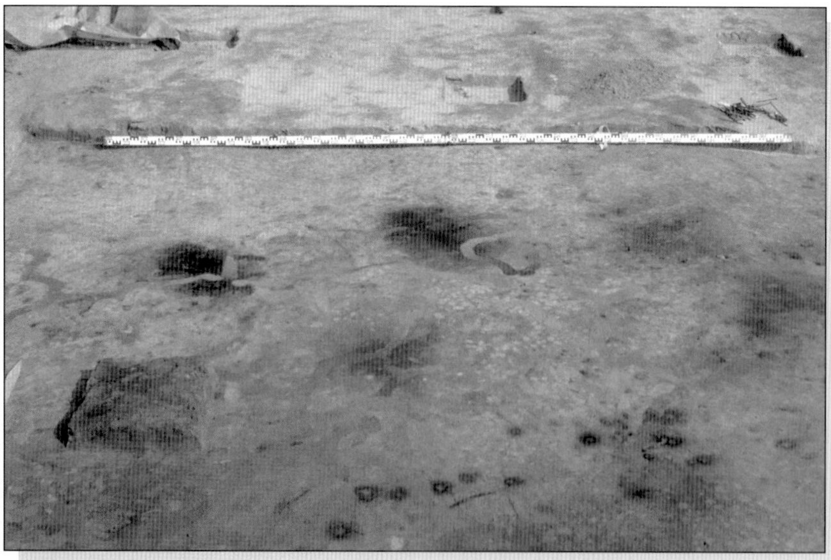

Figure 56: Verrebroek "Dok 1"; view of a small cluster of typical hearth-pits.

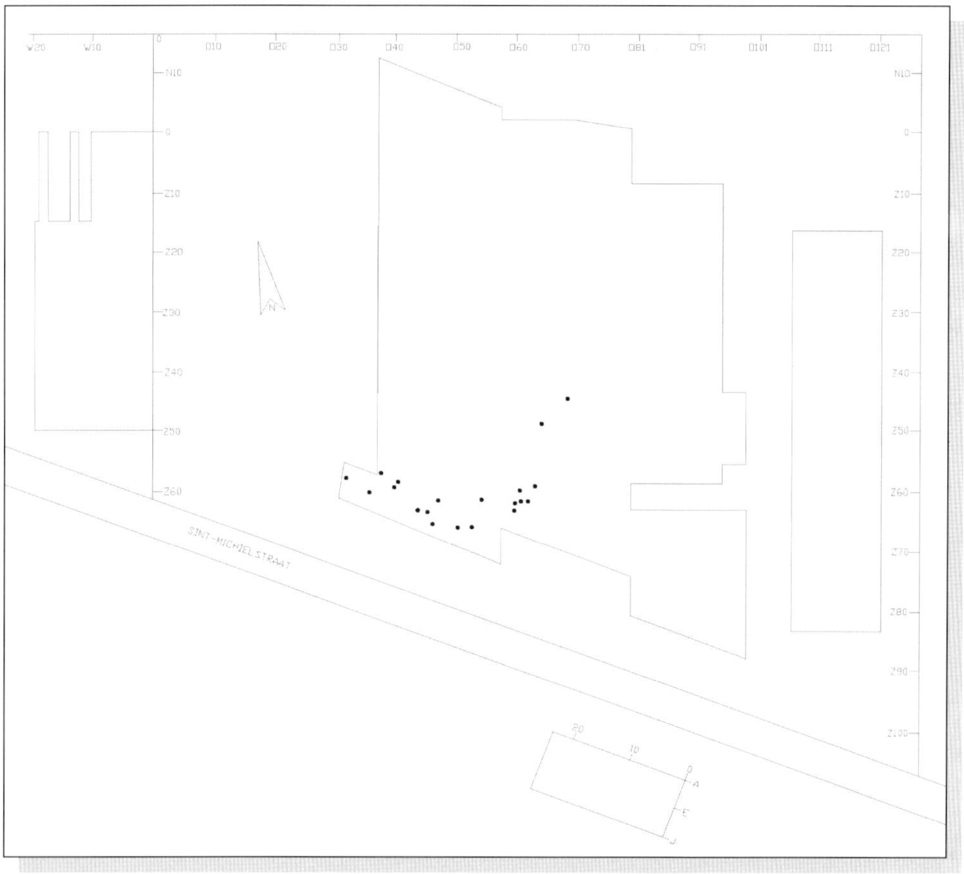

Figure 57: Verrebroek "Dok 1"; general distribution map of the typical hearth-pits.

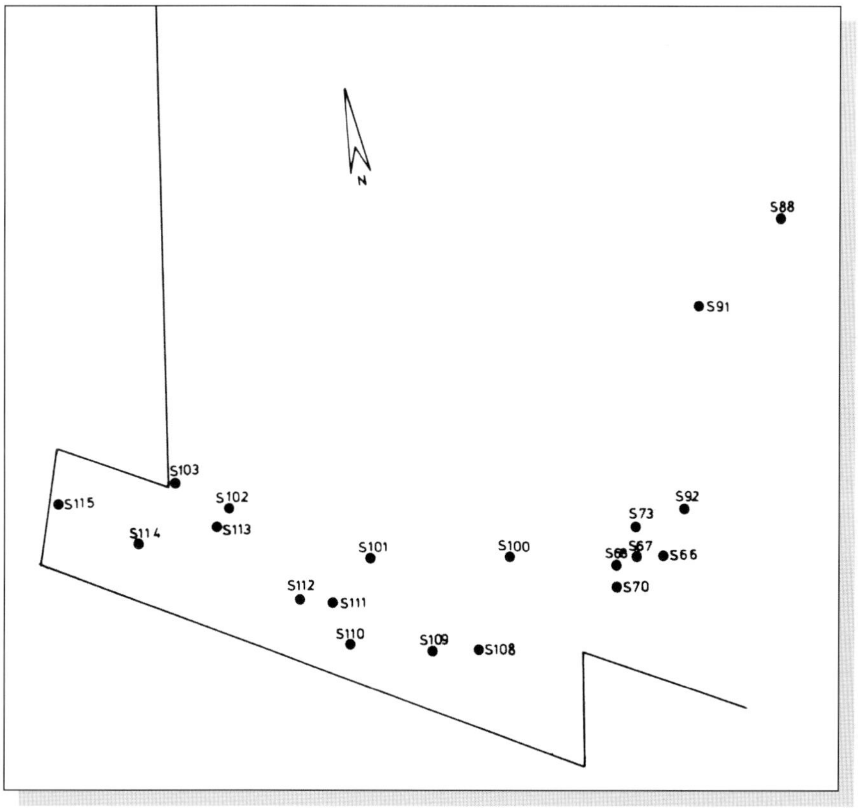

Figure 58: Verrebroek "Dok 1"; detailed distribution map of the typical hearth-pits.

The excavations revealed a total of 20 pits (fig. 56) which correspond to Groenendijk's (1987) definition of hearth-pits. These are small circular to elliptical features, characterised by a homogeneous dark grey to black fill coloured by large amounts of ash and charcoal fragments. These features only occur in a very restricted area of the site. They have been found clustered in the south-western part of the excavations, which corresponds to one of the highest parts of the former coversand landscape (fig. 57, 58). As a result of this position, all hearth-pits are severely truncated; generally they are perturbed into the B/C or C-horizon of the podzol. Hence, the original dimensions of these features can no longer be reconstructed (fig. 59-61).

feature code	length (cm)	width (cm)	relative depth (cm)	position (artefact unit)	archaeological content (N of lithic finds)			
					unburnt	moderately burnt	overheated	total
S66	56	56	25	22	22	10	0	32
S67	80	60	20	22	28	8	14	50
S68	71	44	11	22	5	3	5	13
S70	62	31	17	22	32	9	4	45
S73	62	57	23	22	8	0	9	17
S88	52	46	15	-	1	0	0	1
S91	>32	43	15	21	7	10	2	19
S92	>38	46	10/13	-	0	0	0	0
S100	100	87	22	-	7	2	64	73
S101	65	50/57	21	-	2	0	5	7
S102	74	54	16	-	0	0	0	0
S103	66	52	19	-	0	0	0	0
S108	65	46	20	-	0	0	0	0
S109	55	47	15	-	0	0	0	0
S110	71	60	10	-	1	0	0	1
S111	72	53	15	-	0	0	71	71
S112	72	67	19	-	26	9	303	338
S113	>50	>12	11	-	0	0	0	0
S114	76/66	64	19	-	5	14	15	34
S115	95	>88	27	-	7	26	8	41

Table 28: Verrebroek "Dok 1"; list of hearth-pit features.

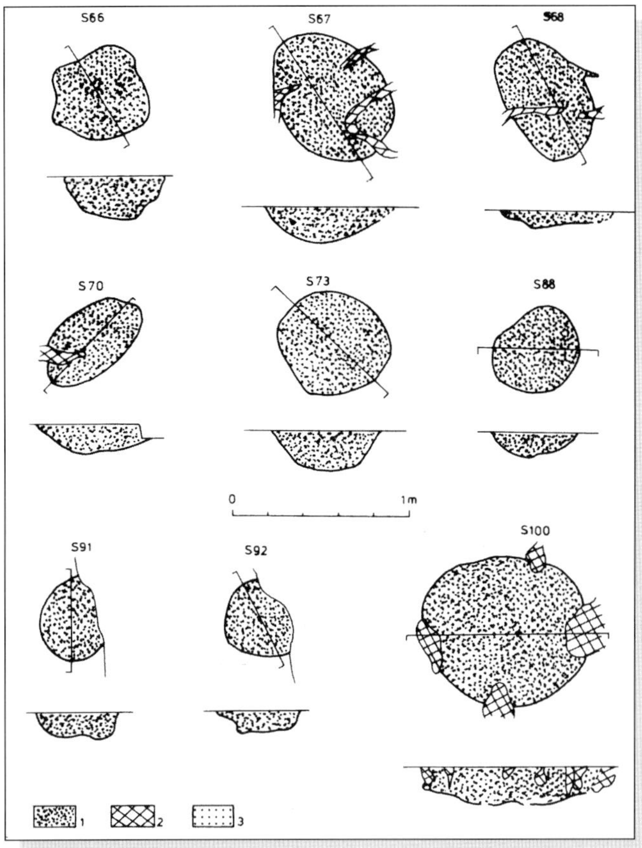

Figure 59: Verrebroek "Dok 1"; plan of typical hearth-pits. Key: (1) homogeneous dark grey to black layer with numerous charcoal fragments; (2) bioturbation/perturbation; (3) light grey layer, containing fewer and generally smaller charcoal fragments.

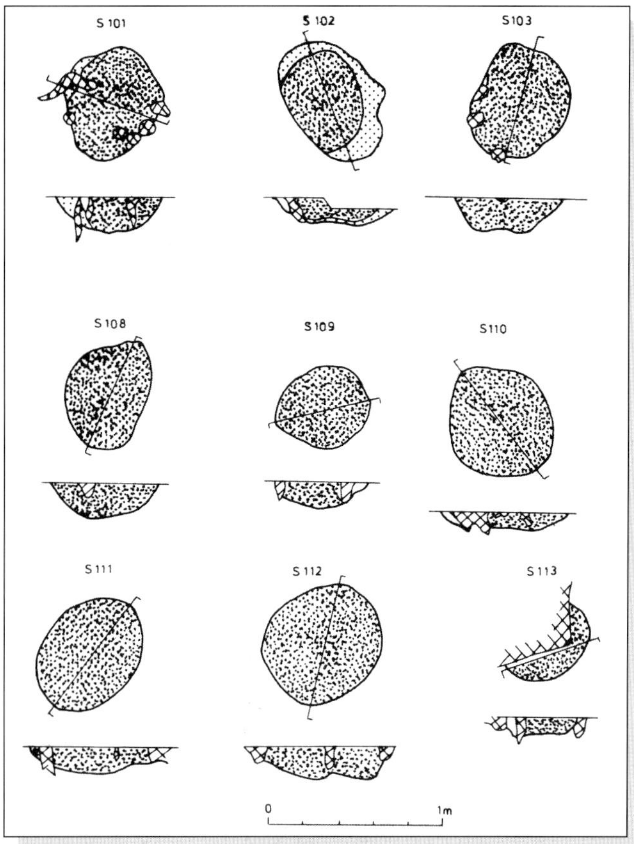

Figure 60: Verrebroek "Dok 1"; plan of typical hearth-pits.

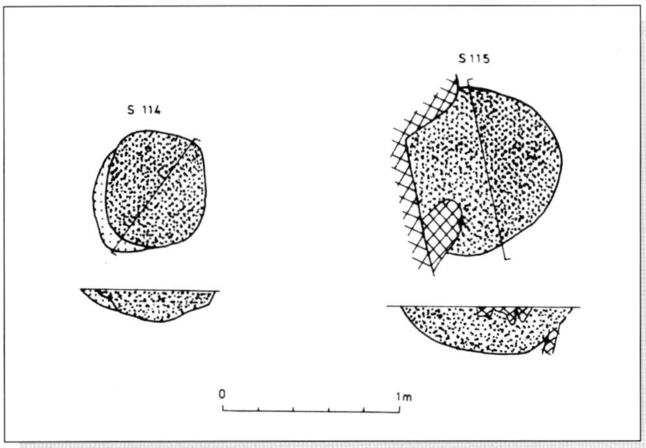

Figure 61: Verrebroek "Dok 1"; plan of typical hearth-pits.

In their present state, the hearth-pits are on average 70.24cm long, 53.89cm wide and 17.65cm deep (tab. 28). The mean length/width ratio amounts to 1.28. The length varies from minimum 52cm to maximum 100cm, yet about half of the hearth-pits cluster between 65cm and 74cm (fig. 62). These features also vary considerably in width - from minimum 31cm to maximum 88cm – though nearly $^{3}/_{4}$ cluster between 45cm and 64cm (fig. 63). The length-width ration fluctuates between 1 and 2, with a clear cluster between 1 and 1.5 (fig. 65). The deepest hearth-pit reaches down 27cm, the most shallow one is barely 10cm deep. Almost half of them are 15cm to 19cm deep (fig. 64). Within the cluster of 20 hearth-pits, S100 and S115 clearly stand alone because of their larger size.

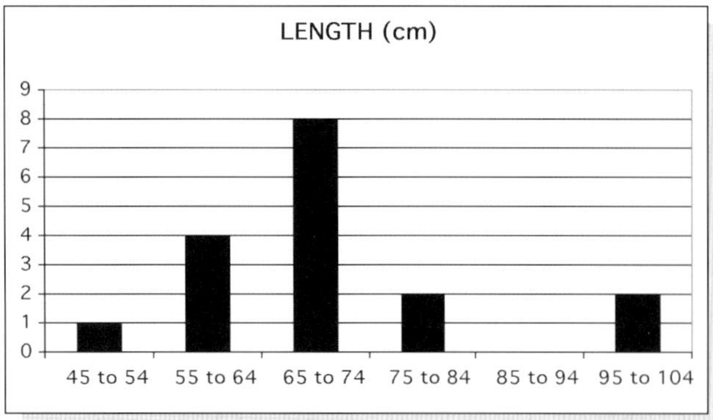

Figure 62: Verrebroek "Dok 1"; distribution chart of the length of typical hearth-pits.

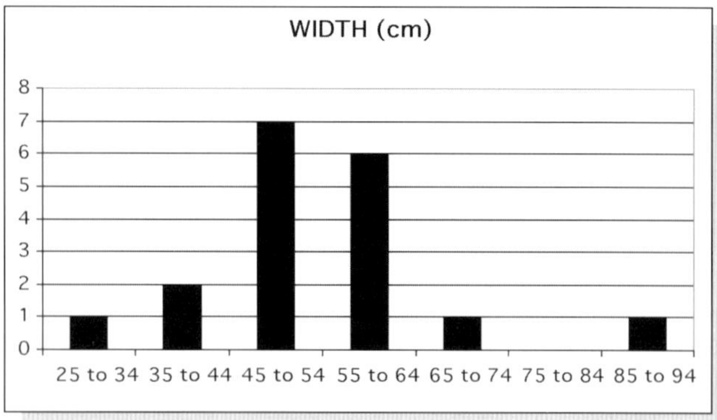

Figure 63: Verrebroek "Dok 1"; distribution chart of the width of typical hearth-pits.

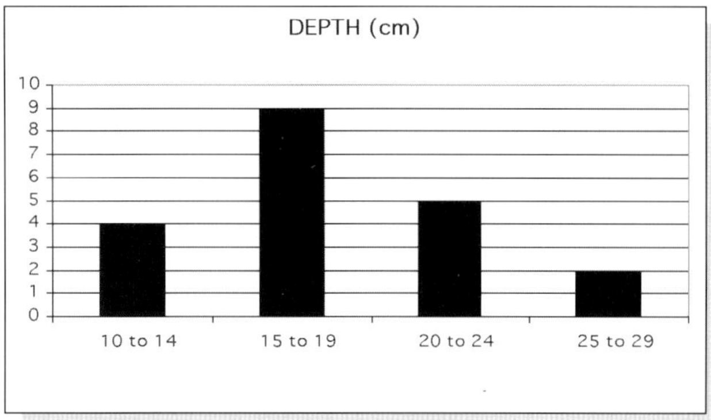

Figure 64: Verrebroek "Dok 1"; distribution chart of the depth of typical hearth-pits.

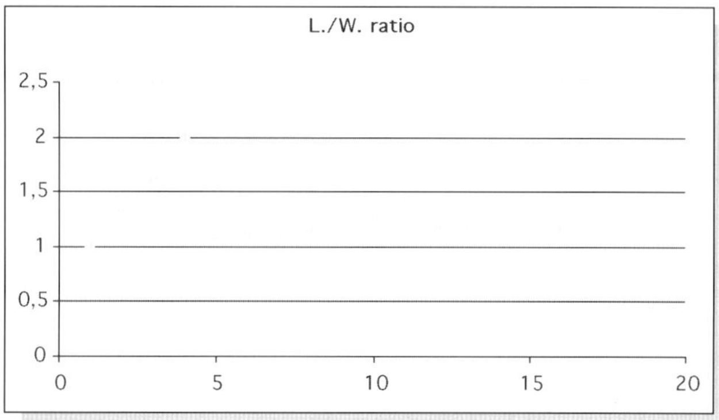

Figure 65: Verrebroek "Dok 1"; distribution chart of the length/width ratio of typical hearth-pits.

Regarding their archaeological content (tab. 28), a considerable variation exists among the hearth-pits of Verrebroek. Only a few of them (e.g. S92, S102, S103, S108, S109) are archaeologically sterile. The majority yielded a certain number of lithic artefacts and/or burnt ecofacts, in particular carbonized hazelnut shells (cf 3.4.2.2.). However, in most hearth-pits the number and size of these finds are rather limited.

The number of lithic artefacts varies between minimum 1 (S110, S88) and maximum 338 specimens (S112), the latter being an exception. In general hearth-pits contain between ca. 10 and ca. 70 artefacts, mostly chips. Neither artefacts larger than 1cm nor retouched artefacts occur often, the number of specimens being 33 and 1 respectively. The only tool – a fragment of an indeterminate microlith fragment – was found in S100. In the absence of diagnostic artefacts these finds do not contribute to a better dating of the hearth-pits. Yet in general these artefacts do not differ in any respect (raw material, technology, etc.) from those recovered from the upper layers of the coversand.

Compared to the surface-hearths (cf further) the number of carbonized hazelnut shells in hearth-pits is very limited. In most cases only few very small fragments were recovered (cf 3.4.2.2.). Apparently these features were not used as roasting-pits for hazelnuts. This is also corroborated by the radiocarbon evidence, which points to an intrusive character of the hazelnut shells within hearth-pits (cf 3.2.3.2.).

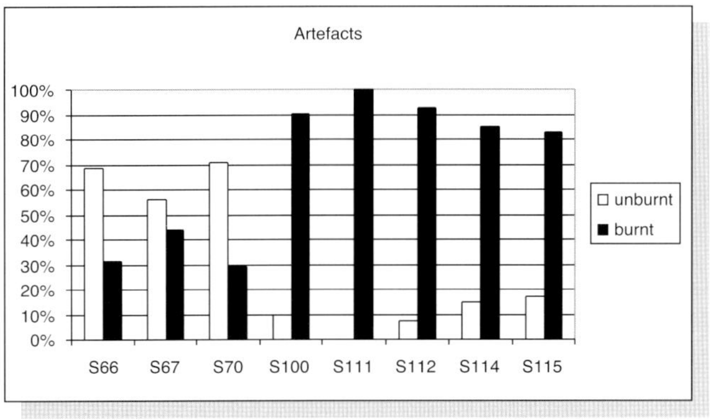

Figure 66: Verrebroek "Dok 1"; ratio burnt/unburnt artefacts from typical hearth-pits.

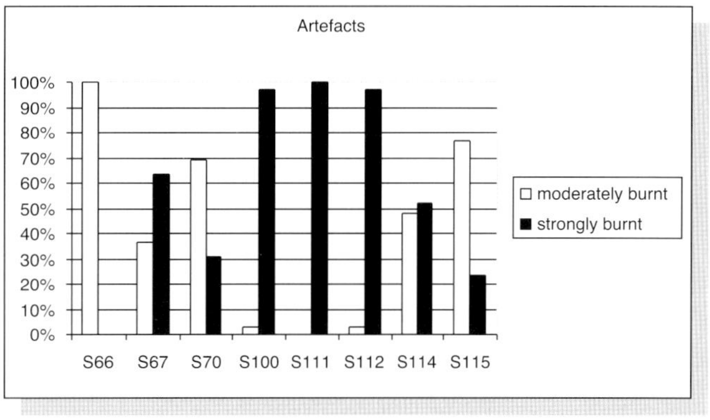

Figure 67: Verrebroek "Dok 1"; ratio moderately/strongly burnt artefacts from typical hearth-pits.

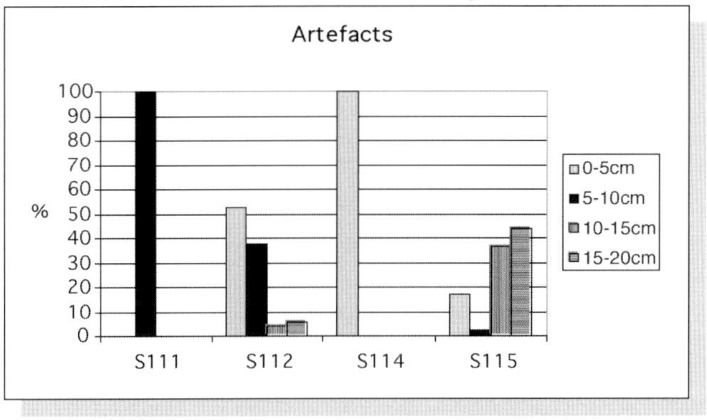

Figure 68: Verrebroek "Dok 1"; vertical distribution of artefacts within some typical hearth-pits.

Variation also exists with regard to the frequency of burnt lithics. Except for features S88 and S110, these are present within all hearth-pits (tab. 28). From figure 66 it can be deduced that in particular in features S100, S111, S112, S114 and S115 a high proportion of burnt items occurs. In all five hearth-pits burnt lithics are absolutely predominant (> 80%). In contrast, burnt artefacts represent less than half of the total industry within pits S66, S67 and S70. Spatially these three features fall within the limits of the intensively truncated artefact concentration 20[1]. Looking at the ratio between moderately and intensively (overheated) burnt artefacts (fig. 67), another clustering emerges. Apparently overheated artefacts are predominant only in three hearth-pits, namely S100, S111 and S112. In contrast, pit S66 yielded exclusively moderately burnt items. In all other hearth-pits both categories of burnt lithics are well represented.

1. Whether the other hearth-pits at the site also make up part of a lithic concentration can no longer be verified, as most of them are situated in areas in which the Mesolithic find-layer is eroded. Exceptions are pits S91 and S88, the former situated within the lithic concentration 21.

Finally, major differences can be observed in the vertical distribution of the lithic finds within these features (fig. 68)[2]. In some pits artefacts cluster within one level (e.g. S111 and S114), while in others there is a gradual increase (e.g. S115) or decrease (S112) of finds from top to bottom. The true meaning of these differences is not at all clear.

Charcoal-pits

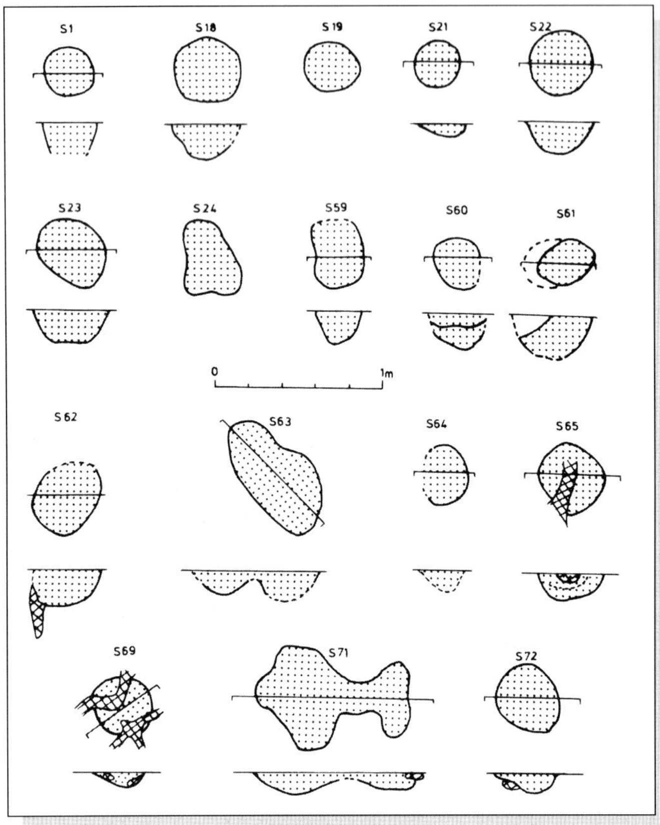

Figure 69: Verrebroek "Dok 1"; plan of some charcoal-pits. Key: cf figure 59.

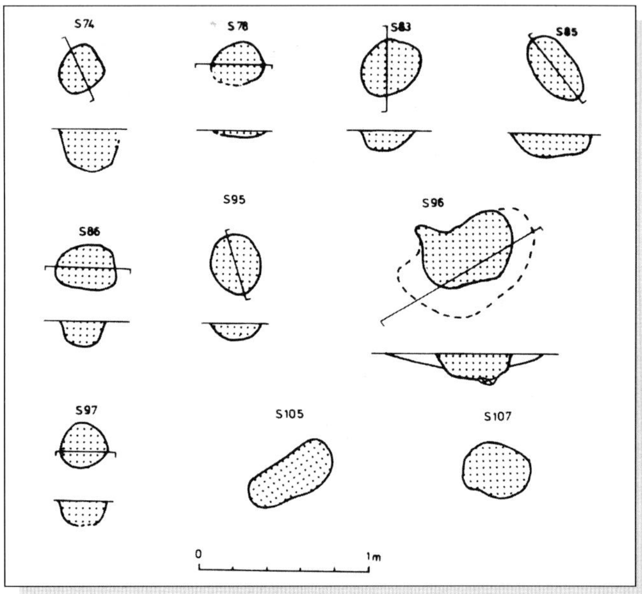

Figure 70: Verrebroek "Dok 1"; plan of some charcoal-pits.

At least forty pits have been classified as charcoal-pits (fig. 69, 70). Despite the fact that they contain lots of charcoal, they are in no way comparable to the hearth-pits described above. The principal differences can be summarized as follows: 1° very diffuse and irregular outlines, especially in vertical section (fig. 71, 72); 2° light greyish sandy fill; 3° the presence of mainly smaller charcoal fragments; 4° smaller dimensions; 5° random spatial distribution.

2. A systematic study of the vertical spread of artefacts is not possible due to the fact that only 8 pits have been excavated in artificial layers, while the remaining were bulk sampled. In addition among the former only 4 hearth-pits (S111, S112, S114, S115) yielded enough artefacts to allow a statistically significant analysis.

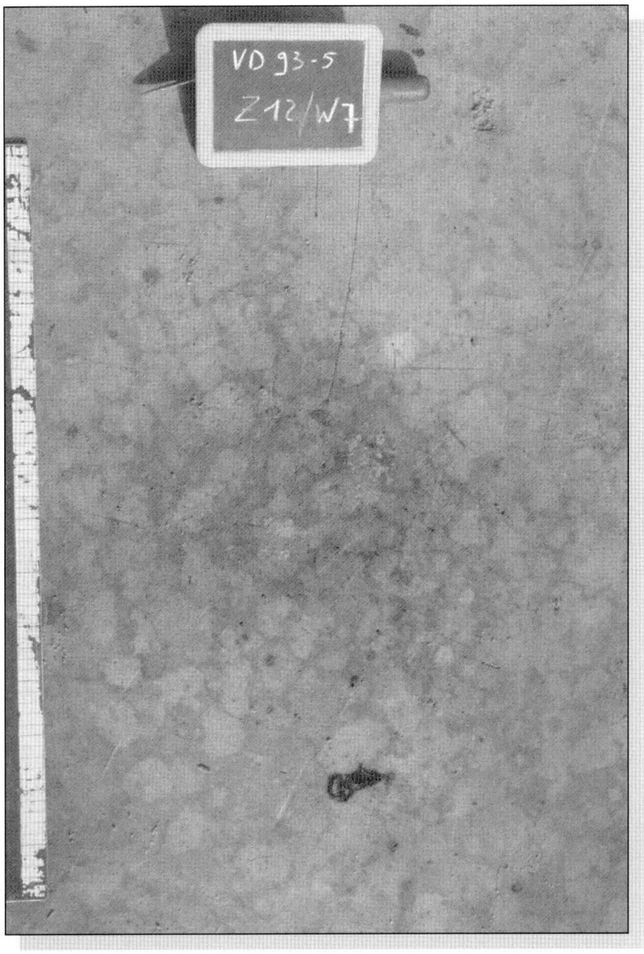

Figure 71: Verrebroek "Dok 1"; view of charcoal-pit S22.

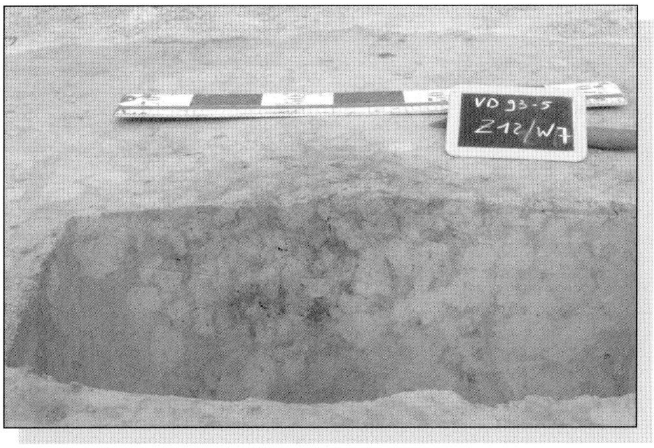

Figure 72: Verrebroek "Dok 1"; view on a cross-section of charcoal-pit S22.

Judging by the mean dimensions (tab. 29), these charcoal-pits are considerably smaller than the hearth-pits. The average length amounts to 44.6cm, the average width to 34.63cm. From the histograms (fig. 73, 74) it can be deduced that the majority of pits are between 25cm and 44cm long (ca. 62%) and between 25cm and 44cm wide (ca. 71%). Judging by the mean depth of 42.8cm, these features seem to be deeper than the hearth-pits. However, one has to take into account the severe truncation of the latter, making a meaningful comparison almost impossible. Except for some examples, the bottom of most charcoal-pits is situated at 35cm to 54cm below the top of the coversand surface (fig. 75).

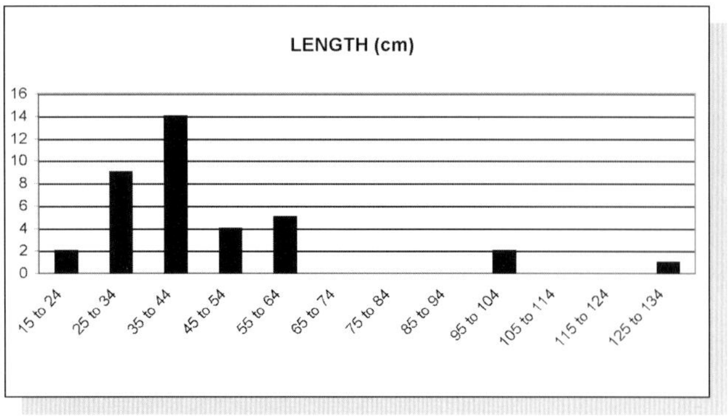

Figure 73: Verrebroek "Dok 1"; distribution chart of the length of charcoal-pits.

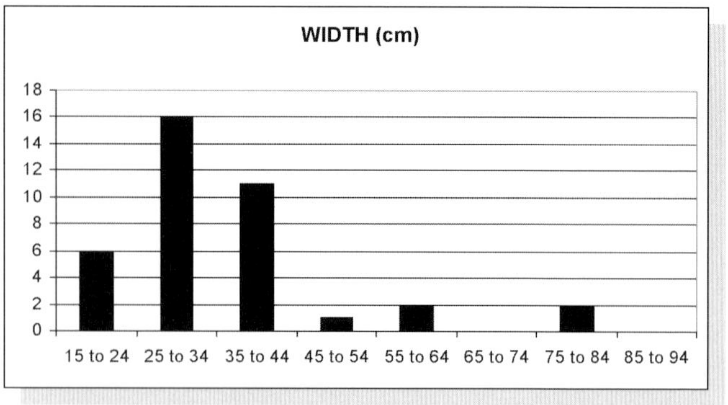

Figure 74: Verrebroek "Dok 1"; distribution chart of the width of charcoal-pits.

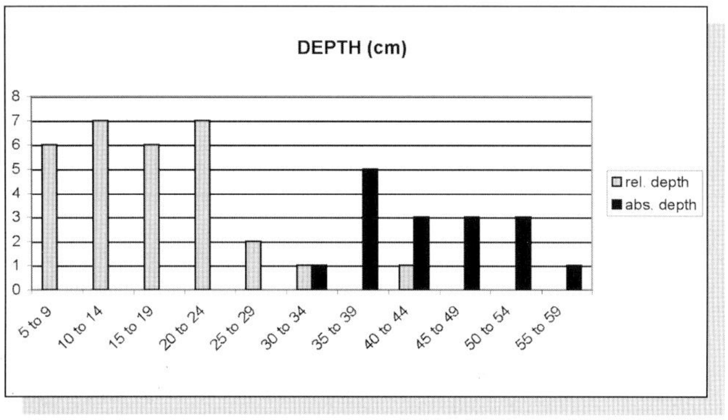

Figure 75: Verrebroek "Dok 1"; distribution chart of the depth of charcoal-pits.

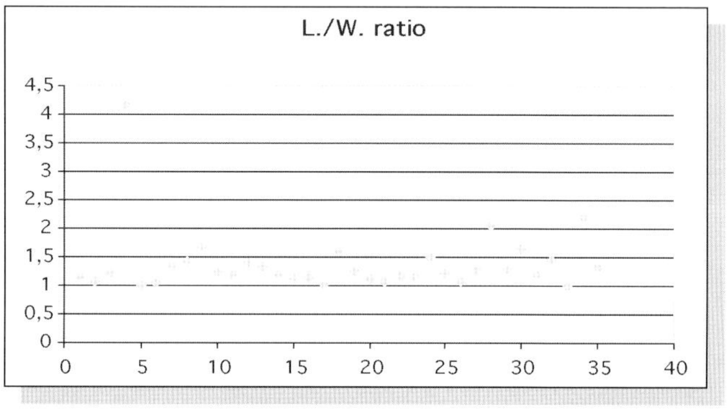

Figure 76: Verrebroek "Dok 1"; distribution chart of the length/width ratio of charcoal-pits.

Another difference from the hearth-pits is the random distribution of the charcoal-pits. Contrary to the former, they are not confined to the highest parts of the contemporary coversand landscape, but occur almost randomly spread over the entire excavated surface. Some have been reported within artefact concentrations, while others occur in areas where little or no artefacts have been found (tab. 29).

With respect to their archaeological content (tab. 29), most charcoal-pits yielded either only few lithic artefacts or none at all. The low number of finds is not surprising since most features are situated at the periphery of or even completely removed from the artefact units. One exception is feature S107, with 90 artefacts.

feature code	length (cm)	width (cm)	relative depth (cm)	absolute depth cm)	horizon	position (artefact unit)	archaeological content (N of lithic finds)			
							unburnt	moderately burnt	overheated	total
S1	33	29	20	35/40	B	1	17	0	2	19
S16	35	-	40	40	A2	6	10	0	8	18
S18	42	39	22	truncated	C	9 (S)	3	0	2	5
S19	35	29	5	truncated	C	9 (centre)	0	0	0	0
S20	125(26)	30	30	truncated	C	9 (centre)	0	0	0	0
S21	28	28	8	truncated	C	9 (N)	0	0	0	0
S22	40	38	20	truncated	C	-	1	0	1	2
S23	47	35	20	truncated	C	10 (W)	5	0	3	8
S24	50	35	5	truncated	C	11(SE)	0	0	0	0
S50	>38	36	-	-	-	14 (N)	not sampled			
S51	30	18	8	-	-	-	0	4	1	5
S58	100	80	-	-	-	15 (SW)	not sampled			
S59	>38	30	20	50	B3	17(NW)	3	1	0	4
S60	32	27	21	51	B3	17(NW)	0	0	0	0
S61	35	25	25	55	B3	17(NW)	1	1	0	2
S62	49	37	22	52	B3	17(N)	1	1	0	2
S63a	36	30	15	35	B3	-	0	0	0	0
S63b	36	40	19	39	B3	-	0	0	1	1
S64	35	>22	15	45	B3	-	1	1	0	2
S65	40	35	16	48	charcoal layer	18(E)	0	0	0	0
S69	34	33	9	truncated	C	-	5	1	1	7
S71	97	61	14	truncated	C	-	0	0	0	0
S72	44	35	11	truncated	C	-	0	0	0	0
S74	30	27	26	truncated	B/C	-	0	0	0	0
S75	26	24	-	-	-	-	not sampled			
S76	22	19	-	-	-	-	not sampled			
S77	21	18	-	-	-	-	not sampled			
S78	33	22	5	35	C	-	2	0	0	2
S79	-	80	12	-	-	-	not sampled			
S81	56	46	-	-	A2	-	not sampled			
S82	32	30	-	-	-	67(W)	not sampled			
S83	40	31	13	-	B3	22(E)	1	1	0	2
S85	47	23	14	44	-	70(S)	2	1	0	3
S86	36	28	16	46	B3	71(E)	0	0	1	1
S94	62	38	-	-	B3	-	not sampled			
S95	36	30	10	30	B	65(N)	0	0	6	6
S96	57	39	15	35	B	-	1	0	0	1
S97	56	56	14	44	B3	29	0	0	0	0
S105	55	25	-	-	-	-	not sampled			
S107	40	30	-	-	-	-	53	11	26	90

Table 29: Verrebroek "Dok 1"; list of charcoal-pits.

Other prehistoric features

Surface-hearths

The delimitation and reconstruction of so-called surface-hearths at Verrebroek is problematic, since none of these features is preserved as real structural features or soil discolorations. This is mainly due to later pedogenetic processes (so called podzolisation), which have faded all shallow prehistoric features, and due to the absence of large accumulations of carbonised organic fuel remains, i.e. charcoal. The latter is most likely to be explained by the fact that most, if not all, charcoal[3] originally lying in these shallow and superficial features has disappeared as a result of wind activity and weathering (rain, frost, etc.)[4]. By contrast, charcoal in the dug out features described earlier (hearth-pits and charcoal-pits) survived thanks to its deeper position.

Despite these problems, the existence of surface-hearths on the site of Verrebroek "Dok 1" is beyond any doubt (Crombé 1998, 52; Sergant 2004). Detailed spatial analyses of various burnt artefacts and ecofacts, in particular carbonized hazelnut shells and burnt bone fragments, have clearly indicated the presence of numerous small clusters, which most likely correspond to former burning places[5]. In the present state of analysis at least 50 possible surface-hearths can be located. This is just a provisional count; the number will certainly increase as the spatial analysis progresses.

Nearly all are situated within the limits of artefact concentrations, and hence can be tentatively interpreted as so-called domestic-hearths. It should however be emphasised that only those hearths which are lying amidst artefacts and ecofacts can be traced; surface-hearths that were possibly constructed in areas without or with only few artefacts, or which did not produce large amounts of carbonized hazelnut shells or bone fragments, can no longer be traced.

Apart from their approximate position, further information on size and internal structure of these surface-hearths can unfortunately no longer be deduced from the available data.

3 Some surface-hearths, which were analysed micro-morphologically, yielded fine carbonised (charcoal) remains (cf 3.3.1.1.).

4 Charcoal in surface-hearths is only preserved if these features are sealed by sediments immediately or shortly after their use. This is only rarely the case on Mesolithic sites in the coversands of Belgium and The Netherlands. Exceptions are some wetland sites in The western Netherlands, e.g. at Almere "Hoge Vaart" (Hamburg *et alii* 2001) and Hardinxveld "De Bruin" (Louwe Kooijmans, Nokkert 2001), where surface-hearths have been quickly covered by peat and/or clay.

5 A detailed report on these analyses, as well as a detailed distribution map of all surface-hearths will be published in a next volume of the "Verrebroek and Doel Excavation Projects".

Hazelnut-pits

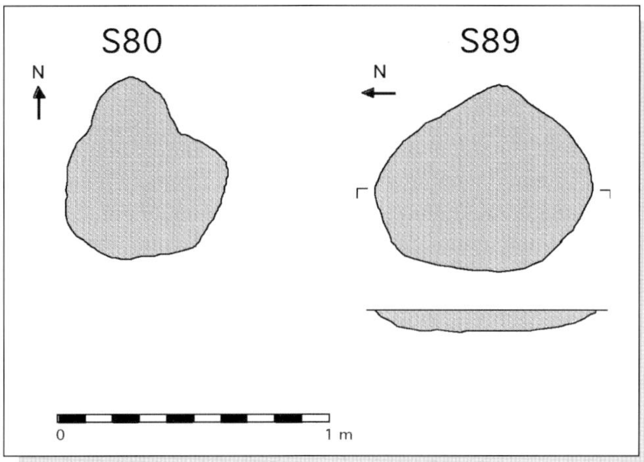

Figure 77: Verrebroek "Dok 1"; plan and cross-section of two hazelnut-pits.

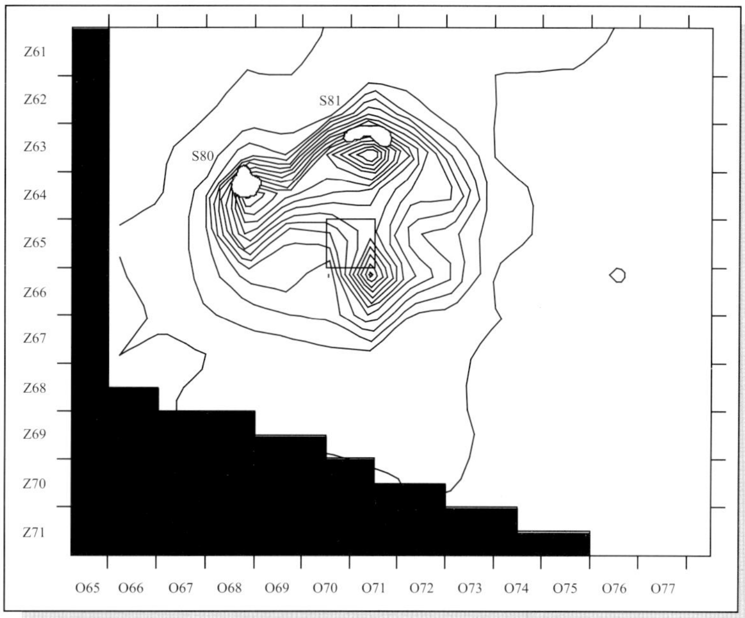

Figure 79: Verrebroek "Dok 1"; artefact concentration 22: distribution map of the burnt hazelnut shells (contour interval 5gr.; lowest level 0.25gr.) in relation to the hazelnut-pit S80 (and S81?) and the surface-hearth (quadrangle).

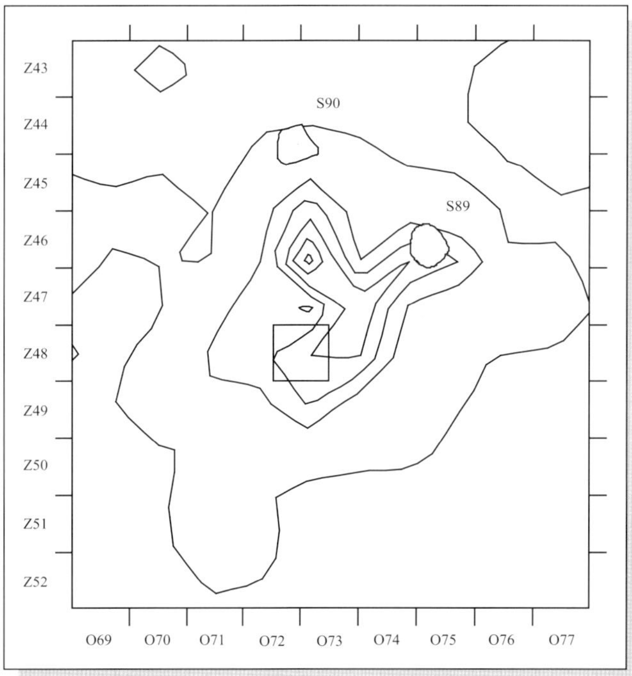

Figure 78: Verrebroek "Dok 1"; artefact concentration 70: distribution map of the burnt hazelnut shells (contour interval 5gr.; lowest level 0.25gr.) in relation to the hazelnut-pit S89, the tree-fall feature S90 and the surface-hearth (quadrangle)

Two enigmatic and vague pit-like features with comparable characteristics have been found within two artefact units (fig. 77). A first feature (S89) was situated in the north-eastern periphery of unit 70, while another (S80) was found to the north-west of a surface-hearth within unit 22[6]. Both features are characterised by a very high density of carbonized hazelnut shells (tab. 30), which seem to coincide with small and shallow pits (fig. 78, 79). Both pits, however, are only very roughly preserved, making the delimitation of their contours extremely difficult. Pit S89 is the more obvious feature. It was first observed at ca. 40cm below the top of the coversand, i.e. after excavation of the heather podzol. At this level the pit was no longer any deeper than 8cm; the size was reduced to 80 x 65cm. The filling of this feature is rather homogeneous and consists of a Brunified Soil (cf 3.3.1.2.). Pit S80 is less clear in outline; it consisted of a light greyish soil discolouration, whose depth could not be delimited even approximately.

6 A third possible hazelnut-pit (S81), containing 221.97gr. of charred hazelnut shells, might be situated to the north of the surface-hearth within unit 22. Because of its larger and irregular size and the presence of large amounts of charcoal fragments, this feature was provisionally classified as a charcoal-pit. A possible natural (tree-fall) origin cannot be excluded either.

feature code	length (cm)	width (cm)	relative depth (cm)	position (artefact unit)	carbonised hazelnut shells (gr.)	charcoal (gr.)
S80	57	49	0.60	22 (NW)	154.95	2.01
S89	80	65	0.48	70 (NE)	77.73	35.09

Table 30: Verrebroek "Dok 1"; list of hazelnut-pits.

Tree-fall and tree-sway features

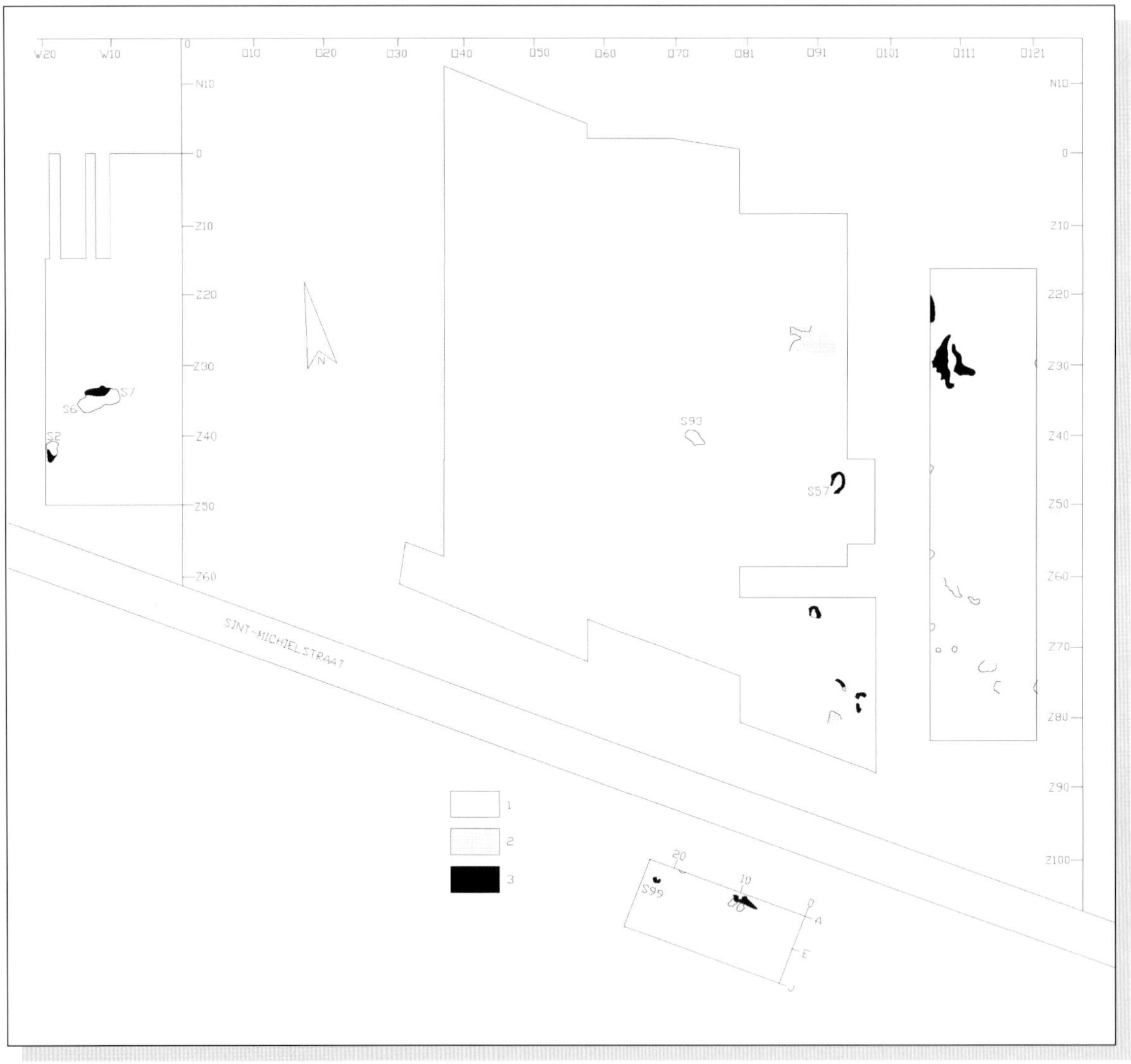

Figure 80: Verrebroek "Dok 1"; general distribution map of the natural features. Key: (1) tree-fall features without peaty filling; (2) tree-sway features; (3) tree-fall features with partial peaty filling (presumed late generation tree-falls).

Compared to other Mesolithic sites in the coversand area of Belgium and The Netherlands (Crombé 1993; 1998), tree-fall features are relatively scarce at Verrebroek "Dok 1". At most, 10 wind-throw features could be observed, belonging to several generations of uprooting (fig. 80). Most of them were found in the lower parts of the site, in particular on the eastern dune slope (trenches 8, 9, 11 and 12) and within the shallow depression in the western sector (trenches 1-7). On the higher part of the sand dune (trenches 10 and 13), only three clear tree-fall features have been reported. S93, situated between artefact concentrations 71 and 24, is the clearest example. A detailed description of this feature is presented in chapter 3.3.1.2.

Most of the tree-falls were obviously created after the formation of the podzol and prior to the peat-growth. Some were still preserved at the coversand surface as a pit and mound micro relief (Schaetzl *et alii* 1989), indicating that the uprooting process happened only shortly before the formation of peat started. Tree-uprooting was most likely induced by the gradual rise of the groundwater level, which might explain the more frequent occurrence of these natural features in the low-lying parts of the dune. This hypothesis is corroborated by the spread of the so-called tree-sway features, which coincides almost perfectly with that of the tree-fall features (fig. 80). As explained in chapter 3.3.1.2., the formation of these features is directly linked to extremely wet soil (even saturation) conditions. Thanks to their limited spread, the impact of both natural disturbances on the original distribution of Mesolithic artefacts and ecofacts has remained very limited.

3.1.2.2. | "Historic" features

Despite the sedimentary cover, the Mesolithic surface has been locally perturbed by later human activities. During Late Medieval times (13th-15th century) and before the deposition of alluvial sediments, the location was used for agriculture (cf 2.1.4.10.) and hence was divided into different parcels by means of large drainage ditches (fig. 81). These ditches are on average 3-3.5m wide and 1m deep and were dug right through the peat beds deep into the coversand deposits (fig. 82). These ditches delimit more or less rectangular parcels of approx. 30m wide and > 60m long. In between two of these parcels – in the western half of trench 10 – several cart tracks were found within the coversand deposits, which point to the existence of a Medieval farm road. Two smaller (0.80m) and less deep ditches (0.12m under top of coversand) situated in the eastern half of trench 13 and trench 10 most likely date back to the Late Medieval period too. This can be deduced from the orientation of the ditches which is more or less similar to the large drainage ditches. These small features are filled with strongly organic clay, parts of which is locally burnt *in situ*.

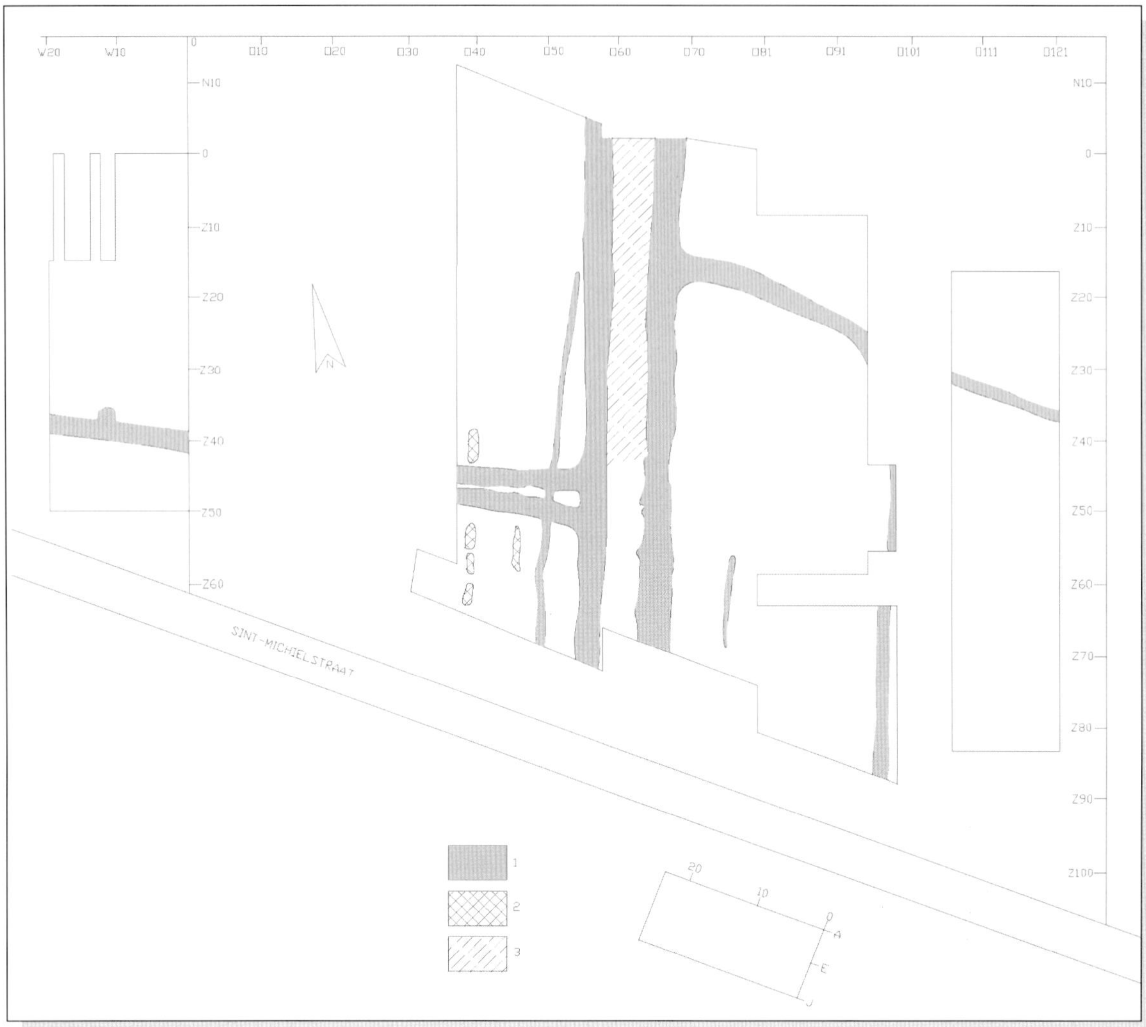

Figure 81: Verrebroek "Dok 1"; general distribution map of the Medieval features. Key: (1) ditches; (2) sand extraction pits; (3) area with cart-tracks.

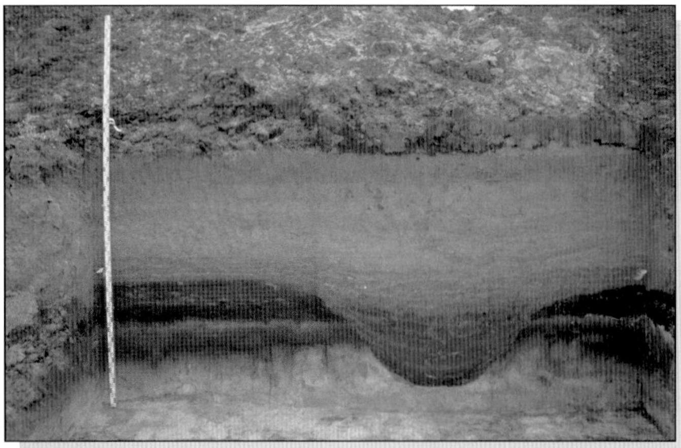

Figure 82: Verrebroek "Dok 1"; cross-section of a Medieval drainage ditch (trench 11, profile 4).

Of more recent origin, are a number of small rectangular pits, situated in the southern part of trench 13 (fig. 82). These pits were dug through the alluvial clay, which is at its thinnest in this part of the site, right into the coversand sediments. These obviously (sub)recent features can probably be interpreted as sand-extraction pits.

3.1.3. | Doel "Deurganckdok-sector B"

At this site a total of more than a hundred structural features were uncovered, all of prehistoric age. All features contain charcoal in their filling and can be classified into different types: 1° typical hearth-pits with lots of charcoal, 2° atypical hearth-pits with fewer and smaller charcoal fragments, and 3° surface-hearths. Stratigraphically the surface-hearths have a higher position than both the other features. They are situated in the upper humiferous layer (A-horizon, cf 2.1.5.), while both the typical and atypical hearth-pits first appeared after removal of this layer or even at a deeper level (cf below).

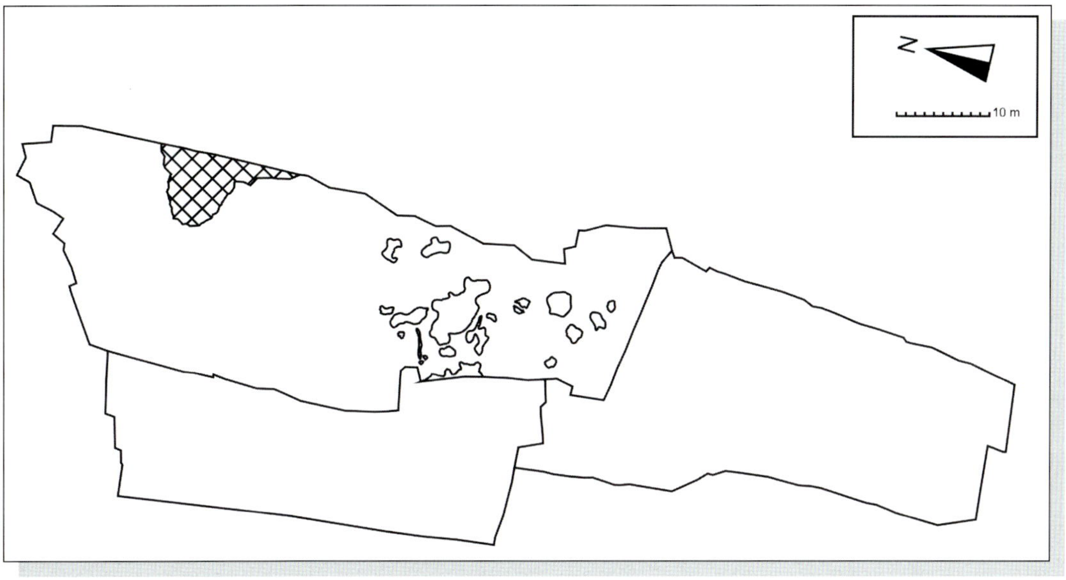

Figure 83: Doel "Deurganckdok-sector B"; general distribution map of local shallow deepenings of the A-horizon, containing Swifterbant settlement waste.

Besides these features, this sector also yielded a number of very irregular shallow deepenings of the A-horizon (fig. 83). Most of these were recorded in the north-eastern zone, where the top of the A-horizon had already been removed by the harbour works. Similar deepenings were observed in the adjacent north-western zone, but could not be recorded as time was too short. Hence it is no longer possible to examine the spatial relation between these deepenings and the archaeological material retrieved from the A-horizon. However, one may assume that some, particularly the smallest ones which yielded numerous burnt bone fragments, are the remains of shallow surface-hearths (cf 3.1.3.3.).

3.1.3.1. | Typical hearth-pits

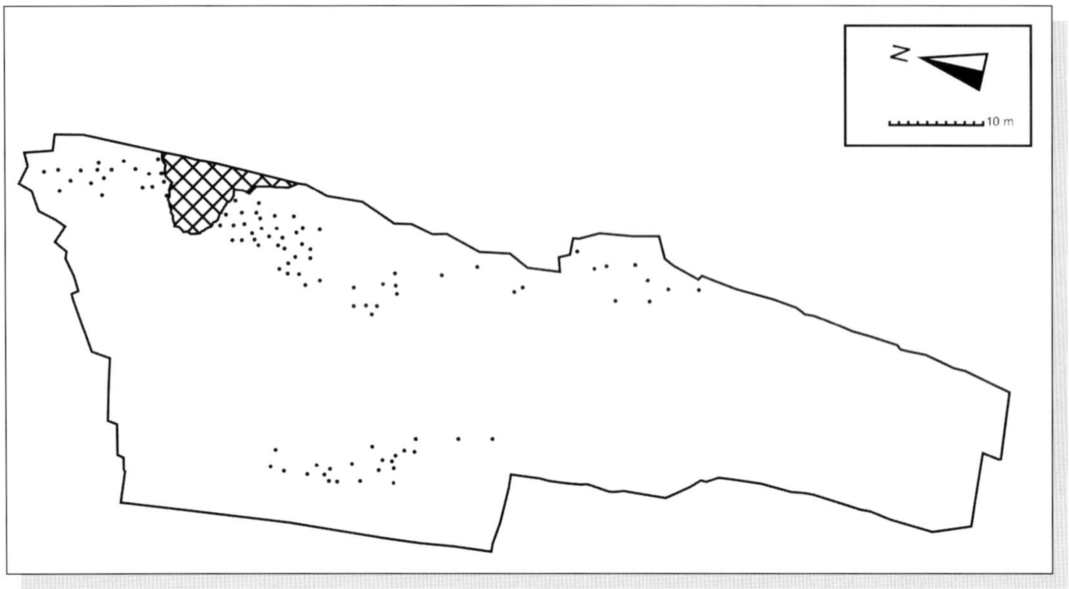

Figure 84: Doel "Deurganckdok-sector B"; general distribution map of typical and atypical hearth-pits.

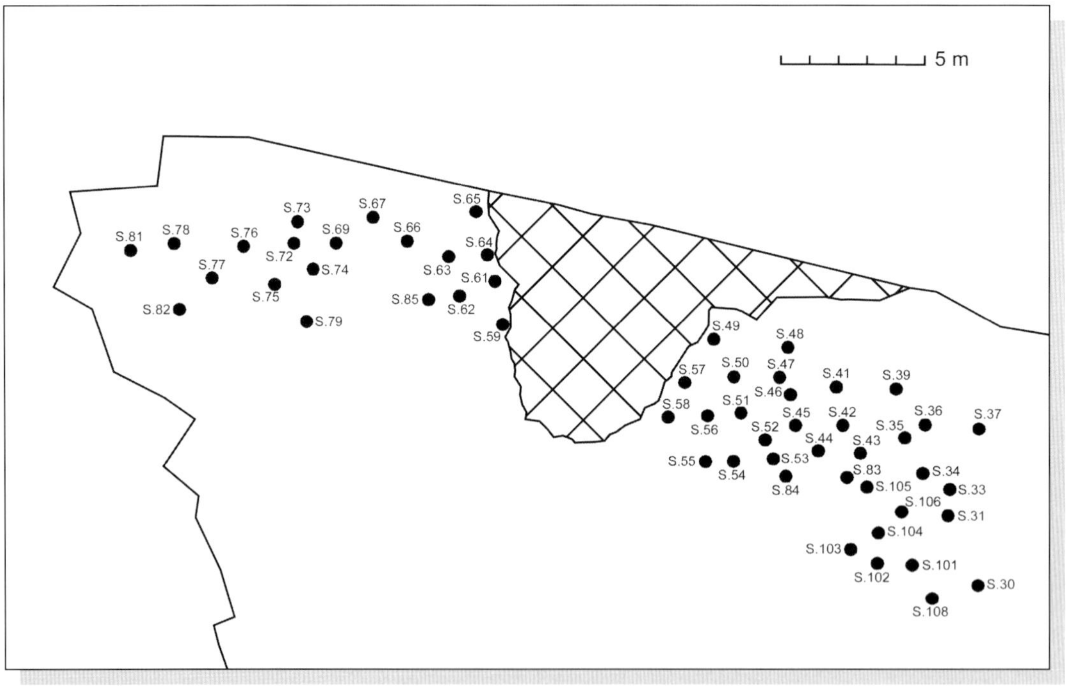

Figure 85: Doel "Deurganckdok-sector B"; detailed distribution map of typical and atypical hearth-pits in part of the north-eastern zone.

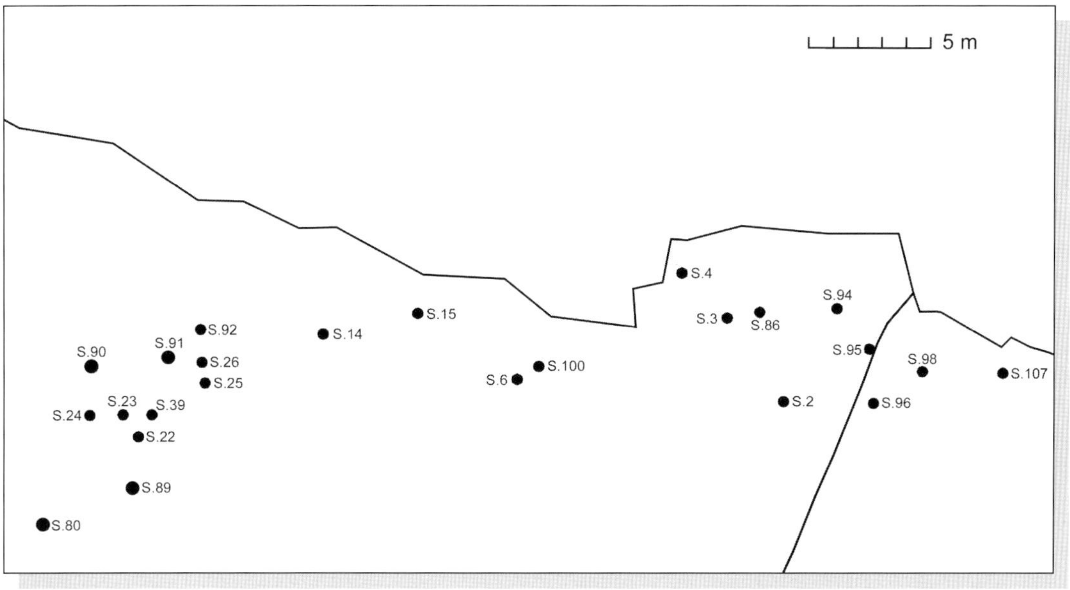

Figure 86: Doel "Deurganckdok-sector B"; detailed distribution map of typical and atypical hearth-pits in part of the north-eastern zone and the south-eastern zone.

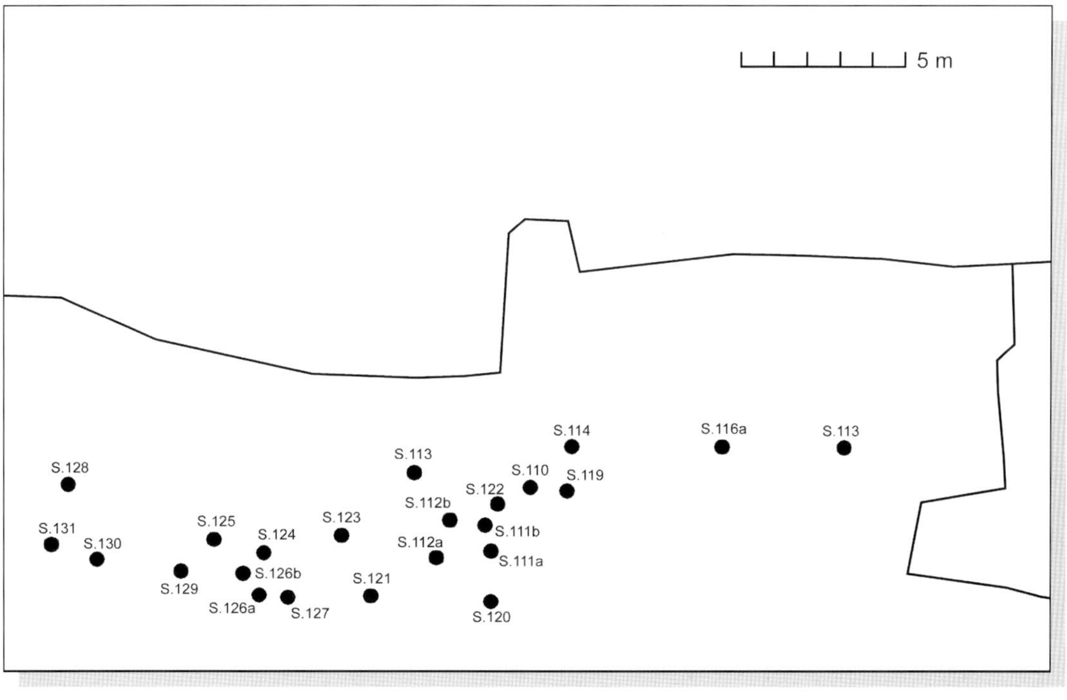

Figure 87: Doel "Deurganckdok-sector B"; detailed distribution map of typical and atypical hearth-pits in the north-western zone.

In total some 79 typical hearth-pits were recorded (fig. 84)[7]. The majority was situated in the north-eastern sector of the site (N=58) (fig. 85, 86); only 17 hearth-pits were found in the north-western sector (fig. 87) and 4 in the south-eastern sector (fig. 86). Due to taphonomic factors and the unfavourable excavation conditions (cf 1.3.4.), it can, however, be assumed that a fairly large number of features were missed. Especially in the north-eastern sector, which was already partially disturbed at the time the excavation started, shallow features may have been destroyed by the harbour works. In addition most of the preserved features were disturbed in their upper part by the digging machines. In the north-western sector, some hearth-pits have certainly been missed simply as a result of shortage of time. Except for the sieved area (cf 1.3.4.), it was not possible to excavate this sector either properly or entirely. Furthermore, in both sectors it could be observed that some hearth-pits were "sealed" by 15 to 20cm of bioturbated sand and consequently became visible only after removal of this layer. Since this removal could not be done systematically[8], probably a lot of these deep-lying pits remained unexplored. The south-eastern sector is the only one that it was possible to investigate in fairly good conditions. Unfortunately here only four hearth-pits were present.

Since the distribution of the hearth-pits at Doel is largely biased and distorted, no reliable spatial analysis can be undertaken. Nevertheless two important observations can be made. Firstly these features apparently cluster in the northern sector of the site (fig. 84), and secondly they never intersect or overlap. Indeed, despite the high density of features, no intersecting or overlapping pits have been observed (fig. 88). In just one case, three aligned hearth-pits (S52, S53, S84) narrowly adjoin (fig. 89).

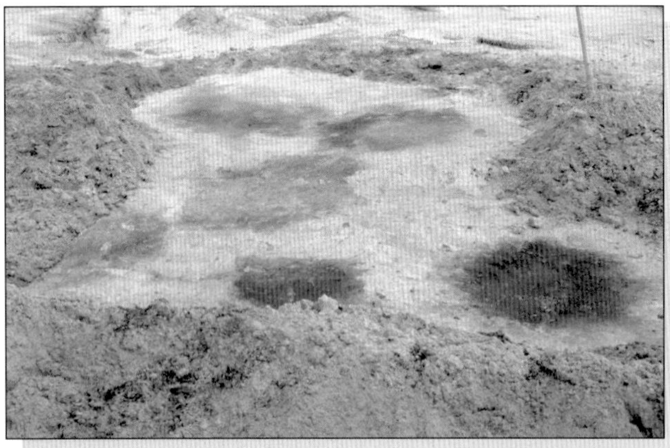

Figure 88: Doel "Deurganckdok-sector B"; view of a cluster of typical hearth-pits in the north-eastern sub-zone.

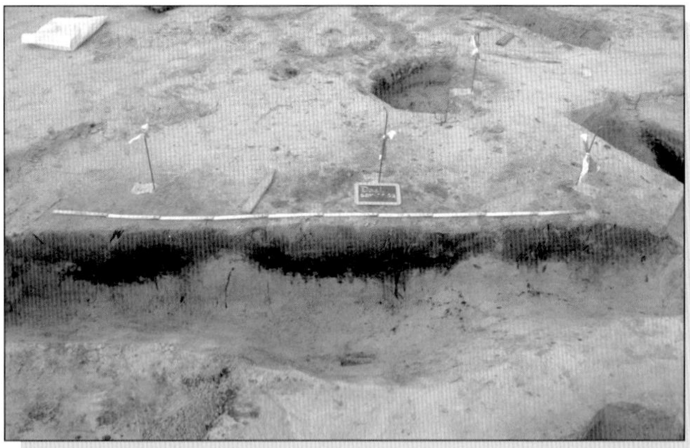

Figure 89: Doel "Deurganckdok-sector B"; cross-section of three bordering hearth-pits (from left to right S52, S53, S84).

7 In addition there is one hearth-pit (S109) which was largely disturbed by a natural feature, i.e. a tree-fall feature.
8 Only locally could this bioturbation layer be removed mechanically in order to check the presence of deep-lying pits.

feature code	length (cm)	width (cm)	relative depth (cm)	absolute depth (cm)	top below TAW (m)	filling layers	bioturbation layer	sector
S2	109	100	12	truncated	1.17	1		NE
S3	90	82	27	truncated	1.17	2		NE
S4	>120	84	22	truncated	1.16	1		NE
S6	101	74	18	truncated	1.04	1		NE
S15	95	71	23	truncated	1.24	2		NE
S22	125	113	55	truncated	1.22	2	+	NE
S23	53	49	15	truncated	1.08	1		NE
S24	94	74	28	truncated	1.14	1	+	NE
S25	87	>52	39	truncated	1.03	1		NE
S26	59	54	16	truncated	1	1	+	NE
S30	92	82	21	truncated	1.26	1	+	NE
S34	123	95	28	truncated	1.42	2		NE
S35	82	70	20	truncated	1.56	1		NE
S36	>89	80	20	truncated	1.62	1		NE
S41	115	102	22	truncated	1.71	2		NE
S42	70	59	11	truncated	1.63	1	+	NE
S44	83	>77	14	truncated	1.55	1		NE
S45	102	89	28	truncated	1.57	1		NE
S46	68	48	9	truncated	1.65	1	+	NE
S47	100	90	25	truncated	1.65	2		NE
S49	>87	>91	30	truncated	1.7	2		NE
S50	81	>64	8	truncated	1.67	1		NE
S51	>132	110	33	truncated	1.6	2		NE
S52	72	65	23	truncated	1.51	2		NE
S53	95	91	29	truncated	1.51	2		NE
S54	71	70	20	truncated	1.52	1		NE
S55	140	106	54	truncated	1.55	2		NE
S57	80	>70	15	truncated	1.65	1		NE
S58	145	80	33	truncated	1.55	2		NE
S59	>39	>93	>30	truncated	1.52	1		NE
S62	70	65	12	truncated	1.87	1		NE
S63	111	113	34	truncated	1.77	2		NE
S65	>61	>73	13	truncated	1.83	1	+	NE
S67	109	>82	40?	truncated	1.84	2		NE
S69	>60	50	17	truncated	1.88	1	+	NE
S72	>38	>67	25	truncated	1.91	1		NE
S74	97	90	32	truncated	1.87	2		NE
S75	111	92	31	truncated	1.92	2		NE
S76	100	92	28	truncated	2.02	2		NE
S77	>82	90	18	truncated	2	1		NE
S78	52	46	17	truncated	2.04	1	+	NE
S81	87	79	20	truncated	2.05	1	+	NE
S82	84	67	30	truncated	2.08	2		NE
S83	>84	93	16	truncated	-	1		NE
S84	74	70	28	truncated	-	2		NE

S86	113	112	33	truncated	1.15	2		NE
S89	90	82	43	truncated	1.29	2		NE
S90	72	66	41	truncated	1.37	1	+	NE
S93	51	51	31	truncated	1.31	1		NE
S94	96	>72	19	truncated	1.19	1		NE
S95	>103	95	31	?	1.04	2		SE
S96	114	91	40	?	1.04	2		SE
S98	>80	71	30	?	0.96	1		SE
S100	100	63	21	truncated	-	2		NE
S101	97	91	29	truncated	1.5	1	+	NE
S102	65	64	35	truncated	1.47	1		NE
S103	97	>45	47	truncated	1.52	2		NE
S104	122	107	36	truncated	1.48	2		NE
S105	109	97	39	truncated	1.46	2		NE
S106	100	85	38	truncated	1.54	2		NE
S107	62	55	-	truncated	-	-		SE
S108	-	-	-	truncated	-	-		NE
S111a	162	-	33	truncated	1.14	2		NW
S111b	95	-	33	truncated	1.14	2		NW
S112a	96	-	41	truncated	1.15	1		NW
S112b	95	-	26	truncated	1.15	2		NW
S113	145	-	20	truncated	1.16	1		NW
S114	130	-	53	truncated	1	2		NW
S116a	115	-	61	truncated	0.98	2		NW
S120	>98	-	56	truncated	-	2		NW
S121	61	-	23	truncated	-	2		NW
S122	>77	-	24	44	-	1		NW
S123	>62	-	18	38	-	1		NW
S124	94	-	29	38	-	1		NW
S125	>116	-	27	48	-	1		NW
S127	97	-	33	truncated	-	2		NW
S128	>105	-	29	60	-	1		NW
S129	143	-	35	truncated	-	2		NW
S131	70	-	20	truncated	-	2		NW

Table 31: Doel "Deurganckdok-sector B"; list of typical hearth-pits.

Most hearth-pits are roughly circular in ground plan (fig. 90-95), although elliptical examples also occur. In section all have a bowl-shaped profile. On the basis of the filling, the hearth-pits can be classified into two groups. About half of the pits (N=38) are filled by two different layers: at the base a lenticular layer of dark grey to black charcoal rich sand of 10 to 25cm thick covered by a homogeneous light greyish sandy packet in which no charcoal occurs. This upper layer is missing in the remaining hearth-pits (N=39). These are entirely filled with black charcoal rich sand. Underneath a limited number of pits (N=12), a light greyish layer with bioturbation galleries was attested.

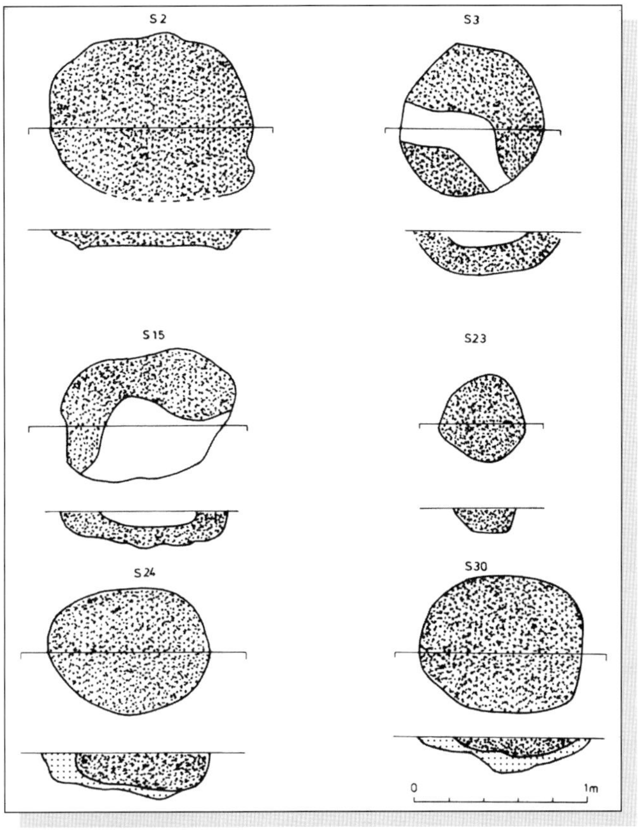

Figure 90: Doel "Deurganckdok-sector B"; plan of some typical hearth-pits. Key: cf figure 59.

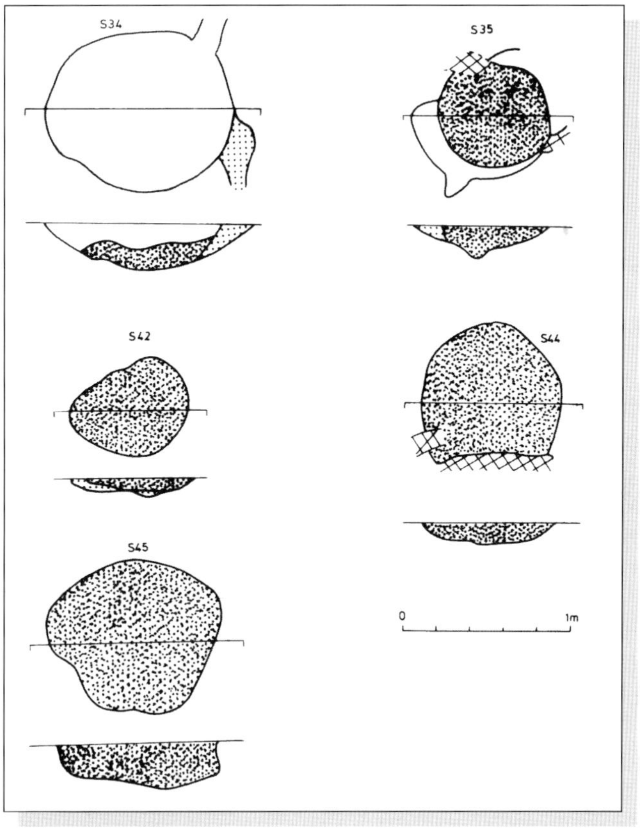

Figure 91: Doel "Deurganckdok-sector B"; plan of some typical hearth-pits.

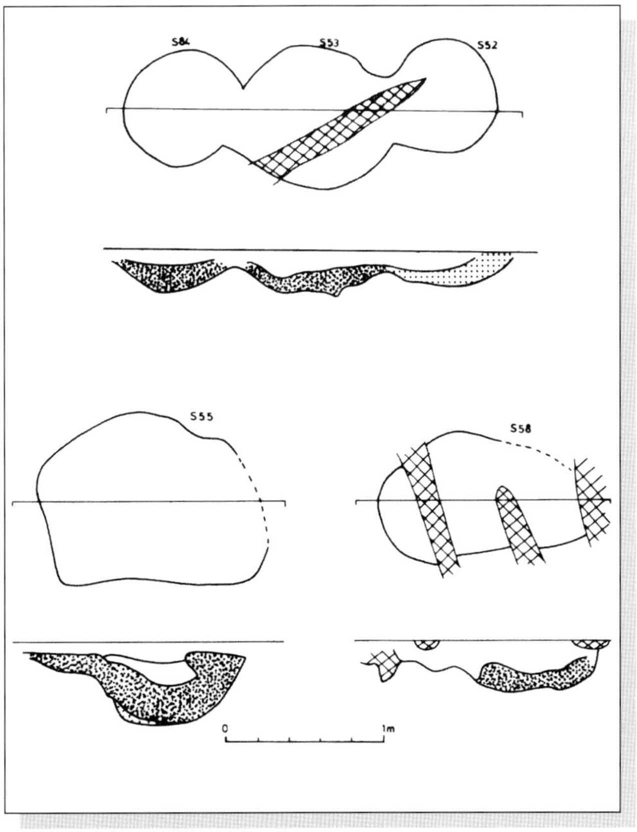

Figure 92: Doel "Deurganckdok-sector B"; plan of some typical hearth-pits.

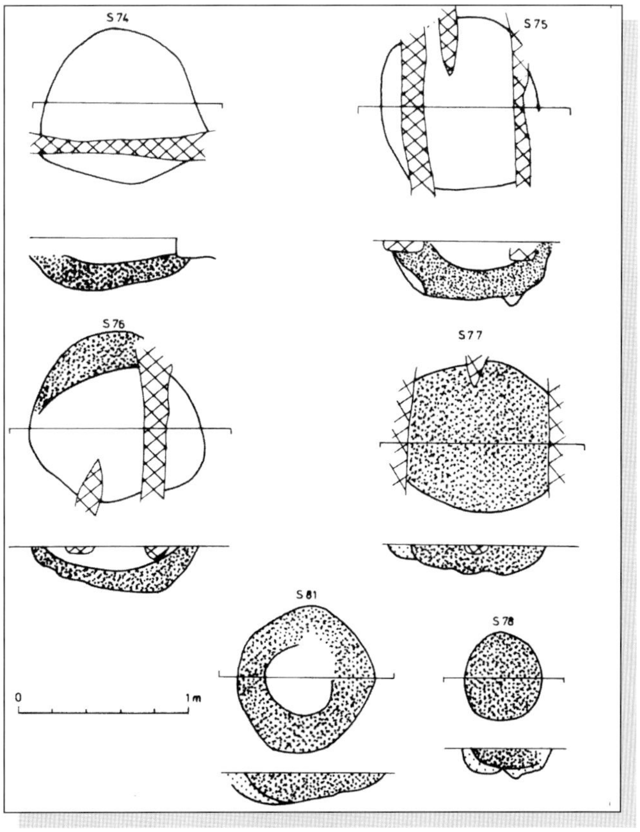

Figure 93: Doel "Deurganckdok-sector B"; plan of some typical hearth-pits.

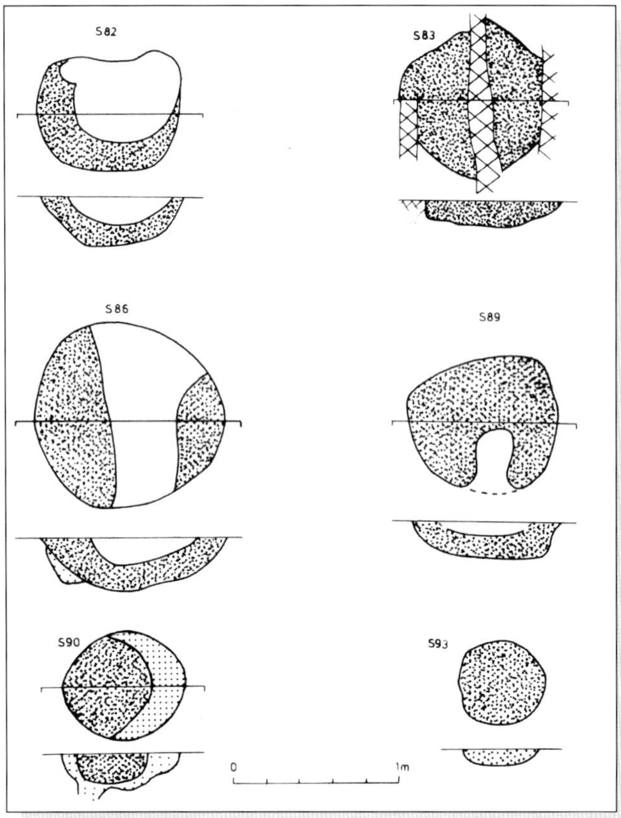

Figure 94: Doel "Deurganckdok-sector B"; plan of some typical hearth-pits.

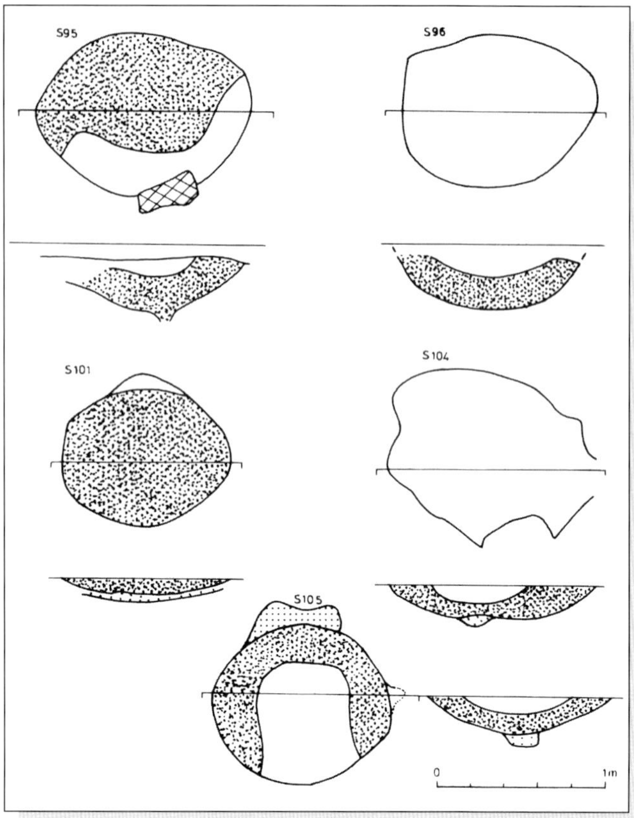

Figure 95: Doel "Deurganckdok-sector B"; plan of some typical hearth-pits.

An analysis of the size of these features, although partly biased by the truncation (especially in the north-eastern sector), reveals that they are on average 95.38cm long, 80.30cm wide and 27.91cm deep. The mean length-width ratio amounts to 1.17. However, from figures 96 to 99 it can be deduced that there is a considerable variation in the dimensions of these features. The length varies from minimum 51cm to maximum 162cm, yet the majority (ca. 72%) have a length between 65cm and 114cm. Furthermore two peaks can be observed in the distribution curve: one around 65-74cm (N=9) and a second around 95-104cm (N=16). The width varies between minimum 46cm and maximum 113cm, with a clear cluster (ca. 58%) between 65cm and 94cm. Here too, two minor peaks can be noted: one around 65-74cm and a second around 85-94cm. Regarding the depth, the majority of hearth-pits (ca. 69.33%) cluster around 15-34cm. Knowing that most features are truncated, their original depth must have been greater. On the evidence of the few complete hearth-pits (N=8), it can be concluded that the original depth must have varied between 30 and 60cm below the top of the Pleistocene sand surface.

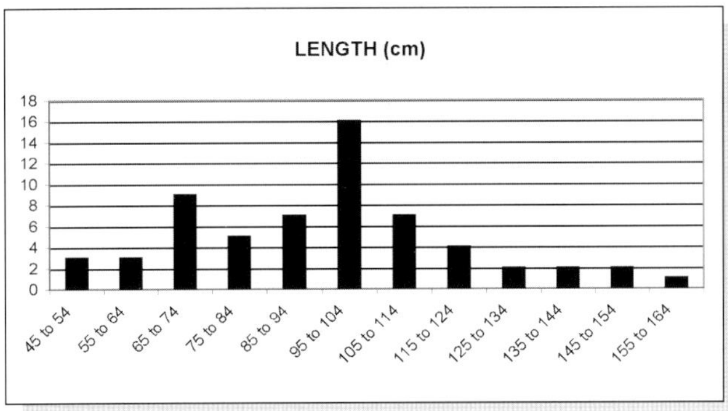

Figure 96: Doel "Deurganckdok-sector B"; distribution chart of the length of typical hearth-pits.

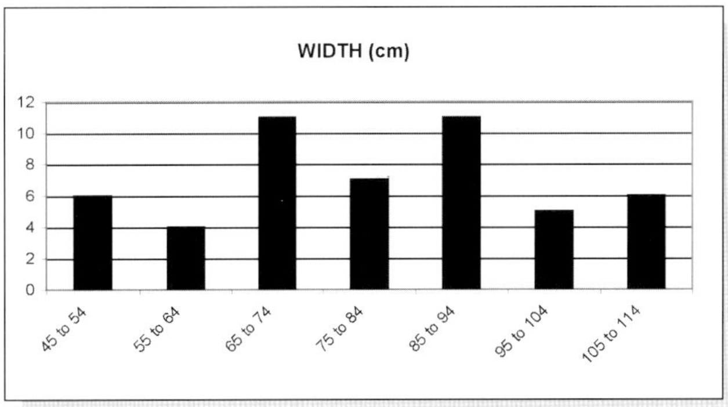

Figure 97: Doel "Deurganckdok-sector B"; distribution chart of the width of typical hearth-pits.

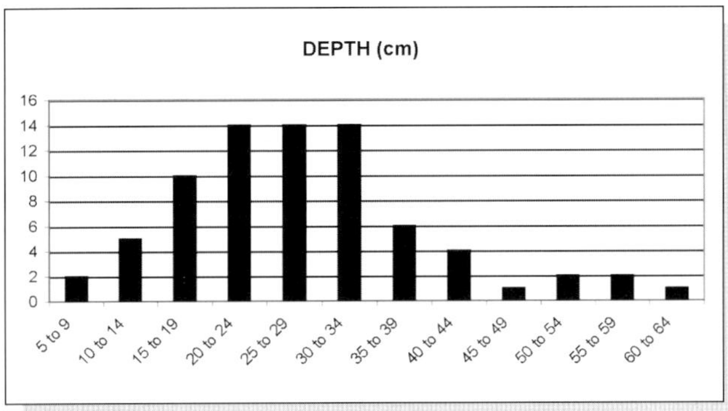

Figure 98: Doel "Deurganckdok-sector B"; distribution chart of the depth of typical hearth-pits.

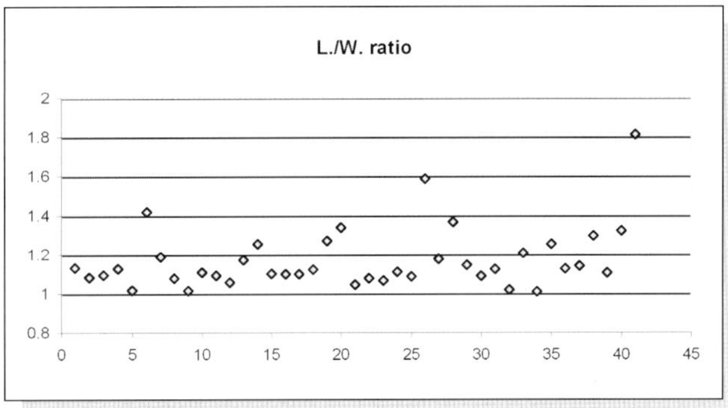

Figure 99: Doel "Deurganckdok-sector B"; distribution chart of the length/width ration of typical hearth-pits.

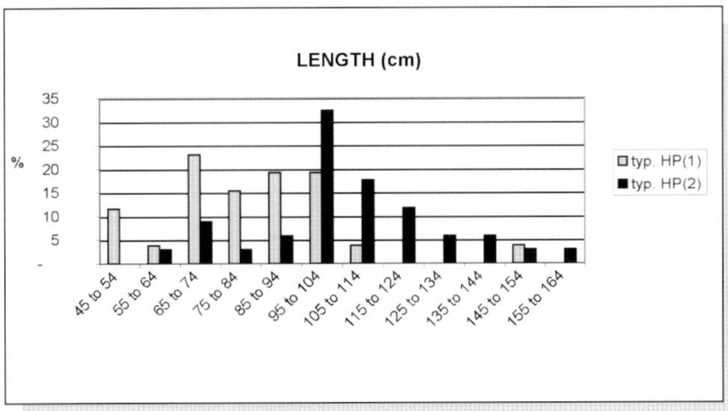

Figure 100: Doel "Deurganckdok-sector B"; comparative distribution chart of the length of typical hearth-pits with one (1) or two (2) filling layers.

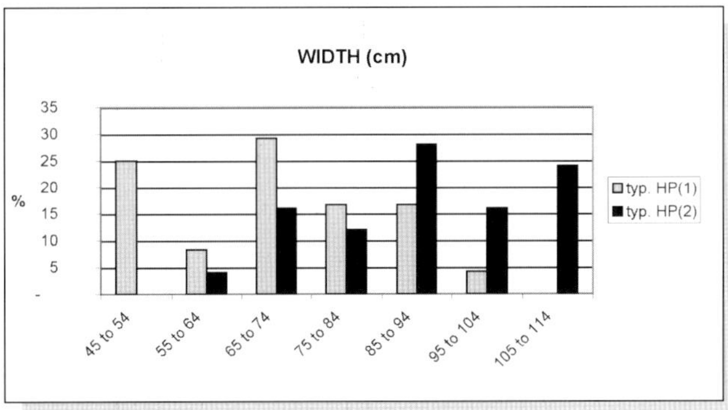

Figure 101: Doel "Deurganckdok-sector B"; comparative distribution chart of the width of typical hearth-pits with one (1) or two (2) filling layers.

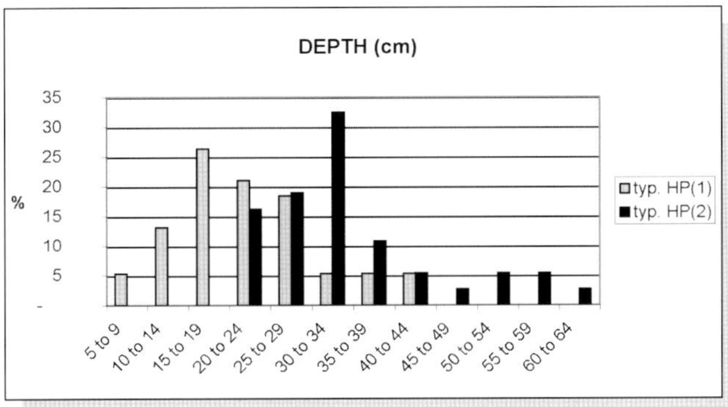

Figure 102: Doel "Deurganckdok-sector B"; comparative distribution chart of the depth of typical hearth-pits with one (1) or two (2) filling layers.

Worthy of note, however, is the existence of quite a significant metric difference between hearth-pits with one filling layer and those with two (fig.100-102). Clearly those with one homogenous filling are generally smaller (length < 100cm) and less deep (depth < 30cm), although there exists some overlap.
The archaeological content of typical hearth-pits will be discussed below together with that of the atypical ones.

3.1.3.2. | Atypical hearth-pits

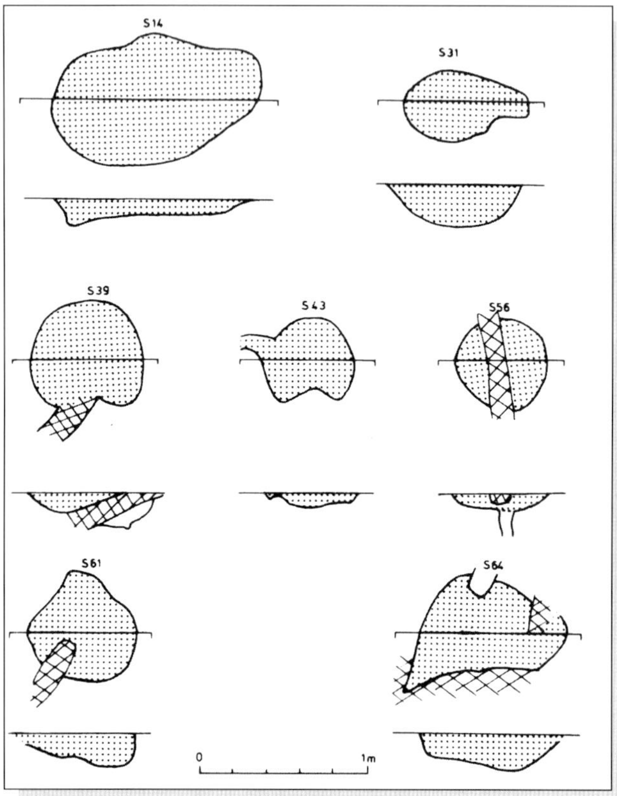

Figure 103: Doel "Deurganckdok-sector B"; plan of some atypical hearth-pits. Key: cf figure 59.

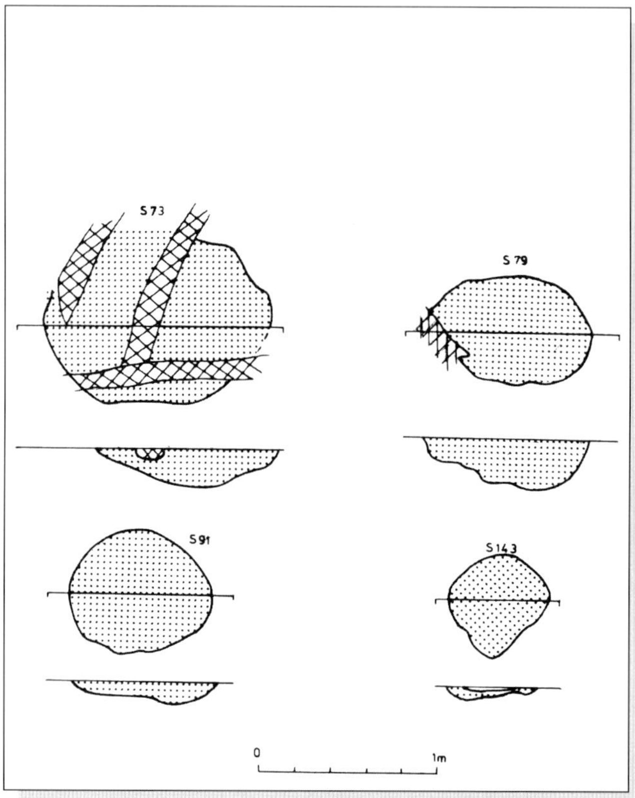

Figure 104: Doel "Deurganckdok-sector B & C"; plan of some atypical hearth-pits.

A series of 23 pits closely resemble the above-described hearth-pits, except for their filling (fig. 103, 104). Within these features the typical black charcoal lens does not occur. On the contrary, the filling generally consists of a homogenous light grey to pale grey sandy matrix mixed with charcoal fragments. Compared to the typical hearth-pits, the amount of charcoal is generally smaller and so are the fragments.

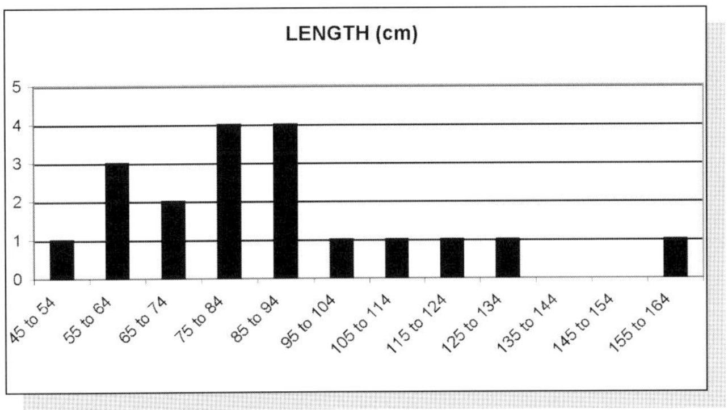

Figure 105: Doel "Deurganckdok-sector B"; distribution chart of the length of atypical hearth-pits.

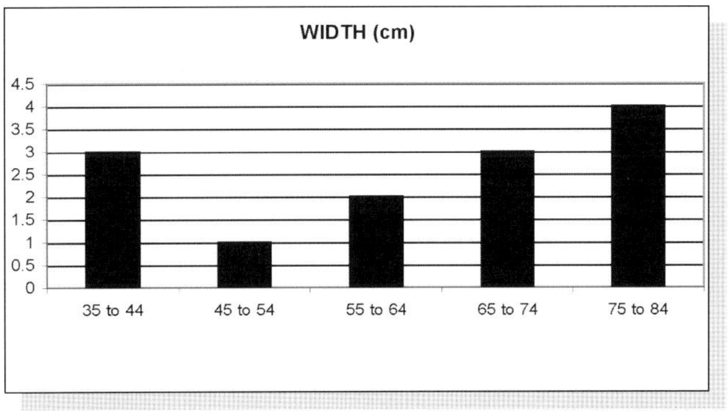

Figure 106: Doel "Deurganckdok-sector B"; distribution chart of the width of atypical hearth-pits.

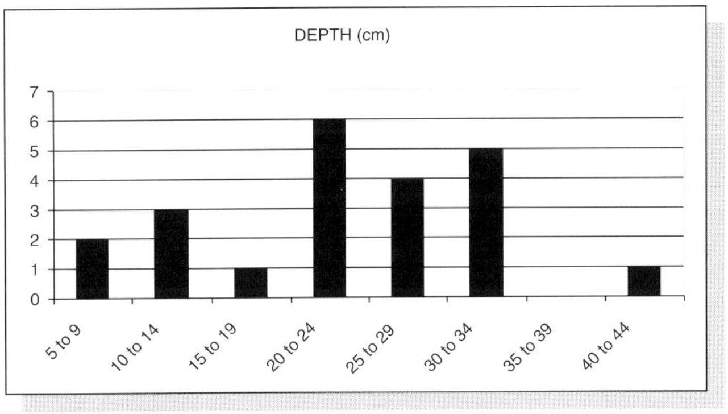

Figure 107: Doel "Deurganckdok-sector B"; distribution chart of the depth of atypical hearth-pits.

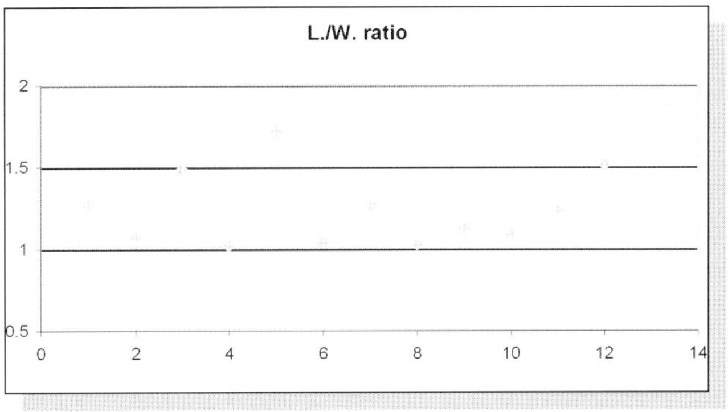

Figure 108: Doel "Deurganckdok-sector B"; distribution chart of the length/width ratio of atypical hearth-pits.

The mean dimensions of these features are: length 86.79cm, width 63.77cm, depth 22.82cm and length/width ratio 1.24. The length varies from minimum 45cm to maximum 164cm, with the majority (68.42%) clustering between 55cm and 94cm (fig. 105). The minimum width amounts to 39cm, the maximum to 84cm (fig. 106). The depth fluctuates between minimum 8cm and maximum 40cm, with a clear cluster around 20-34cm (68.18%) (fig. 107). According to the length/width ratio most pits are circular to weakly elliptical in ground plan (fig. 108). Compared to the typical hearth-pits, the size of these features fits perfectly with those of the hearth-pits with one single fill.

feature code	length (cm)	width (cm)	relative depth (cm)	absolute depth (cm)	top below TAW (m)	filling layers	sector
S14	121	80	17	truncated	-	1	NE
S31	74	43	25	truncated	-	1	NE
S33	76	73	25	truncated	1.41	1	NE
S37	45	>38	?	truncated	-	1	NE
S39	76	60	22	truncated	1.66	1	NE
S43	56	44	9	truncated	1.56	1	NE
S48	>34	>13	8	truncated	1.84	1	NE
S56	58	54	11	truncated	1.58	1	NE
S61	66	65	20	truncated	1.83	1	NE
S64	89	>58	23	truncated	1.77	1	NE
S66	94	76	21	truncated	1.77	1	NE
S73	127	>94	23	truncated	1.97	1	NE
S79	>82	63	30	truncated	1.91	1	NE
S80	58	39	12	truncated	1.5	1	NE
S85	91	84	13	truncated	-	1	NE
S91	80	71	33	truncated	1.25	1	NE
S92	79	77	40	truncated	1.25	1	NE
S110	90	-	30	truncated	-	1	NW
S118	>80	-	30	truncated	-	1	NW
S119	100	-	25	truncated	-	1	NW
S126a	>80	-	24	truncated	-	1	NW
S126b	105	-	27	truncated	-	1	NW
S130	164	-	34	truncated	-	1	NW

Table 32: Doel "Deurganckdok-sector B"; list of atypical hearth-pits.

On a spatial level no specific pattern can be observed to distinguish the typical from the atypical hearth-pits. Apparently both kinds occur randomly spread over the excavated surface.

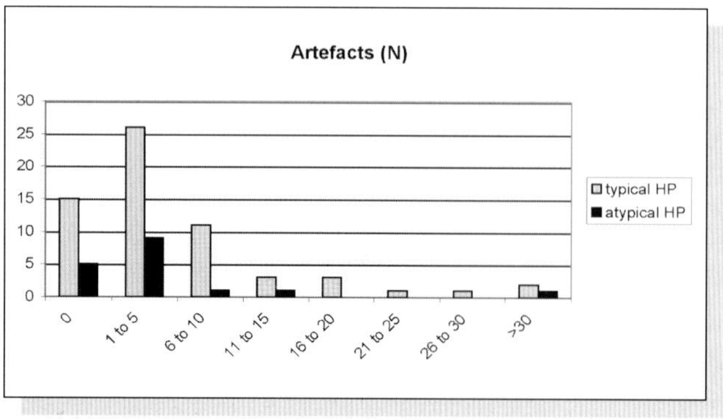

Figure 109: Doel "Deurganckdok-sector B"; comparative distribution chart of the number of artefacts from typical and atypical hearth-pits.

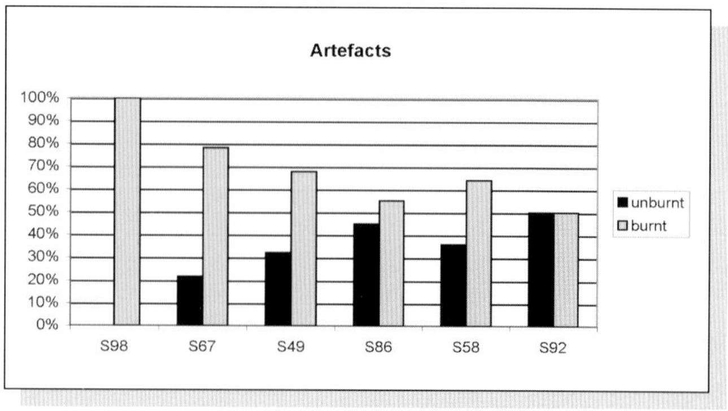

Figure 110: Doel "Deurganckdok-sector B"; ratio burnt/unburnt artefacts within some typical and atypical hearth-pits.

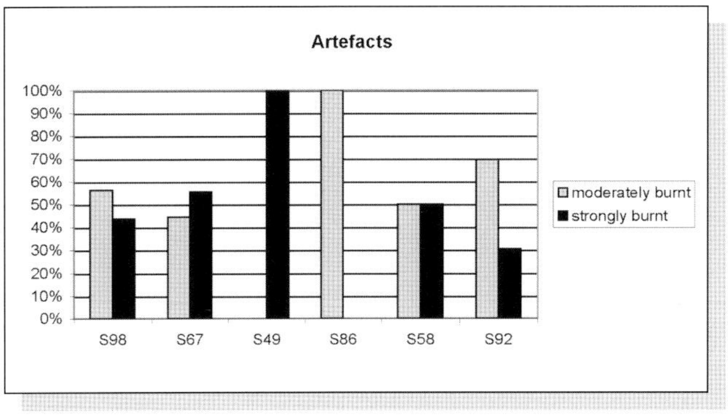

Figure 111: Doel "Deurganckdok-sector B"; ratio moderately/strongly burnt artefacts within some typical and atypical hearth-pits.

Regarding their archaeological content, the typical and atypical hearth-pits from Doel "Deurganckdok-sector B" are as heterogeneous as those from the nearby Early Mesolithic site of Verrebroek "Dok 1". The frequency of lithic artefacts varies considerably, as does the ratio between burnt and unburnt artefacts. The majority of hearth-pits clearly didn't yield many lithic artefacts (tab. 33, 34). As shown in the diagram (fig. 109) in most features, fewer than five to ten artefacts or none were found. Three pits stand outside this overall picture, yielding a substantial number of lithic implements. They are hearth-pits S58, S86 and S92, with 100, 40 and 46 artefacts respectively.

Except for eight typical hearth-pits and one atypical one, in all artefact bearing features some burnt items were found, either weakly and/or strongly (overheated) burnt. The ratio between burnt and unburnt items, calculated only for those features that contained enough lithics (fig. 110), is very variable, ranging from 50% (S92) to 100% (S98) of burning. The ratio between the moderately burnt and overheated artefacts too is not constant (fig. 111), with extremes of 100% of the former (S86) or of the latter (S49).

feature code	unburnt	moderately burnt	strongly burnt	total	potlids	tools	patinated artefacts	carbonised hazelnuts (gr.)	burnt bone (gr.)	potsherds
S2	0	1	0	1						
S3	2	8	1	11			1			
S4	0	0	0	0						
S6	1	0	2	3					0.08	
S15	1	3	1	5					0.01	
S22	12	2	0	14			3	0.03	0.10	
S23	0	0	0	0						
S24	0	0	0	0						
S25	7	2	0	9		retouched blade				1
S26	2	0	1	3						
S30	4	0	0	4			1			
S34	1	1	1	3						
S35	1	0	0	1						
S36	0	3	3	6						
S41	4	0	0	4			4			
S42	1	0	0	1						
S44	1	0	0	1						
S45	10	1	6	17			1			
S46	0	0	0	0						
S47	2	3	2	7		chip with retouche				
S49	9	0	19	28						
S50	0	0	0	0						
S51	8	2	0	10			7			
S52	5	0	2	7		undet. fragment	1			
S53	3	1	0	4			2			
S54	1	0	0	1						
S55	2	2	4	8						
S57	9	3	1	13						
S58	36	32	32	100		undet. fragment	2	0.05		
S59	0	0	0	0						
S62	0	0	1	1						
S63	2	3	0	5						
S65	0	0	2	2						
S67	5	8	10	23			1			
S69	2	0	0	2			1			
S72	3	0	0	3			1			
S74	1	1	2	4						
S75	8	0	0	8		backed point	2			
S76	0	0	0	0						
S77	0	0	0	0						
S78	2	0	0	2						
S81	1	0	0	1						
S82	0	0	0	0						
S83	1	1	0	2					0.03	
S84	6	1	1	8			1			
S86	18	22	0	40						

feature code	unburnt	moderately burnt	strongly burnt	total	potlids	tools	patinated artefacts	carbonised hazelnuts (gr.)	burnt bone (gr.)	potsherds
S89	3	3	0	6		trapeze				
S90	1	0	0	1						
S93	0	0	0	0						
S94	3	3	2	8					0.12	
S95	3	6	7	16	51	microlith fragment				
S96	0	2	6	8	60					
S98	0	9	7	16						
S100	0	0	0	0						
S101	0	0	0	0						
S102	0	1	1	2						
S103	0	0	0	0						
S104	0	0	1	1						
S105	2	0	0	2						
S106	0	0	0	0						
S107	0	0	0	0						
S108	0	3	2	5						
S111a	not sampled									
S111b	not sampled									
S112a	not sampled									
S112b	not sampled									
S113	not sampled									
S114	not sampled									
S116a	not sampled									
S120	not sampled									
S121	not sampled									
S122	not sampled									
S123	not sampled									
S124	not sampled									
S125	not sampled									
S127	not sampled									
S128	not sampled									
S129	not sampled									
S131	not sampled									

Table 29: Doel "Deurganckdok-sector B"; list of the archaeological content of the typical hearth-pits (expressed in number of artefacts).

feature code	unburnt	moderately burnt	strongly burnt	total	potlids	tools	patinated artefacts	carbonised hazelnuts (gr.)	burnt bone (gr.)	potsherds
S14	1	0	0	1			1	0.05	0.03	1
S31	0	0	0	0						
S33	0	2	1	3						
S37	0	0	0	0					0.02	
S39	3	0	0	3			1			
S43	0	0	0	0						
S48	0	0	0	0						
S56	3	0	0	3						
S61	2	2	0	4		triangle				
S64	0	0	0	0						
S66	2	0	1	3						
S73	0	1	0	1						
S79	4	8	3	15				0.41		
S80	1	0	0	1						
S85	7	1	2	10						
S91	0	2	0	2						
S92	23	16	7	46						
S110	not sampled									
S118	not sampled									
S119	not sampled									
S126a	not sampled									
S126b	not sampled									
S130	not sampled									

Table 34: Doel "Deurganckdok-sector B"; list of the archaeological content of the atypical hearth-pits (expressed in number of artefacts).

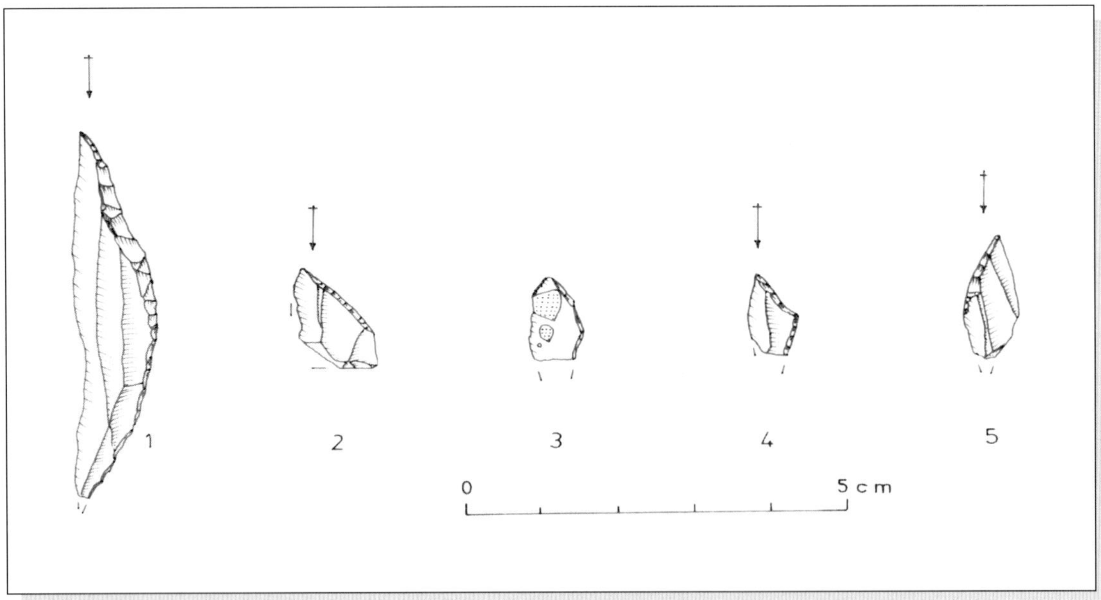

Figure 112: Doel "Deurganckdok-sector B & C"; flint tools found within some typical and atypical hearth-pits.

Only eight features yielded tools (fig. 112). Strangely enough these tools date back to different prehistoric stages. In a few hearth-pits artefacts were collected which can be attributed to the Final Palaeolithic and/or Early Mesolithic rather securely on the basis of weathering pattern (presence of patina), raw material (fine grained blackish flint), technology and especially typology. Most convincing is the find of a typical arced-backed, double-pointed projectile in S75 (fig. 112-1) and a blade fragment with facetted butt in S67, typical of the *Federmesser* culture. Furthermore, some pits yielded flakes and blade fragments covered by a weakly developed white to blue-white patina (30 artefacts), which also occurs on typical Final Palaeolithic tools, such as burins and backed points, found in the coversand deposits outside features (Crombé *et alii* 2000). From the distribution map it may be deduced that the majority of Final Palaeolithic finds cluster at the northern limit of the north-eastern sector of the site (fig. 113). This area corresponds to the zone where a lot of Final Palaeolithic artefacts were found in the coversand deposits (Crombé *et alii* 2000).

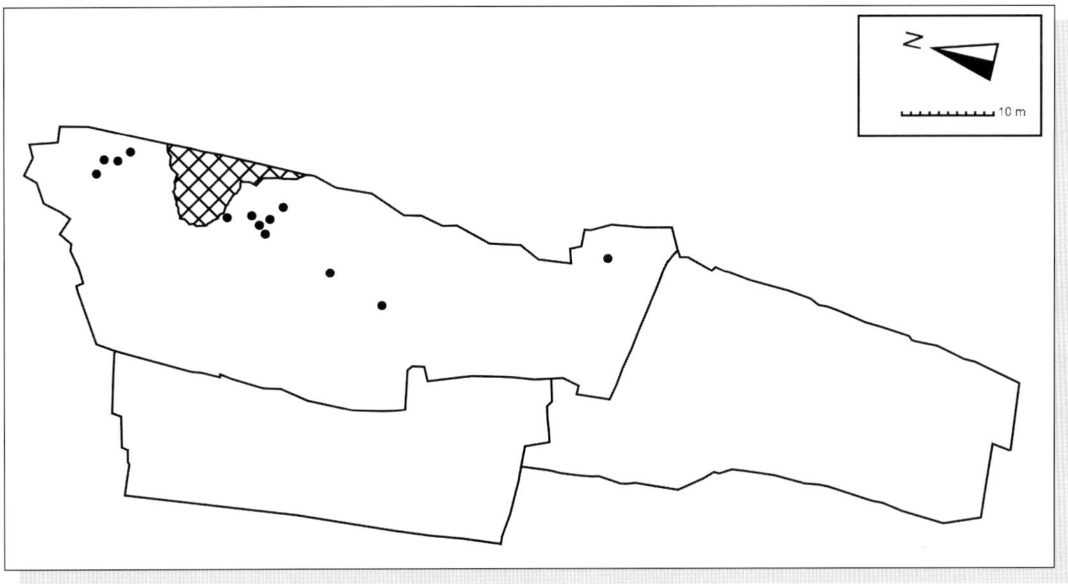

Figure 113: Doel "Deurganckdok-sector B"; distribution map of hearth-pits containing patinated flint artefacts.

Other hearth-pits yielded artefacts of clearly Mesolithic age. Altogether, three pits contained microliths, namely a small rectangular trapeze (S89) (fig. 112-2), a broken triangle (S61) (fig. 112-4) and an undetermined (overheated) microlith fragment[9] (S95) (fig. 112-3). A finer chronological attribution of these few tools is difficult, especially for the latter two specimens because they occur throughout the Mesolithic period (Vermeersch 1984; Crombé, Cauwe 2001). The trapeze, however, can be dated to the Late or Final Mesolithic, with a preference for the latter. Indeed, this specimen is much more comparable with the small trapezes in Wommersom quartzite and flint found in the A-horizon (Swifterbant level) at Doel "sector B" than with the broad trapezes found on many (surface)sites in the Belgian coversand region (Vermeersch 1984). Two other Final Mesolithic finds, each a very small pottery fragment, were found in surrounding hearth-pits (S14 and S25)[10]. Technologically both potsherds are perfectly similar to the Swifterbant pottery from the above-lying A-horizon (Crombé *et alii* 2000, 2002).

9 Although an exact determination of this find is no longer possible, it is clearly not a fragment of a trapeze, but rather of a microlith with a backed edge (crescent or unilaterally backed point or ...).

10 One was a rim fragment of hardly 13x14mm large (S14) and the other a wall fragment of ca. 15x15mm (S25).

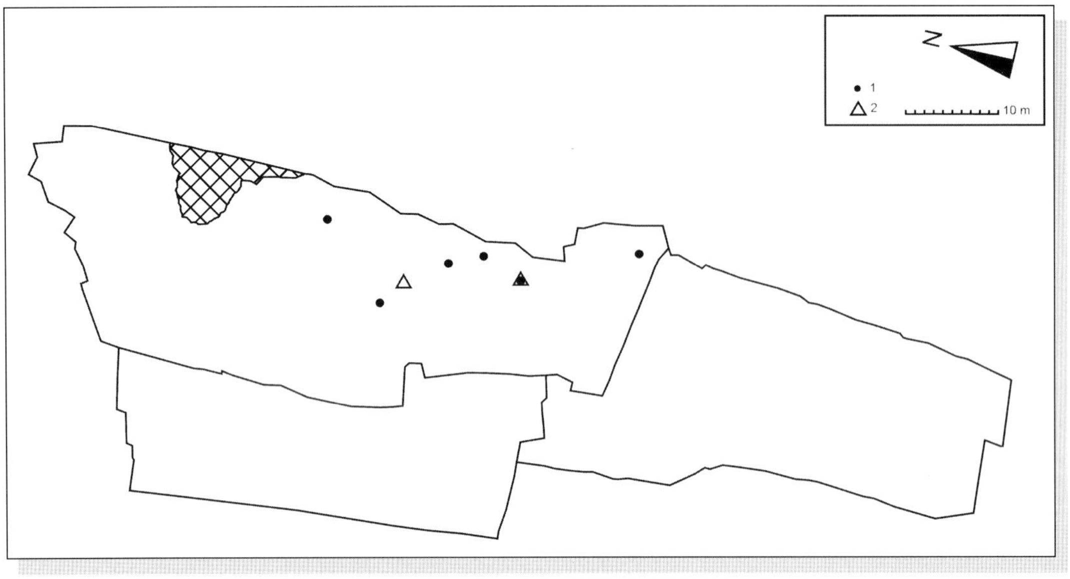

Figure 114: Doel "Deurganckdok-sector B"; distribution map of hearth-pits containing fragments of burnt bone (black dots) and pottery (triangles).

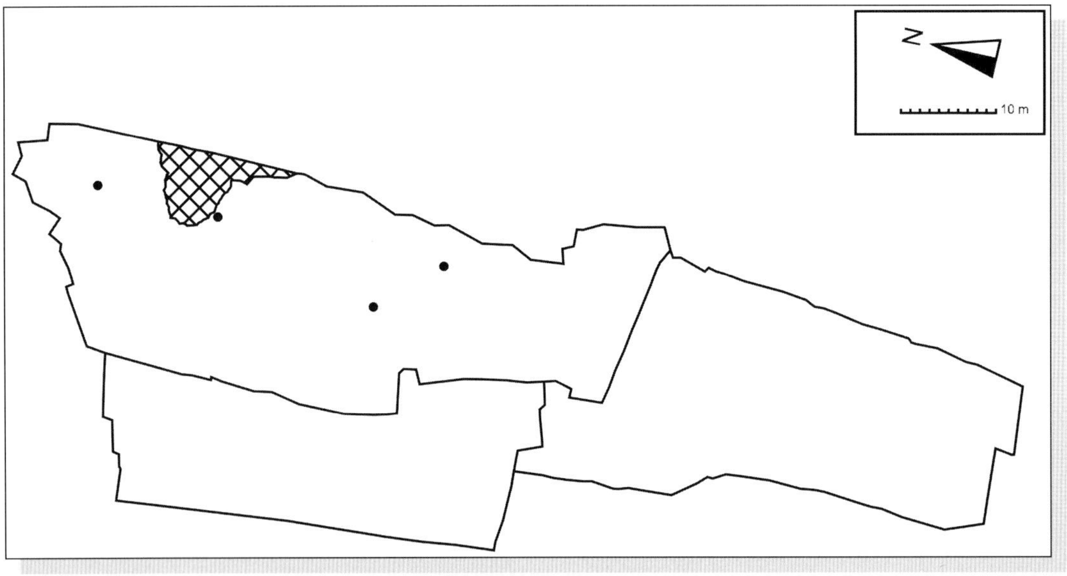

Figure 115: Doel "Deurganckdok-sector B"; distribution map of hearth-pits containing carbonised hazelnut shells.

Some hearth-pits - typical hearth-pits S58, S83, S94 and atypical hearth-pits S6, S15, S22 - also contained a minor quantity of burnt ecofacts other than charcoal, such as burnt bone fragments and/or carbonised hazelnut shells. In most cases only a few small fragments were recovered (tab. 33, 34). The distribution pattern of the pits with burnt bones (fig. 114) corresponds perfectly with that of the above-lying A-horizon, which contains large quantities of Swifterbant settlement waste, including burnt bone fragments (cf 3.5.4.). The hearth-pits with carbonized hazelnut shells, on the other hand, are more randomly spread over the site (fig. 115).

3.1.3.3. | Surface-hearths

At Doel several small clusters of burnt bones have been excavated, especially in the north-western (sieved) area of the site. Most of the retrieved bones belong to mammals and fish (cf 3.5.4.). Although at present a reliable spatial analysis of these finds cannot be undertaken due to the fact that only part of the sieved area has so far been analysed[11], it may be assumed that these bone clusters are the remains of former surface-hearths or hearth-dumps. A detailed analysis of these latent features will be discussed in a later volume of the "Verrebroek and Doel Excavation Projects".

3.1.4. | Doel "Deurganckdok-sector C"

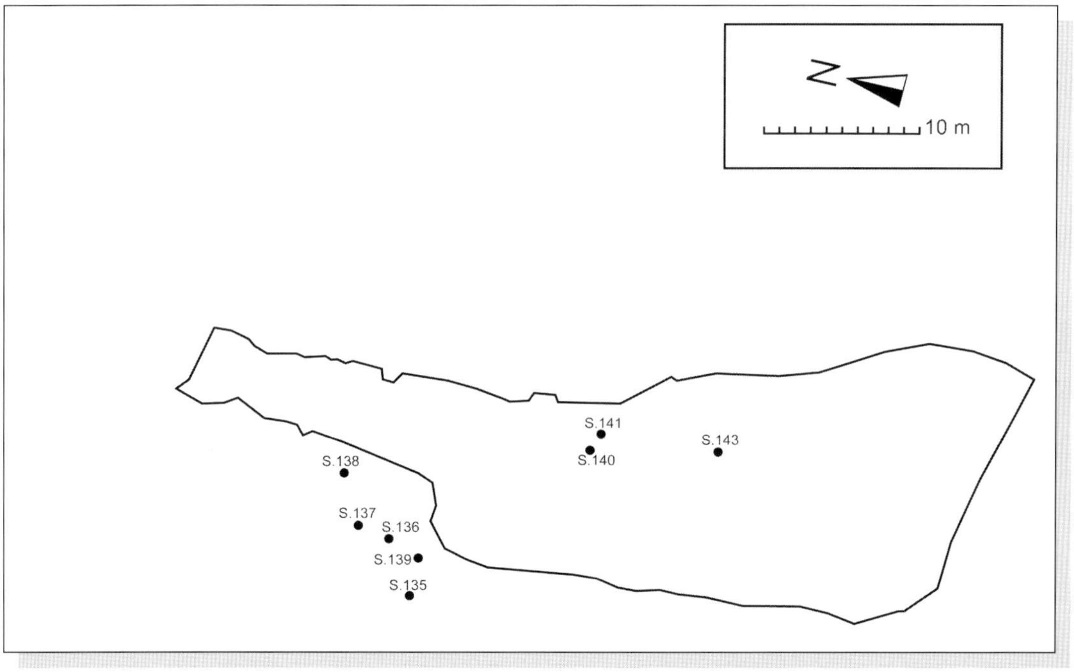

Figure 116: Doel "Deurganckdok-sector C"; distribution map of the hearth-pits.

At this site at least 10 to 12 features with a charcoal-rich fill could be observed (fig. 116). Unfortunately due to time pressure and the unfavourable working conditions (cf 1.3.5.) only one (S143) (fig. 104) could be properly investigated[12]. Judging by the scant available information, most features are comparable with the hearth-pits excavated in sector B. Feature S143 (110 x 120cm; 15cm depth) yielded a microlithic point, namely an obliquely truncated point in greyish flint (fig. 112-5).

11 Due to a lack of time and funding, so far only the squares with the highest bone density have been analysed in detail (cf 3.5.4.1.).

12 Nearly all features were situated in areas which were already largely disturbed by the harbour works. In addition, shortly after their discovery, they were dug away before any archaeological investigation could take place.

3.1.5. | Conclusions

3.1.5.1. | Hearth-pits

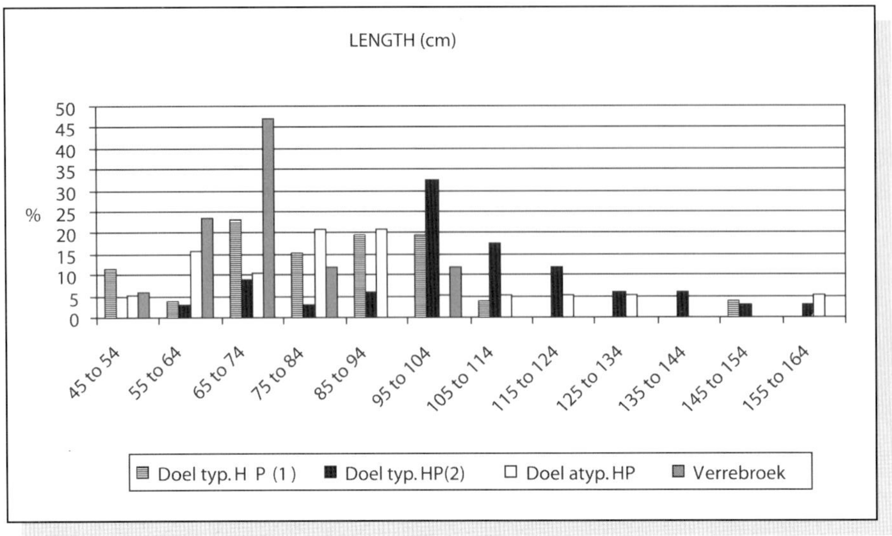

Figure 117: comparative chart of the length of hearth-pits from Doel and Verrebroek.

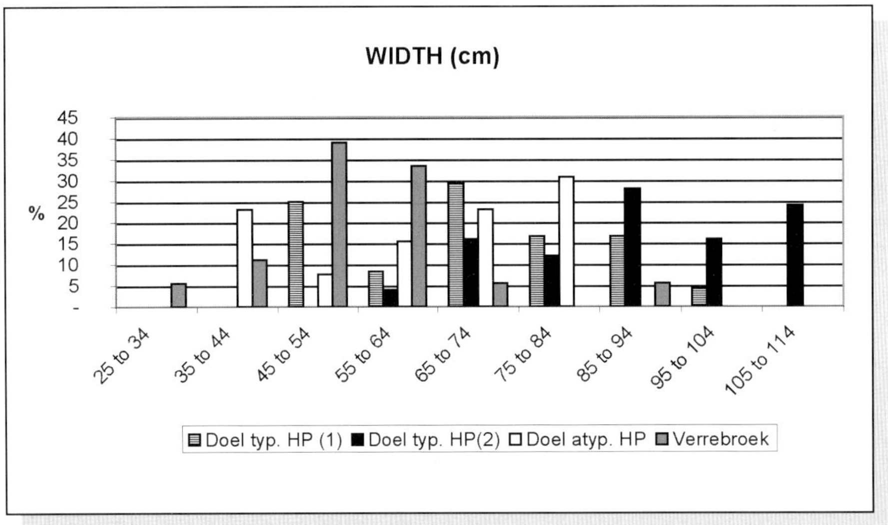

Figure 118: comparative chart of the width of hearth-pits from Doel and Verrebroek.

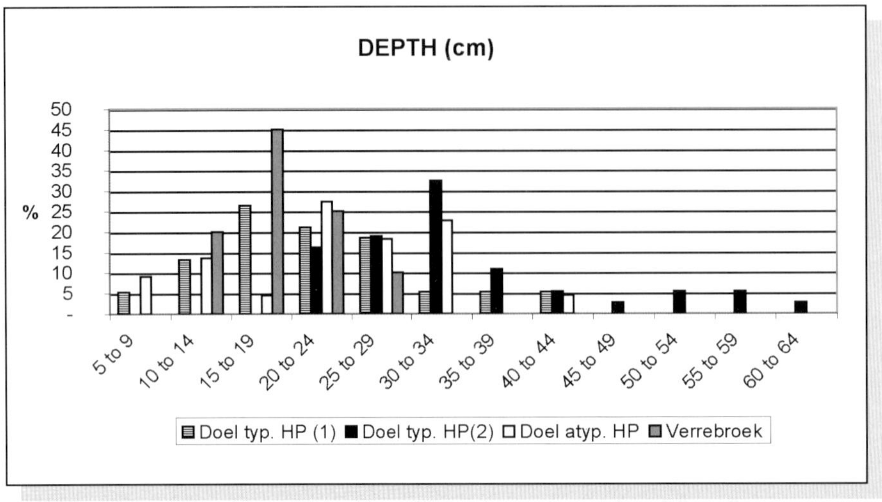

Figure 119: comparative chart of the depth of hearth-pits from Doel and Verrebroek.

A comparison of these pit features excavated at Verrebroek "Dok 1" and Doel "Deurganckdok-sector B" (fig. 117-119) shows a close similarity in size, shape and filling between the hearth-pits at the former site and both the typical hearth-pits with one single (charcoal) filling and the atypical hearth-pits from the latter site. On average these features are smaller than 0.90/1.0m, bowl-shaped and filled with a homogenous charcoal-rich layer or lens. In this respect these features are very comparable to the hearth-pits found at other Mesolithic sites in the coversand area of Belgium and The Netherlands[13]. The hundreds of hearth-pits excavated in the Veenkoloniën area of Groningen in the northern part of The Netherlands, for example, vary between 0.40 and 0.80m in diameter. The majority, however, cluster between 0.40 and 0.60m (Groenendijk 1987). At the Final Mesolithic/Early Neolithic (Swifterbant) sites of Almere "Hoge Vaart" (Hamburg *et alii* 2001)[14] and Urk-E4 (Peters, Peeters 2001) the hearth-pits, of which there are 111 and 35 examples respectively, are of similar sizes. On both sites these features do not generally exceed 0.70/0.80m in diameter. The average depth of hearth-pits on all the above-mentioned Dutch sites amounts to 0.40 to 0.60m. This is very much in agreement with the observations at Doel on the basis of the few hearth-pits which were not truncated by erosion or the harbour works. The hearth-pits of Verrebroek and Doel do not differ at all from the Dutch examples with respect to content either. In general hearth-pits do not contain large amounts of artefacts or ecofacts, except for charcoal. At Almere one exception to the rule was found: a deep hearth-pit containing ca. 725gr. of flint.

Large hearth-pits exceeding 1.00m in diameter as found at Doel "Deurganckdok-sector B" hardly occur on the Dutch Mesolithic sites. At Almere pits of this size are exceptional, whereas in the Veenkoloniën they are non-existent (as far as we can deduce from the published data).

Similarities between the Belgian and Dutch sites also exist as regards the spatial distribution of the hearth-pits. On all sites, hearth-pits are found clustered in a restricted area of the site, mostly on the higher and drier parts of the coversand landscape. Furthermore, in most cases they are situated away from artefact scatters and surface-hearths. At Almere the hearth-pits concentrate on the western slope and the highest part of the sand ridge, whereas the surface-hearths are situated on the top and the eastern slope of the dune. In the lower lying periphery of the site, only isolated hearth-pits occur. At the NP3 site in the Veenkoloniën too, hearth-pits are confined to the top of the coversand ridge, where they occur in small clusters of 2 to 3 pits (Groenendijk 1987, 2004; Smit 1995).

13 Hearth-pits are also known from sandy sites in other countries, such as Germany (Vollbrecht 2003), Denmark (Brinch Petersen 1989) and France (Hinout 2002). Unfortunately most of these features are not (yet) published in detail.

14 On this site the deep pits classified as type a and b (Hamburg et alii 2001, 11) can be compared with the typical hearth-pits of Verrebroek and Doel, whereas the deep pits of type c are the counterparts of the atypical hearth-pits of Doel.

3.1.5.2. | Charcoal-pits

Compared to the hearth-pits these features form a less homogeneous group in terms of size and morphology. Another important difference, in particular at Verrebroek, is the spatial distribution, which is apparently random, whereas hearth-pits tend to cluster in specific parts of the landscape and/or site. Comparable features are reported on other Mesolithic coversand sites, such as different sites in the Veenkoloniën (Groenendijk 1987, footnote 9), Bergumermeer (so-called "pits", in Newell 1981, 257, fig. 13), Havelte (so-called "small pits", in Price *et alii* 1974, 22), Merselo-Haag (Verhart 2000, 78) and many others. In most cases these features are considered anthropogenic, the interpretations ranging from "hearths" to refuse or storage pits. In our opinion there is not enough proof, on any of the above-mentioned sites, for such interpretations. While awaiting further and more detailed analyses, such as micromorphology and phosphate analysis[15], we would not fully exclude a possible natural origin for this type of pit-feature.

3.1.5.3. | Hazelnut-pits

The interpretation of these vague features is controversial. According to the pedologists (cf 3.3.1.2.) at least one of these features (S89) might be considered as a pre-podzol tree-fall structure. Their main arguments are: 1° the presence of a tilted A-horizon; 2° the absence of fine carbonised organic matter, excluding *in situ* burning; 3° the asymmetric distribution of the artefacts. However, in our opinion the third argument is not valid, since the distribution pattern is biased due to the excavation grid[16]. Furthermore, the absence of *in situ* burning evidence does not necessarily exclude an anthropogenic origin for this feature. Indeed, it might also indicate that this feature was not used for roasting hazelnuts, but rather to store carbonised hazelnuts or to dump waste of carbonising hazelnuts. Bearing in mind the results of burning experiments conducted by David Score and Steven Mithen (2000), it is questionable whether the rather deep hazelnut-pits found at Verrebroek would have been appropriate for roasting hazelnuts. Indeed, from these experiments it appears that the most efficient way to roast hazelnuts in terms of minimizing fuel outlay and the charring of nuts is to construct a shallow oven over which fires burn for a relatively short period. In the experimental pit 4, dug out at a depth of 0.30m, it was found that roasting had only taken place to a depth of 0.21m, leaving 25% of the nuts uncooked. Better results were obtained in three shallow pits, which were only 0.12m deep. Here the percentage of unroasted nuts varied between 0% and 8%. This makes our assumption that the surface-hearths at Verrebroek are most likely hazelnut roasting places much more plausible.

15 At the site of Havelte some phosphate determinations have been conducted (Price et alii 1974, 55-57). The results indicate a clear difference in phosphate values between hearth-pits (mean 123.6 ppm) and charcoal-pits (mean 70.8 ppm).

16 Indeed, this feature was excavated within the general grid of the trench. As a result its filling was not evenly spread over the quadrates O75/Z46(3) and O75/Z46(4). As the largest pit volume corresponds to quadrate O75/Z46(3), it is not surprising that it yielded the largest number of artefacts and ecofacts.

Comparable features with concentrations of carbonised hazelnut shells are reported on other Mesolithic sites such as Staosnaig on the island of Colonsay in Scotland (Mithen, Finlay 2000), Stadskanaal in The Netherlands (Groenendijk 1987) as well as on the neighbouring site of Verrebroek "Aven Ackers" (Van Roeyen 1990, 67-69).

At Staosnaig two large pits - F24 and F41 - were excavated, yielding >3000gr. and >500gr. respectively of carbonised nut remains. In light of the rarity of burnt stone, fire-cracked stone and wood charcoal, the excavators interpret these pits tentatively as dumps. It should, however, be emphasised that these Staosnaig pits are considerably larger than those found at Verrebroek. Feature F24 is sub-circular with a diameter of 4.5m and a depth of ca. 0.70m; feature F41 is guitar-shaped measuring ca. 3 x 1.25 x 0.45m.

Henny Groenendijk reports the find of a possible storage-pit full of complete, carbonised hazelnuts near Stadskanaal. According to the observations of J.H. Keizer, the pit had the same shape as the hearth-pits uncovered at various Mesolithic sites in the Veenkoloniën. Unfortunately detailed data on this feature are not available.

The best parallels are to be found at the nearby Early Mesolithic site of Verrebroek "Aven Ackers". Here a partially excavated pit with a diameter of more than 0.70m and a depth of ca. 0.60m was vaguely visible as a greyish soil discolouration. The delimitation was facilitated by the large quantities of hazelnut shells. The only difference with respect to the pits excavated at Verrebroek "Dok 1" is the presence in the southern half of the pit of numerous charcoal fragments.

In conclusion the presently available data is not sufficient to determine with certainty the exact function of either pit defined as hazelnut-pits, at Verrebroek "Dok 1". Unlike the pedologists we are, however, more in favour of an anthropogenic origin for these features, although hard and irrefutable evidence is still lacking. Based on parallels with other Mesolithic sites, a tentative interpretation as storage or dumping pits is proposed.

3.2. | Radiocarbon dating

Mark Van Strydonck & Philippe Crombé

3.2.1. | Sample selection and testing

All dated samples were cleaned using the Acid Alkali Acid method, converted into graphite and measured by AMS (Van Strydonck, van der Borg 1990/91). Sample pre-treatment and preparation took place at the Royal Institute for Cultural Heritage (Brussels). The measurements were carried out either at the Van de Graaff Laboratory, State University of Utrecht (UtC), or at Rafter Radiocarbon Laboratory, Institute of Geological and Nuclear Sciences, Lower Hutt (NZA), or at the Leibniz Labor für Altersbestimmung und Isotopenforschung, Christian-Albrechts-Universität, Kiel (KIA).

At all archaeological sites only single entities were dated. Each dated sample consisted of one single organic fragment, either a fragment of a charred hazelnut shell, one piece of charcoal or one seed or pip. Wherever possible, surface-hearths were dated on samples of charred hazelnut shells (Verrebroek) or carbonized seeds (Doel), while charcoal samples were commonly used to date hearth-pits. Most features were dated once.

3.2.1.1. | Test 1

Some surface-hearths however were dated by more than one sample to obtain more precision. At the site of Verrebroek "Dok 1", for example, all features dated between 1992 and 1994 (Crombé 1998, 20-22) were re-dated in 2000, because some of the earliest AMS dates had very large standard deviations. This was the case for the surface-hearths within the following artefact units:

Unit 2.1.
UtC-3445 9100±130 BP
UtC-8398 9265±40 BP
A T-test shows a more than 95% probability that the two samples represent the same true age. The mean of the two measurements is 9250±38BP

Unit 2.2.
UtC-3436 9130±170 BP
UtC-8391 8850±50 BP
A T-test shows a more than 95% probability that the two samples represent the same true age. The mean of the two measurements is 8872±47BP

Unit 10
UtC-2744 8920±130 BP
UtC-8394 8835±40 BP
A T-test shows a more than 95% probability that the two samples represent the same true age. The mean of the two measurements is 8842±38BP

Unit 5.2.
UtC-2743 9000±190 BP

UtC-8396 8795±40 BP
A T-test shows a more than 95% probability that the two samples represent the same true age. The mean of the two measurements is 8803±39BP

It is obvious that due to the large standard deviation of the dates obtained between 1992-1994, the means do not deviate much from the recently obtained radiocarbon dates. As a consequence the contribution of these older dates to the problem under consideration is nil.

3.2.1.2. | Test 2

Since some of the charred hazelnut samples at Verrebroek "Dok 1" were badly preserved and some dates deviated significantly from the bulk of the measurements (see below) it was feared that humic acid infiltration from the overlying peat layer had changed the ^{14}C content of some of the charred hazelnut samples. To test this possibility, the efficiency of the pre-treatment was tested in two ways. First a duplo sample from artefact unit 17 was dated after a repeated pre-treatment (tab. 35). Then a larger sample, containing different individual hazelnut shells, from artefact unit 22 (SH.1) was pre-treated several times and the different residues as well as the soluble fractions were dated (tab. 36). All treatments were performed in hot 1% solutions for 20 to 30 min. All samples were rinsed with hot distilled water after treatment and dried.

	lab. reference	BP date	$\delta^{13}C$ ‰
After 1 pretreatment	UtC 7047	6260±50	-25.5
NaOH extract	UtC 7940	5470±90	N.A.
After 2 pretreatments	UtC 7941	6290±40	-25.5

Table 35: Verrebroek "Dok 1"; repeated Acid Alkali Acid (AAA) pre-treatment on a hazelnut sample from a surface-hearth within artefact unit 17 (test 2).

#	UtC	pre-treatment	BP date	$\delta^{13}C$ ‰
1	9692	No	9120±60	-24.4
HCl				
2	9928	soluble	5790±50	-25.4
3	9436	insoluble fraction neutralisation	9200±90	-23.9
# 3 + NaOH				
4	9926	soluble	9060±50	-25.4
5	9437	residue fraction after HCl neutralisation	9370±80	-24.3
#5 + NaOH				
6	9927	soluble	8920±50	-25.3
7	9438	residue fraction after HCl neutralisation	9290±80	-26.0

Table 36: Verrebroek "Dok 1"; repeated Acid Alkali Acid (AAA) pre-treatment on a hazelnut sample from surface-hearth 2 (SH.1) within artefact unit 22 (test 2).

The results in tables 35 and 36 show very clearly that almost all contamination was removed after the first wash and that the quantitative contribution of the humic acids to the date was very small. There is more than 95% probability that samples #1 and #2 in table 35 and #1 and #7 in table 36 have the same true age. These tests demonstrate that the AAA method was sufficient to clean the samples.

3.2.1.3. | Test 3

One of the surface-hearths at Verrebroek "Dok 1"which yielded burnt bone fragments – artefact unit 72 – was also dated on a sample consisting of a single bone fragment. The burnt bone no longer contained any collagen. Therefore the ^{14}C date of the bone carbonate was measured and compared to the ^{14}C date of associated charred hazelnut shell (tab. 37 and fig. 120).

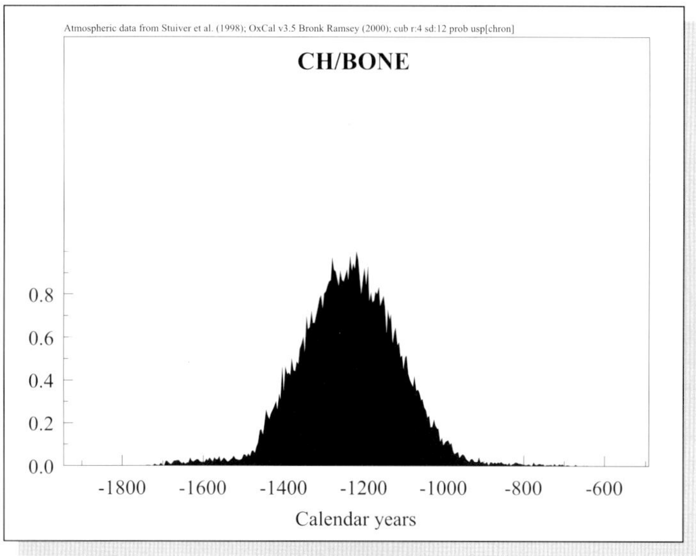

Figure 120: Verrebroek "Dok 1"; difference in ^{14}C age between a burnt bone and a charred hazelnut shell from a surface-hearth within artefact unit 72.

dating material	lab. reference	BP date
bone	UtC-7252	7745±65
charred hazelnut	NZA-11938	8750±40

Table 37: Verrebroek "Dok 1"; radiocarbon date of a burnt bone and a charred hazelnut shell from a surface-hearth within artefact unit 72 (test 3).

This test shows that the difference between the two dates is more than a millennium (1360 -1110 at 68.2% probability range; 1470-970 at 95.4% probability range). This was not unexpected. Cremated bones, heated at temperatures above 600°C, lose their organic carbon and the bio-apatite recrystallizes and produces larger and better-structured crystals. As a result of this recrystallization the bone carbonate is protected from exchanges with bi-carbonate from the groundwater and the ^{14}C date of cremated bones reflect their true age (Lanting *et alii* 2001; Van Strydonck *et alii* in preparation). This is not the case for burnt bones where carbon exchange between the bones matrix and the environment still exists.

3.2.1.4. | Test 4

It was possible that the dates obtained on the charcoal samples from hearth-pits deviated from the real age of those features due to "old-wood-effects". This old wood effect can arise when an older (or inner) part of a tree is dated. Judging by the small size of most hearth-pits, however, only a limited "old-wood-effect" on the charcoal was expected. This was confirmed by dating identified branch and trunk wood from S66 (tab. 38, fig. 121) and S91 (tab. 39, fig. 122) excavated at Verrebroek "Dok 1". It must, however, be admitted that anthracologically the term "trunk wood" had already been applied to all wood with a diameter exceeding 5cm (Crombé *et alii* 1999a).

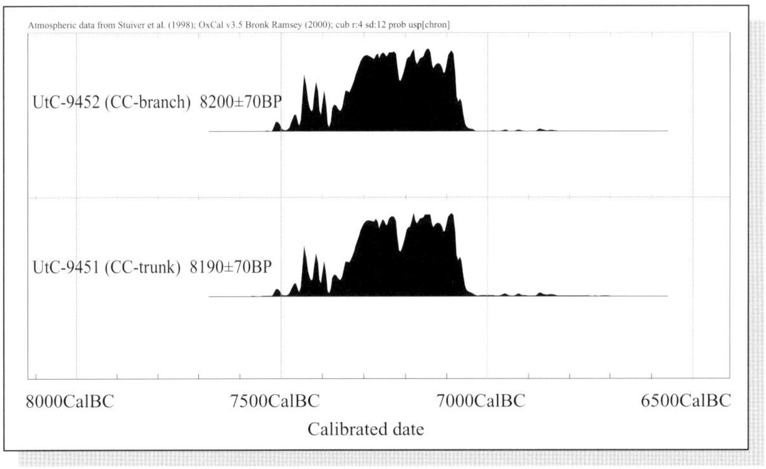

Figure 121: Verrebroek "Dok 1"; calibrated radiocarbon dates from a test on different materials from hearth-pit S66 within artefact unit 20.

lab. reference	dating material	BP date	$\delta^{13}C$ ‰	68.2% confidence	95.4% confidence
UtC-9452	Charcoal *Pinus* branchwood	8200±70	-25.3	7320BC (68.2%) 7080BC	7460BC (8.1%) 7390BC 7380BC (87.3%) 7060BC
UtC-9451	Charcoal *Pinus* trunkwood	8190±70	-24.5	7310BC (29.3%) 7220BC 7200BC (38.9%) 7080BC	7460BC (6.7%) 7390BC 7380BC (88.7%) 7050BC
DIFFERENCE					
				-170 (68.2%) 0	-350 (95.4%) 10

Table 38: Verrebroek "Dok 1"; calibrated radiocarbon dates of different charcoal samples from hearth-pit S66 within artefact unit 20 (test 4).

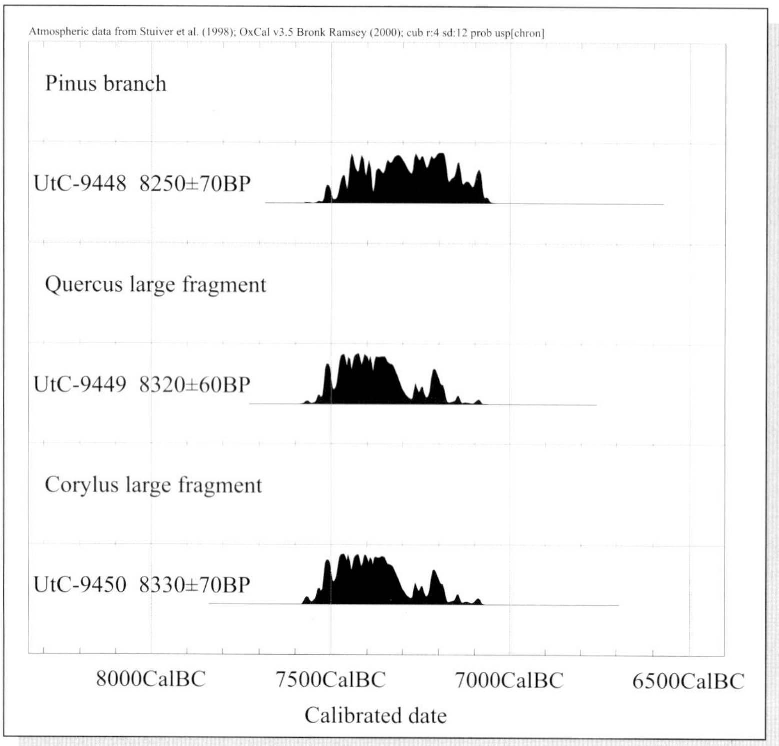

Figure 122: Verrebroek "Dok 1"; calibrated radiocarbon dates from a test on different charcoal fragments from hearth-pit S91.

lab. reference	dating material	BP date	$\delta^{13}C$ ‰	68.2% confidence	95.4% confidence
UtC-9448	*Pinus* branch	8250±70	-24.5	7450BC (12.8%) 7390BC 7380BC (51.5%) 7170BC 7160BC (3.0%) 7140BC 7100BC (0.8%) 7080BC	7520BC (1.0%) 7500BC 7500BC 7480BC (94.4%) 7080BC
UtC-9449	*Quercus* large fragment	8320±60	-25.3	7520BC (4.4%) 7500BC 7480BC (61.4%) 7320BC 7220BC (2.4%) 7200BC	7540BC (95.4%) 7180BC
UtC-9450	*Corylus* large fragment	8330±70	-25.1	7520BC (65.8%) 7320BC 7220BC (2.4%) 7200BC	7550BC (95.4%) 7170BC

Table 39: Verrebroek "Dok 1"; calibrated radiocarbon dates of charcoal fragments from different wood species from hearth-pit S91 (test 4).

Both tests show very clearly that there is no old-wood effect. The differences are even smaller than one would expect based on the standard deviations. Dating different wood species within hearth-pit S91 also revealed that the hearth-pits were, as expected, most probably only used once within a limited period and that no intrusive or residual charcoal was present in the pits.

3.2.1.5. | Test 5

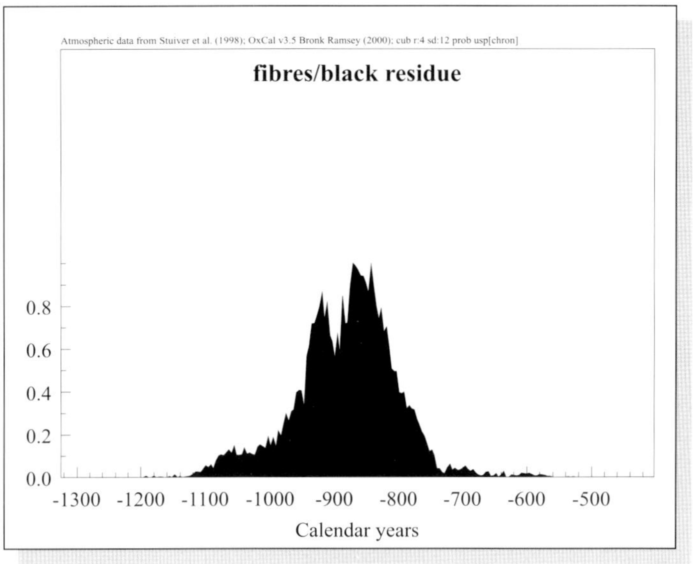

Figure 123: Doel "Deurganckdok-sector B"; difference in ^{14}C age between the black residue and the fibres from a potsherd.

lab. reference	dating material	BP date	$\delta^{13}C$ ‰	68.2% confidence	95.4% confidence
KIA-12258	Plant fibres from black residue	5165±35	-28.1	4040BC (13.9%) 4020BC 3995BC (54.3%) 3955BC	4050BC (85.7%) 3930BC 3860BC (9.7%) 3800BC
KIA-12259	NaOH extract from black residue	5545±35	-28.8	4450BC (28.0%) 4420BC 4400BC (16.7%) 4380BC 4370BC (23.5%) 4340BC	4460BC (95.4%) 4330BC
KIA-12260	Black residue after AAA treatment	5980±35	-28.0	4910BC (27.9%) 4870BC 4860BC (40.3%) 4800BC	4950BC (93.3%) 4770BC 4750BC (2.1%) 4730BC

Table 40: Doel "Deurganckdok-sector B"; calibrated radiocarbon dates of black residue from potsherd WKG 109 (test 5).

The AAA method was also applied to the black, charred food remains found on the inner side of a Swifterbant potsherd from the site of Doel "Deurganckdok-sector B". First the organic coating was sampled and mechanically cleaned. No fatty acids could be extracted from the food remains so the black residue had to be dated. Microscopic examination showed that the black organic residue contained fibres from a different colour. The fibres were removed and dated separately (tab. 40). Figure 123 represents the difference between the black residue and the fibres. The difference between the two (950-800 at 68.2% probability and 1080-740 at 95.4% probability) shows very clearly that the fibres are intrusive and probably derive from the peat covering the site.

3.3.2. | Verrebroek "Dok 2"

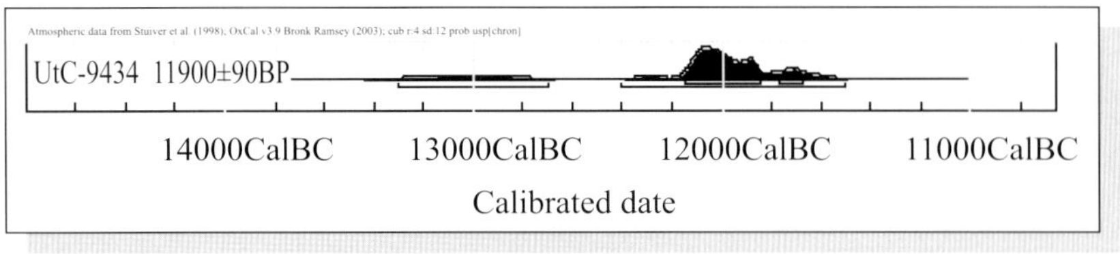

Figure 124: Verrebroek "Dok 2"; calibrated radiocarbon date of a burned bark fragment.

lab. reference	dating material	BP date	$\delta^{13}C$ ‰	68.2% confidence	95.4% confidence
UtC-9434	burned bark	11900±90	-27.2	12150BC (60.2%) 11840BC 11770BC (8.0%) 11670BC	13300BC (12.4%) 12700BC 12400BC (83.0%) 11500BC

Table 41: Verrebroek "Dok 2"; calibrated radiocarbon date of a burned bark fragment.

A sample of partly burned bark, originating from a charcoal patch which was found associated with a small cluster of lithic artefacts (cf 3.1.1.), was radiocarbon dated (tab. 41, fig. 124). Due to the fact that the "bulb" in the calibration curve around 13000 cal BC is artificial (cf 2.3.1.), only the range between 12400 and 11500 cal. BC (95.4% confidence level) can be considered valid. If, however, we anticipate the publication of the new calibration curve (INCAL 04), a calibrated date range between 11900 and 11400 BC can be expected. The obtained date confirms the age expected on the basis of the artefact techno-typology (Final Palaeolithic), litho-stratigraphy and bio-stratigraphy (end of Bølling/Dryas II). Furthermore the date agrees very well with the peat dates obtained on samples from the nearby profile 1 (cf 2.3.2.).

3.2.3. | Verrebroek "Dok 1"

3.2.3.1. | Surface-hearths

The radiocarbon dates performed on carbonized hazelnut shells are divided into a coherent group of 48 dates (tab. 42, fig. 125) and 3 outliers (tab. 43, fig. 126). All samples were taken from within or nearby the presumed centre of these latent features, as defined on the basis of detailed spatial analysis (cf 3.1.2.1.). Although the majority of surface-hearths are dated by one sample, some features were selected for further dating. This was the case for 4 surface-hearths which were excavated between 1992 and 1994 (cf 3.2.1.1.). Another two surface-hearths – one within artefact unit 17(4 dates), another in unit 22 (3 dates) - were dated by several samples in order to verify the chronological integrity of these features and associated lithic industry. Both surface-hearths were selected because they differ from the other surface-hearths in terms of size and density. Indeed, according to detailed spatial analyses both of them consist of a much wider and denser accumulation of overheated artefacts and burnt ecofacts. In order to check whether this is the result of a longer or repeated use of these fire-places, several samples were dated.

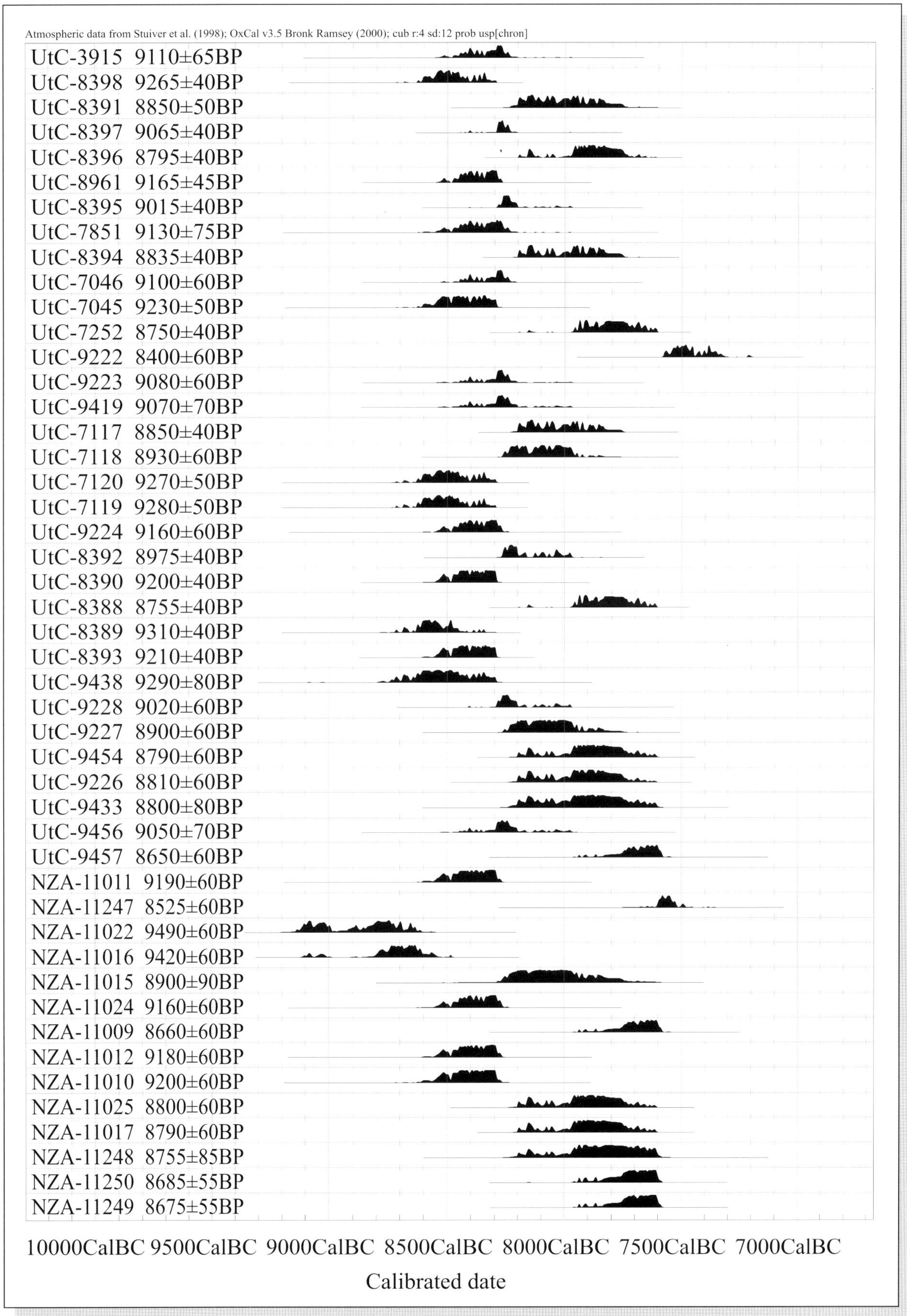

Figure 125: Verrebroek "Dok 1"; calibrated dates of 48 hazelnut samples from surface-hearths.

arte-fact unit	sub unit	feature reference	sample location	lab. reference	BP date	$\delta^{13}C$ ‰	dating material	68.2% probability	95.2% probability
1		SH	Z45/W17(1)	UtC-3915	9110±65	-23.0	hazelnut	8430BC (22.3%) 8350BC 8340BC (43.6%) 8260BC 8250BC (2.3%) 8240BC	8530BC (3.6%) 8490BC 8480BC (91.8%) 8210BC
2.1		SH	Z37/W5(4)	UtC-8398	9265±40	-23.2	hazelnut	8600BC (3.7%) 8570BC 8560BC (54.1%) 8430BC 8420BC (0.8%) 8410BC 8390BC (2.4%) 8380BC 8370BC (2.7%) 8360BC 8350BC (4.6%) 8330BC	8630BC (95.4%) 8290BC
2.2		SH	Z38/W2(20)	UtC-8391	8850±50	-23.7	hazelnut	8200BC (23.3%) 8100BC 8090BC (11.3%) 8040BC 8020BC (22.5%) 7910BC 7900BC (5.9%) 7870BC 7860BC (5.2%) 7830BC	8210BC (95.4%) 7760BC
4		SH	Z32/W11(4)	UtC-8397	9065±40	-21.6	hazelnut	8288BC (56.8%) 8259BC 8250BC (11.4%) 8242BC	8430BC (3.6%) 8390BC 8360BC (1.2%) 8350BC 8320BC (90.7%) 8200BC
5.2		SH	Z25/W8(3)	UtC-2743	9000±190	-23.7	hazelnut	8450BC (68.2%) 7800BC	8700BC (95.4%) 7600BC
5.2		SH	Z25/W7(4)	UtC-8396	8795±40	-23.8	hazelnut	8160BC (1.2%) 8140BC 7970BC (67.0%) 7760BC	8200BC (95.4%) 7650BC
6		SH	Z21/W1(3)	UtC-8961	9165±45	-24.2	hazelnut	8450BC (68.2%) 8280BC	8530BC (3.8%) 8500BC 8480BC (91.6%) 8270BC
7		SH	Z20/W10(4)	UtC-8395	9015±40	-26.4	hazelnut	8275BC (68.2%) 8210BC	8300BC (86.4%) 8160BC 8130BC (1.2%) 8120BC 8110BC (1.7%) 8080BC 8050BC (4.1%) 8010BC 7990BC (1.9%) 7970BC
9		SH	Z12/W2(4)	UtC-7851	9130±75	-26.2	hazelnut	8450BC (68.2%) 8260BC	8560BC (95.4%) 8210BC
10		SH	Z10/W10(3)	UtC-8394	8835±40	-27.1	hazelnut	8200BC (2.6%) 8180BC 8170BC (15.1%) 8100BC 8090BC (8.1%) 8040BC 8010BC (42.4%) 7820BC	8210BC (24.1%) 8100BC 8090BC (71.3%) 7750BC
14		SH1	081/Z45(1A)	UtC-7046	9100±60	-25.6	hazelnut	8430BC (9.4%) 8390BC 8380BC (3.6%) 8370BC 8360BC (3.0%) 8350BC 8340BC (52.2%) 8240BC	8530BC (1.3%) 8500BC 8480BC (94.1%) 8210BC
14		SH2	084/Z43(2C)	UtC-7045	9230±50	-24.3	hazelnut	8540BC (16.6%) 8490BC 8480BC (27.5%) 8410BC 8400BC (24.1%) 8320BC	8610BC (3.1%) 8570BC 8560BC (92.3%) 8290BC
14	72	SH	088/Z41(1A)	UtC-7252	8750±40	-23.9	hazelnut	7950BC (12.4%) 7890BC 7880BC (50.3%) 7710BC 7700BC (5.5%) 7680BC	7970BC (95.4%) 7600BC
14	68	SH	068/Z46(2B)	UtC-9222	8400±60	-27.4	hazelnut	7580BC (2.8%) 7560BC 7550BC (51.5%) 7450BC 7410BC (13.9%) 7350BC	7580BC (95.4%) 7320BC
14	70	SH	072/Z48(2B)	UtC-9223	9080±60	-23.9	hazelnut	8420BC (6.3%) 8390BC 8380BC (2.2%) 8370BC 8360BC (2.0%) 8350BC 8330BC (57.7%) 8230BC	8550BC (94.3%) 8200BC 8050BC (1.1%) 7950BC
14	71	SH	076/Z44(1A)	UtC-9419	9070±70	-24.6	hazelnut	8420BC (5.3%) 8390BC 8380BC (2.1%) 8370BC 8360BC (1.8%) 8350BC 8340BC (59.1%) 8200BC	8550BC (95.4%) 7950BC

arte-fact unit	sub unit	feature reference	sample location	lab. reference	BP date	$\delta^{13}C$ ‰	dating material	68.2% probability	95.2% probability
16		SH	088/Z25(3B)	UtC-7117	8850±40	-23.4	hazelnut	8200BC (5.2%) 8180BC 8170BC (18.7%) 8100BC 8090BC (11.7%) 8040BC 8020BC (20.1%) 7940BC 7930BC (3.3%) 7910BC 7900BC (5.7%) 7870BC 7860BC (3.6%) 7840BC	8210BC (95.4%) 7760BC
17		SH	087/Z17(1A)	UtC-7118	8930±60	-25.4	hazelnut	8240BC (22.9%) 8160BC 8140BC (45.3%) 7970BC	8270BC (94.3%) 7940BC 7860BC (1.1%) 7840BC
17		SH	085/Z17(1B)	UtC-7120	9270±50	-25.2	hazelnut	8610BC (11.6%) 8570BC 8560BC (48.2%) 8440BC 8390BC (2.2%) 8380BC 8370BC (2.4%) 8360BC 8350BC (3.9%) 8330BC	8630BC (95.4%) 8290BC
17		SH	089/Z17(4B)	UtC-7119	9280±50	-23.3	hazelnut	8610BC (63.8%) 8440BC 8370BC (1.5%) 8360BC 8350BC (2.9%) 8340BC	8690BC (2.1%) 8660BC 840BC (93.3%) 8290BC
18		SH	090/Z67(4A)	UtC-9224	9160±60	-27.1	hazelnut	8450BC (68.2%) 8280BC	8540BC (8.7%) 8490BC 8480BC (86.7%) 8260BC
19		SH	092/Z76(4A)	UtC-8392	8975±40	-22.8	hazelnut	8270BC (48.3%) 8190BC 8180BC (3.0%) 8160BC 8130BC (1.1%) 8120BC 8100BC (4.8%) 8080BC 8050BC (9.5%) 8020BC 7980BC (1.5%) 7970BC	8270BC (58.4%) 8160BC 8140BC (37.0%) 7960BC
21		SH	061/Z50(4B)	UtC-8390	9200±40	-23.5	hazelnut	8460BC (68.2%) 8290BC	8540BC (11.1%) 8490BC 8480BC (84.3%) 8280BC
22	67	SH2	075/Z60(4B)	UtC-8388	8755±40	-26.0	hazelnut	7950BC (64.0%) 7730BC 7700BC (4.2%) 7680BC	8000BC (95.4%) 7600BC
22		SH1	069/Z64(4B)	UtC-8389	9310±40	-23.1	hazelnut	8630BC (54.6%) 8520BC 8510BC (13.6%) 8470BC	8730BC (7.3%) 8660BC 8650BC (84.3%) 8430BC 8400BC (1.0%) 8380BC 8370BC (2.8%) 8330BC
22		SH1	071/Z64(3B)	UtC-8393	9210±40	-25.2	hazelnut	8520BC (3.3%) 8510BC 8460BC (64.9%) 8290BC	8540BC (14.3%) 8490BC 8480BC (81.1%) 8290BC
22		SH1		UtC-9438	9290±80	-26.0	hazelnut	8690BC (1.4%) 8670BC 8630BC (59.2%) 8410BC 8400BC (2.3%) 8380BC 8370BC (5.3%) 8330BC	8740BC (95.4%) 8290BC
23		SH	067/Z37(4A)	UtC-9228	9020±60	-26.0	hazelnut	8290BC (62.0%) 8200BC 8100BC (1.5%) 8090BC 8050BC (4.8%) 8020BC	8420BC (1.1%) 8390BC 8300BC (68.1%) 8150BC 8140BC (26.3%) 7960BC
26		SH	059/Z28(2A)	UtC-9227	8900±60	-23.4	hazelnut	8210BC (14.5%) 8150BC 8140BC (53.7%) 7960BC	8270BC (95.4%) 7820BC
27		SH	060/Z16(3B)	UtC-9454	8790±60	-24.4	hazelnut	8160BC (5.1%) 8130BC 7970BC (63.1%) 7740BC	8200BC (95.4%) 7600BC
28	99	SH	063/Z11(2A)	UtC-9226	8810±60	-26.0	hazelnut	8170BC (7.3%) 8130BC 8080BC (1.6%) 8070BC 8060BC (1.9%) 8040BC 7970BC (57.4%) 7750BC	8250BC (95.4%) 7650BC
28		SH	073/Z12(1A)	UtC-9433	8800±80	-25.8	hazelnut	8200BC (68.2%) 7650BC	8250BC (95.4%) 7600BC

arte-fact unit	sub unit	feature reference	sample location	lab. reference	BP date	$\delta^{13}C$ ‰	dating material	68.2% probability	95.2% probability
33		SH	069/Z20(3A)	UtC-9456	9050±70	-26.1	hazelnut	8420BC (3.5%) 8390BC 8380BC (1.1%) 8370BC 8360BC (1.0%) 8350BC 8330BC (59.7%) 8200BC 8040BC (2.9%) 8020BC	8460BC (79.2%) 8160BC 8140BC (16.2%) 7960BC
35		SH	074/Z31(2B)	UtC-9457	8650±60	-28.4	hazelnut	7740BC (68.2%) 7590BC	7920BC (1.7%) 7890BC 7880BC (1.5%) 7850BC 7840BC (92.2%) 7570BC
36		SH1	A11(1C)	NZA-11011	9190±60	-24.7	hazelnut	8460BC (68.2%) 8290BC	8560BC (95.4%) 8270BC
36		SH2	A15(2B)	NZA-11247	8525±60	-23.2	hazelnut	7600BC (68.2%) 7535BC	7680BC (95.4%) 7470BC
37		SH	E19(1B)	NZA-11022	9490±60	-27.3	hazelnut	9120BC (26.7%) 8990BC 8900BC (1.8%) 8880BC 8840BC (32.0%) 8720BC 8710BC (5.0%) 8690BC 8670BC (2.7%) 8640BC	9150BC (95.4%) 8600BC
38		SH	I16(4A)	NZA-11016	9420±60	-25.7	hazelnut	8790BC (68.2%) 8600BC	9150BC (10.3%) 8950BC 8900BC (85.1%) 8450BC
39		SH	052/Z56(4B)	NZA-11015	8900±90	-22.5	hazelnut	8240BC (66.9%) 7960BC 7950BC (1.3%) 7940BC	8300BC (95.4%) 7750BC
40	97B	SH	047/Z57(3A)	NZA-11024	9160±60	-23.7	hazelnut	8450BC (68.2%) 8280BC	8540BC (8.7%) 8490BC 8480BC (86.7%) 8260BC
40		SH	041/Z53(4)	NZA-11009	8660±60	-26.7	hazelnut	7750BC (68.2%) 7590BC	7940BC (1.0%) 7930BC 7920BC (2.1%) 7890BC 7880BC (1.8%) 7850BC 7840BC (90.5%) 7580BC
41		SH	046/Z44(4A)	NZA-11012	9180±60	-25.2	hazelnut	8460BC (68.2%) 8280BC	8550BC (95.4%) 8260BC
44		SH1	045/Z35(4C)	NZA-11010	9200±60	-23.0	hazelnut	8520BC (2.9%) 8510BC 8480BC (65.3%) 8290BC	8560BC (95.4%) 8270BC
44		SH2	047/Z36(2A)	NZA-11025	8800±60	-26.4	hazelnut	8160BC (6.2%) 8130BC 8080BC (0.7%) 8070BC 8060BC (1.0%) 8050BC 7970BC (60.3%) 7740BC	8250BC (95.4%) 7650BC
44	92	SH	051/Z36(2A)	NZA-11017	8790±60	-26.5	hazelnut	8160BC (5.1%) 8130BC 7970BC (63.1%) 7740BC	8200BC (95.4%) 7600BC
44	91	SH4	045/Z31(1A)	NZA-11248	8755±85	-24.4	hazelnut	7970BC (67.1%) 7650BC 7620BC (1.1%) 7610BC	8200BC (95.4%) 7550BC
45		SH	052/Z26(1B)	NZA-11250	8685±55	-25.7	hazelnut	7760BC (68.2%) 7600BC	7940BC (4.6%) 7890BC 7880BC (90.8%) 7580BC
49		SH	042/Z15(4A)	NZA-11249	8675±55	-23.7	hazelnut	7750BC (68.2%) 7600BC 7920BC (2.4%) 7890BC 7880BC (2.0%) 7850BC 7840BC (89.9%) 7580BC	7940BC (1.1%) 7930BC

Table 42: Verrebroek "Dok 1"; calibrated radiocarbon dates of 48 hazelnut samples from surface-hearths (SH: surface-hearths).

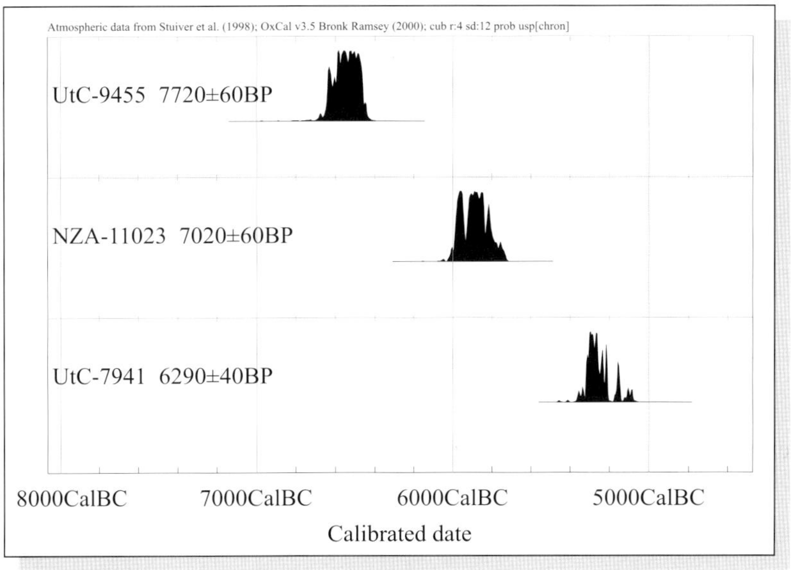

Figure 126: Verrebroek "Dok 1"; calibrated dates of 3 outliers on hazelnut shells from surface-hearths.

arte fact unit	feature reference	sample location	lab. reference	BP date	$\delta^{13}C$ ‰	dating material	68.2% probability	95.2% probability
34	SH	073/Z23(4B)	UtC-9455	7720±60	-24.2	hazelnut	6600BC (68.2%) 6460BC	6650BC (95.4%) 6440BC
46	SH	051/Z22(2A)	NZA-11023	7020±60	-24.9	hazelnut	5990BC (18.9%) 5940BC 5930BC (46.0%) 5830BC 5820BC (3.3%) 5810BC	6010BC (95.4%) 5740BC
17	SH	088/Z20(1A)	UtC-7941	6290±40	-25.5	hazelnut	5320BC (65.5%) 5210BC 5160BC (2.7%) 5150BC	5370BC (81.7%) 5200BC 5180BC (9.2%) 5140BC 5120BC (4.6%) 5080BC

Table 43: Verrebroek "Dok 1"; calibrated outliers of three hazelnut samples from surface-hearths.

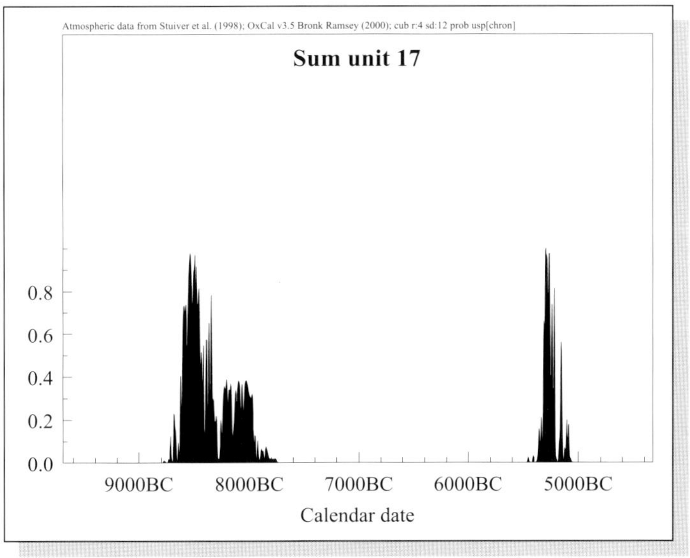

Figure 127: Verrebroek "Dok 1"; sumprobability of four dates from a surface-hearth within unit 17.

The obtained results from the surface-hearth within unit 17 confirm the hypothesis of multiple or repeated use of this area as burning place. The sumprobability from the dated samples (UtC-7118: 8930±60BP; UtC-7120: 9270±50BP; UtC-7119: 9280±50BP; UtC-7941: 6290±40BP) is depicted in figure 127. The graph shows three related dates (interquartile range: 8530-8190 BC; 95% probability range: 8640-7960 BC) and one outlier. We can define UtC-7941 as an outlier because the result differs by more than 1.5 IQR from the middle 50% of the results. Although the dates situated in the 9th millennium form a coherent series, they cannot be synchronous as is shown by a χ^2 test. A χ-Test fails at 5% (χ^2-Test: df=2 T=24.043(5% 6.0)

The three dates from artefact unit 22 (UtC-8389: 9310±40BP; UtC-8393: 9210±40BP; UtC-9438: 9290±80BP), on the other hand, clearly overlap and form a consistent group. Hence it is possible that the three samples date the same event (χ^2-Test: df=2 T=3.2(5% 6.0). If so the real date must be situated between 8610 BC (80.8%) and 8410 BC or 8400 BC (14.6%) and 8330 BC at 95.4% probability. The dates, however, could also represent three independent events. If so, they span a period of 350 years at a 95.4% confidence level, with an average around 200 years.

The 48 hazelnut dates were grouped in a density diagram and the interquartile range and the 95 % probability range was calculated. The use of an interquartile range involves the assumption that all samples are related or belong to the same cultural phase. The interquartile range (8405-7890 BC) and the 95 % probability range (8740-7560 BC) of the bulk and the 95% probability range of each outlier are represented in figure 140. Although there seems to be no relationship between the bulk and the outliers, these results must be considered valid, as was proven by cross dating on one sample. The difference between the two is within statistical expectation (cf 3.2.1.2.).

It thus appears that the majority of surface-hearths, and thus of associated artefact units, were in use during the latest quarter of the Pre-boreal and the first half of the Boreal according to the biozonation of C. Verbruggen *et alii* (1996)[1]. This is in agreement with the relative dating, based on the tool-typology. Indeed, the microlith composition found in spatial association with these surface-hearths perfectly fits with the expected Early Mesolithic range. The dominant types of microliths at Verrebroek "Dok 1" are points with un-retouched base (including mainly obliquely truncated points and unilaterally backed points), points with retouched base, (scalene) triangles and crescents. These microlith types occur in variable frequencies within the different artefact units dated to the Pre-boreal and Boreal[2].

Although associated with Early Mesolithic assemblages, the three outlier dates on hazelnut samples must be regarded as reliable, as proven by the cross dating on a sample from unit 17. These late dates, however, are difficult to explain since none of the dated surface-hearths (artefact units 17, 34 and 46) yielded typical Late Mesolithic artefacts, e.g. trapezes or Montbani blades. The only possible exception is a triangle with flat retouch made of Wommersom quartzite which was retrieved from artefact unit 17. Although this type of microlith is typical of the Middle Mesolithic, the occurrence of it in younger, Late Mesolithic contexts is not exceptional in the coversand area of Belgium and The Netherlands (Crombé, Cauwe 2001; Arts 1989). However, in the absence of other Late Mesolithic finds within these younger-dated artefact units, the three outlier dates should probably be interpreted as reflecting rather ephemeral activities on the site by Late Mesolithic hunter-gatherers[3].

1 C. Verbruggen *et alii* date the Pre-boreal within Sandy Flanders between 10300 BP and 9250 BP, the Boreal between 9250 BP and 7500 BP.

2 A detailed typo-chronological analysis of the assemblages from the many artefact units will be published in the second volume of the "Verrebroek and Doel Excavation Projects".

3 The presence of Late Mesolithic man on the site of Verrebroek "Dok 1" is confirmed by the discovery of a handful of typical trapezes on other locations within the excavated area, especially in the southernmost sector. Unfortunately detailed information on the spatial distribution of these Late Mesolithic trapezes won't be available before the typological analysis of the entire lithic inventory of the site is finished.

3.2.3.2. | Hearth-pits

The radiocarbon dates performed on hearth-pit samples form a coherent group of 14 dates (tab. 44, fig. 128). This series also contains two outliers: one sample (UtC-9453) is much older and one is much younger (NZA-11941) than the bulk. The difference is, however, less conspicuous than in the case of the surface-hearths. The interquartile range (7510-7225 BC) and the 95% probability range (7600-6990 BC) are calculated without the two outlier results and represented in figure 140 (cf Van Strydonck *et alii* 2001). Clearly most hearth-pits were used within the second half of the Boreal as defined in the biozonation of C. Verbruggen.

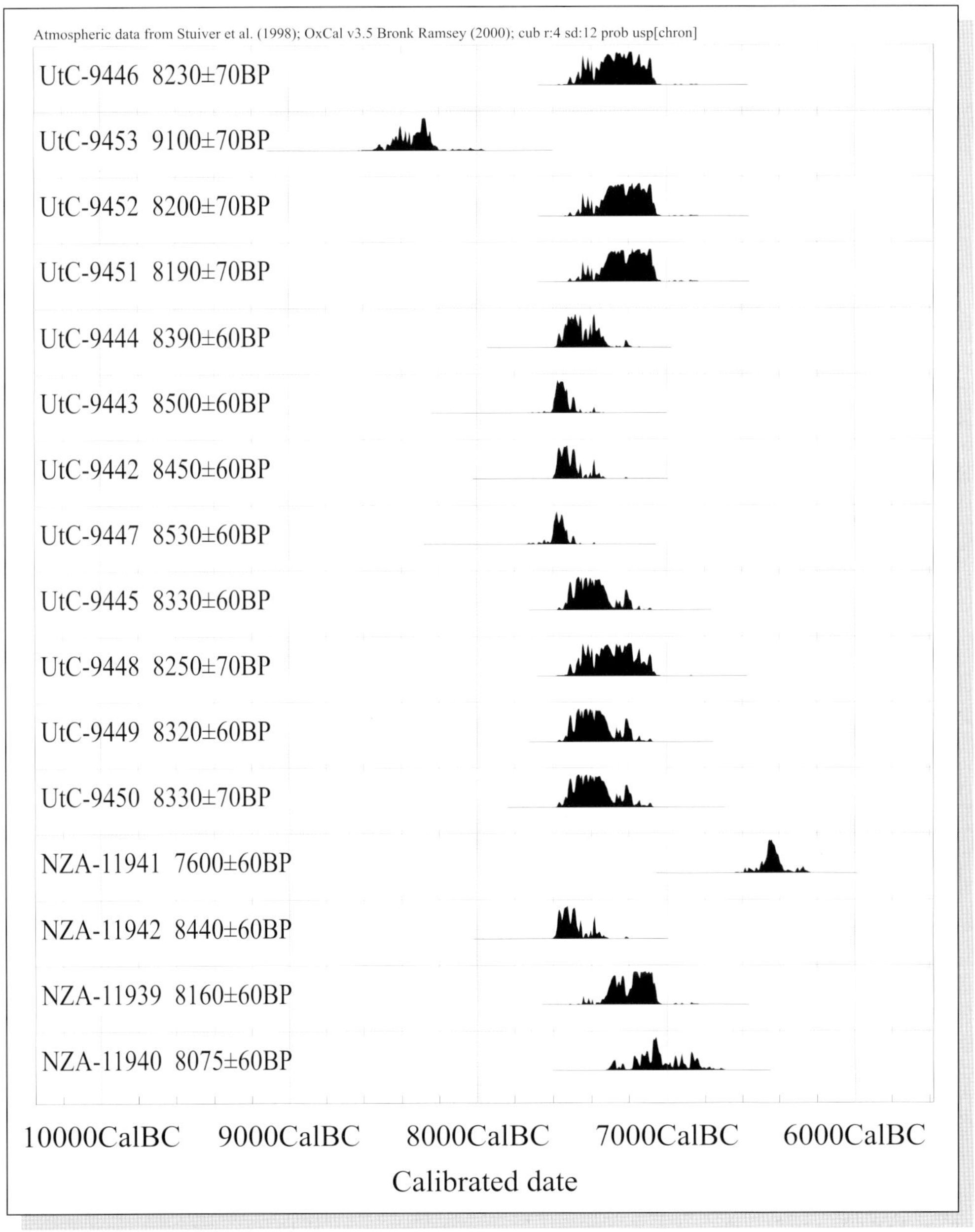

Figure 128: Verrebroek "Dok 1"; calibrated dates of 16 charcoal samples from hearth-pits.

arte-fact unit	sub unit	feature reference	lab. reference	BP date	$\delta^{13}C$ ‰	dating material	68.2% probability	95.2% probability
14	68	S88	UtC-9446	8230±70	-25.5	charcoal	7450BC (3.0%) 7430BC 7420BC (2.4%) 7410BC 7350BC (58.9%) 7130BC 7100BC (3.9%) 7080BC	7480BC (95.4%) 7070BC
20		S66	UtC-9453	9100±70	-27.8	hazelnut	8430BC (19.2%) 8350BC 8340BC (49.0%) 8230BC	8550BC (95.4%) 8200BC
20		S66	UtC-9452	8200±70	-25.3	charcoal	7320BC (68.2%) 7080BC	7460BC (8.1%) 7390BC 7380BC (87.3%) 7060BC
20		S66	UtC-9451	8190±70	-24.5	charcoal	7310BC (29.3%) 7220BC 7200BC (38.9%) 7080BC	7460BC (6.7%) 7390BC 7380BC (88.7%) 7050BC
20		S67	UtC-9444	8390±60	-24.8	charcoal	7550BC (51.2%) 7420BC 7410BC (17.0%) 7350BC	7580BC (94.0%) 7320BC 7220BC (1.4%) 7200BC
20		S68	UtC-9443	8500±60	-25.6	charcoal	7590BC (68.2%) 7520BC	7610BC (93.5%) 7450BC 7410BC (1.9%) 7370BC
20		S70	UtC-9442	8450±60	-25.6	charcoal	7580BC (67.0%) 7470BC 7390BC (1.2%) 7380BC	7600BC (85.2%) 7420BC 7410BC (10.2%) 7350BC
20		S73	UtC-9447	8530±60	-26.7	charcoal	7600BC (68.2%) 7535BC	7680BC (95.4%) 7470BC
20		S92	UtC-9445	8330±60	-22.6	charcoal	7520BC (6.0%) 7500BC 7490BC (62.2%) 7320BC	7540BC (83.2%) 7290BC 7280BC (12.2%) 7180BC
21		S91	UtC-9448	8250±70	-24.5	charcoal	7450BC (12.8%) 7390BC 7380BC (51.5%) 7170BC 7160BC (3.0%) 7140BC 7100BC (0.8%) 7080BC	7520BC (1.0%) 7500BC 7480BC (94.4%) 7080BC
21		S91	UtC-9449	8320±60	-25.3	charcoal	7520BC (4.4%) 7500BC 7480BC (61.4%) 7320BC 7220BC (2.4%) 7200BC	7540BC (95.4%) 7180BC
21		S91	UtC-9450	8330±70	-25.1	charcoal	7520BC (65.8%) 7320BC 7220BC (2.4%) 7200BC	7550BC (95.4%) 7170BC
		S100	NZA-11941	7600±60		charcoal	6500BC (2.2%) 6490BC 6480BC (66.0%) 6390BC	6600BC (88.9%) 6340BC 6320BC (6.5%) 6250BC
		S101	NZA-11942	8440±60		charcoal	7580BC (65.7%) 7470BC 7390BC (2.5%) 7380BC	7590BC (95.4%) 7330BC
		S102	NZA-11939	8160±60		charcoal	7300BC (20.0%) 7220BC 7190BC (48.2%) 7070BC	7350BC (95.4%) 7050BC
		S103	NZA-11940	8075±60		charcoal	7290BC (2.4%) 7270BC 7180BC (6.1%) 7150BC 7140BC (40.4%) 7020BC 6970BC (3.8%) 6950BC 6940BC (3.6%) 6910BC 6880BC (11.9%) 6830BC	7350BC (95.4%) 6750BC

Table 44: Verrebroek "Dok 1"; calibrated radiocarbon dates of 16 charcoal samples from hearth-pits.

Regarding the outliers, it has to be noted that the oldest of these is not a charcoal but a charred hazelnut shell and its date coincides very well with the bulk of surface-hearth dates (cf 3.2.3.1.). The differences in age between this charred hazelnut shell sample and associated charcoal samples from the same context, i.e. hearth-pit S66, are represented in table 45 and figures 129 and 130. It is clear from the data that, although the two charcoal samples are identical and even perhaps from the same tree, the difference from the charred hazelnut shell is in the order of one millennium (68.2% probability: 1240-970; 95.4% probability: 1400-800). Similar chronological deviations have been observed with respect to some charcoal-pits (cf 3.2.3.3.). The most likely explanation for this anomaly is that the hazelnut sample within hearth-pit S66 probably consists of residual material from an older occupation phase, possibly represented by the artefact cluster 20. It seems not unlikely that burnt hazelnut shells - probably from an adjacent older surface-hearth[4] - slipped into the hearth-pit at the moment of its digging or filling (Crombé *et alii* 1999a). This probability is further strengthened by the observation that hearth-pits were apparently not used for the roasting of hazelnuts, since, compared to the surface-hearths, only very small amounts of shells are usually found within these pits (cf 3.1.2.1.). If we accept the above reasoning, this implies that the use and/or filling of hearth-pits can only be accurately dated by means of charcoal samples and not by carbonized hazelnut shells.

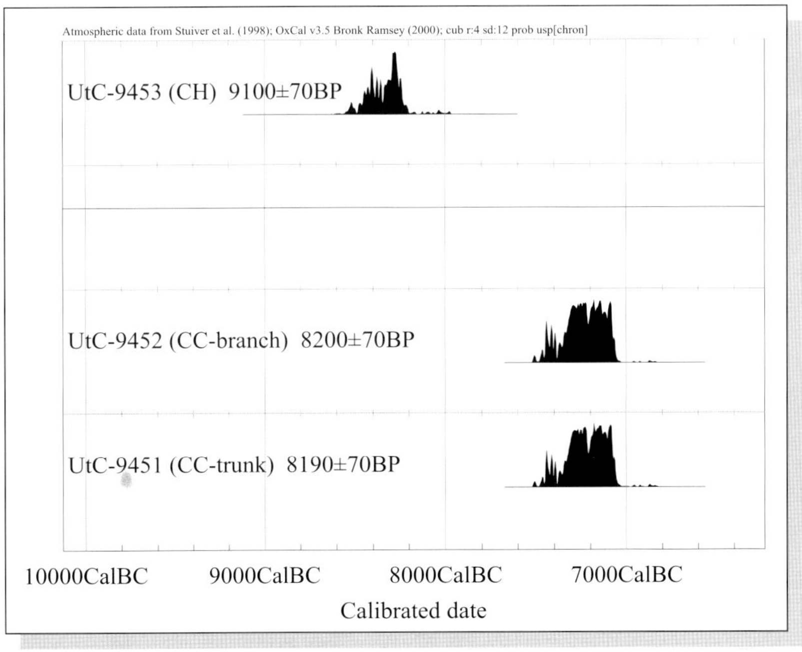

Figure 129: Verrebroek "Dok 1"; calibrated dates of three samples from hearth-pit S66 within artefact unit 20 (CH: carbonised hazelnut shell; CC: charcoal)

4 This presumed surface-hearth is difficult to locate due to the advanced perturbation (soil truncation) of artefact unit 20.

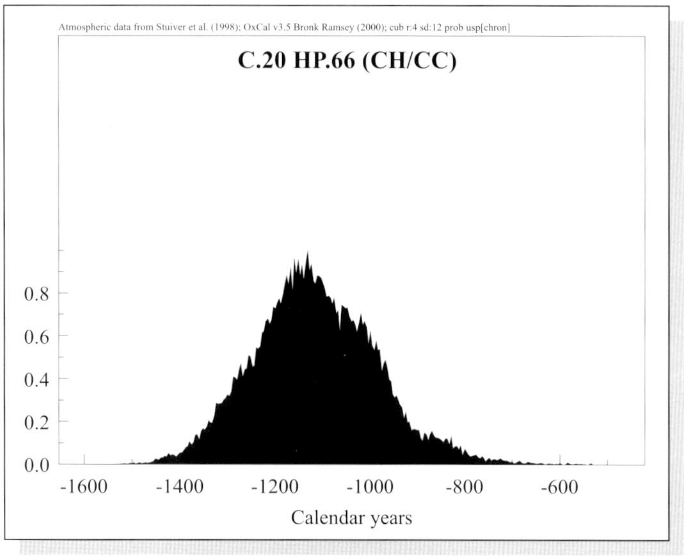

Figure 130: Verrebroek "Dok 1"; difference in ^{14}C age between the average date of two charcoal samples and a charred hazelnut sample from hearth-pit S66 within artefact unit 20.

lab. reference	dating material	BP date	$\delta^{13}C$ ‰	68.2% confidence	95.4% confidence
UtC-9453	charred hazelnut	9100±70	-27.8	8430BC (29.3%) 7220BC 8340BC (49%) 8230BC	7460BC (8.1%) 7390BC 7380BC (87.3%) 7060BC
UtC-9452	charcoal *Pinus* branchwood	8200±70	-25.3	7320BC (68.2%) 7080BC	7460BC (8.1%) 7390BC 7380BC (87.3%) 7060BC
UtC-9451	charcoal *Pinus* trunkwood	8190±70	-24.5	7310BC (29.3%) 7220BC 7200BC (38.9) 7080BC	7460BC (6.7%) 7390BC 7380BC (88.7%) 7050BC
Difference between hazelnut date and average of both charcoal dates					
				-1260 (1.2%) -1250 -1240 (67.0%) -980	-1400 (95.4%) -800

Table 45: Verrebroek "Dok 1"; calibrated radiocarbon dates from hearth-pit S66 within artefact unit 20.

With respect to the younger outlier, it must be noticed that this sample originates from a hearth-pit (S100) which deviates in size from all other hearth-pits at Verrebroek "Dok 1" (cf 3.1.2.1.). S100 is clearly the largest within the cluster of 20 features. In this respect it corresponds better with the hearth-pits excavated at the nearby site of Doel "Deurganckdok-sector B". Coincidentally or not, the ^{14}C-date of feature S100 also fits very well with those of Doel, which cluster between 7865±35 BP and 7580±35 BP (cf 3.2.4.2.).

3.2.3.3. | Charcoal-pits

Only three charcoal-pits were selected for radiocarbon dating. Altogether five samples, some on carbonized hazelnut shells, others on charcoal, were dated. Two pits were double dated on both charcoal and hazelnut.

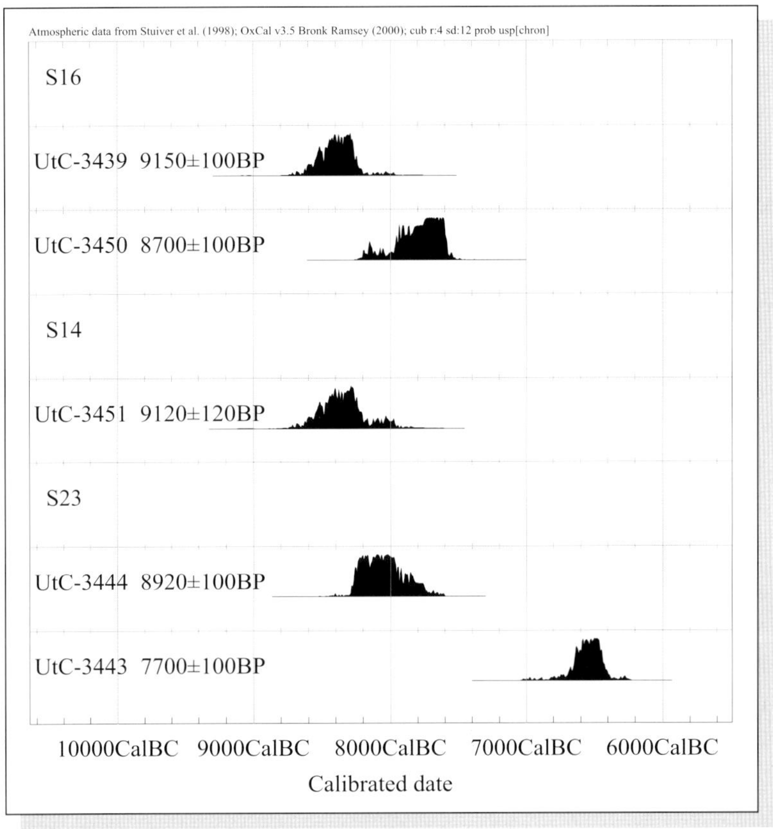

Figure 131: Verrebroek "Dok 1"; calibrated dates of 5 samples from charcoal-pits.

arte fact unit	feature reference	sample location	lab. reference	BP date	$\delta^{13}C$ ‰	dating material	68.2% probability	95.2% probability
6	S16	Z21/W2(4)	UtC-3439	9150±100	-26.9	hazelnut	8530BC (2.7%) 8510BC 8480BC (65.5%) 8260BC	8700BC (95.4%) 7950BC
6	S16	Z21/W2(4)	UtC-3450	8700±100	-26.4	charcoal	7940BC (1.7%) 7930BC 7920BC (3.8%) 7890BC 7880BC (2.6%) 7850BC 7840BC (60.1%) 7590BC	8200BC (95.4%) 7500BC
7	S14	Z21/W10(2)	UtC-3451	9120±120	-25.0	hazelnut	8540BC (7.2%) 8490BC 8480BC (61.0%) 8230BC	8700BC (95.4%) 7950BC
10	S23	Z9/W9	UtC-3444	8920±100	-24.3	hazelnut	8260BC (67.3%) 7960BC 7950BC (0.9%) 7940BC	8300BC (95.4%) 7700BC
10	S23	Z9/W9	UtC-3443	7700±100	-24.8	charcoal	6640BC (68.2%) 6440BC	6900BC (95.4%) 6250BC

Table 46: Verrebroek "Dok 1"; calibrated radiocarbon dates from charcoal-pits.

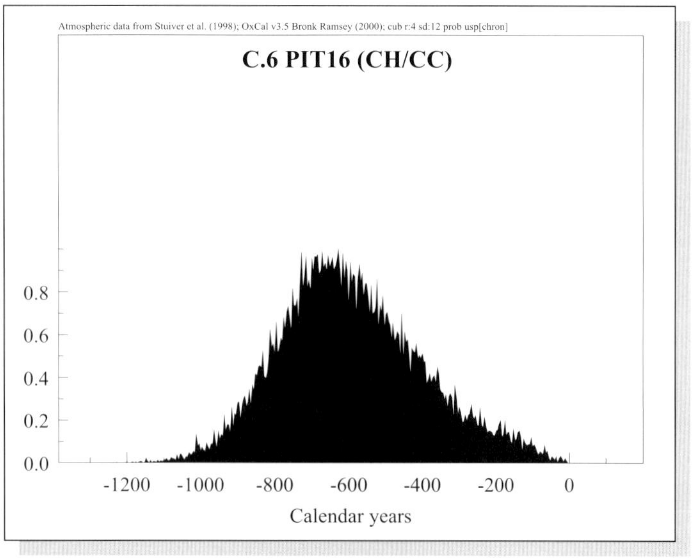

Figure 132: Verrebroek "Dok 1"; difference in ^{14}C age between a charcoal sample and a hazelnut shell sample found in charcoal-pit S16 within artefact unit 6.

The obtained results (tab. 46, fig. 131) once again form a coherent set (interquartile 8385-7920 BC; 95% probability 8590-7630 BC), with one outlier. The three hazelnut dates perfectly fit with the bulk of those from the surface-hearths (cf 3.2.3.1.). The oldest charcoal date also falls within the same time range.

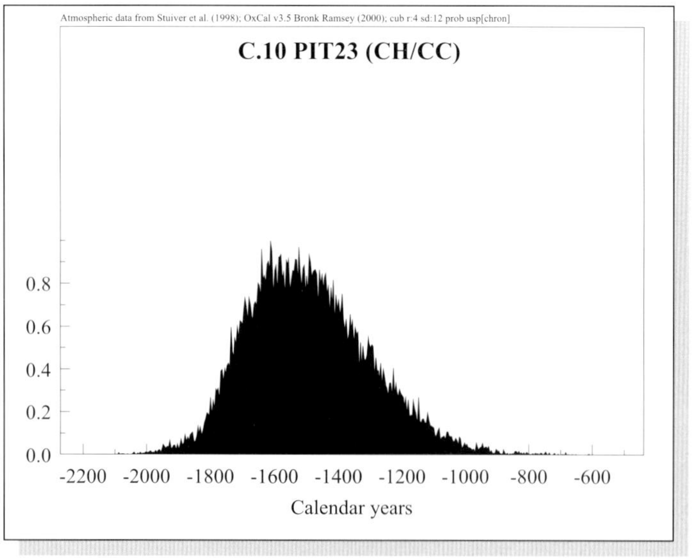

Figure 133: Verrebroek "Dok 1"; difference ^{14}C age between a charcoal sample and a hazelnut shell sample found in charcoal-pit S23 within artefact unit 10.

If we look at the double dated features, a pattern similar to that observed in one of the dated hearth-pits can be observed (cf 3.2.3.2.). Both features S16 and S23 were dated on a sample of carbonized hazelnut shell as well as on a charcoal sample. In both features these dates are not compatible at all, with the hazelnut dates always being the older ones. The difference in age within S16 at 68.2% probability is 840-430; the difference at 95.4% probability is 950-100 (fig. 132). In feature S23 the difference in age at 68.2% probability amounts to 1730-1320; the difference at 95.4% probability to 1850-1050 (fig. 133). As with the abovementioned hearth-pit, the most plausible explanation here is that the dated hazelnut shells are residual, coming from an adjacent older occupation surface. This hypothesis is corroborated by the fact that the hazelnut dates from both features correspond perfectly with those on hazelnut shells from surface-hearths lying in the immediate vicinity of the charcoal-pits (unit 6: UtC-8961=9165±45 BP; unit 10: UtC-2744=8920±130 BP).

In conclusion it is clear that the use and/or filling of these charcoal-pits are dated only by the charcoal samples and not by the hazelnut samples. This means that neither charcoal-dated feature forms a coherent cluster.

3.2.3.4. | Other features

arte fact unit	sub unit	feature reference	sample location	lab. reference	BP date	$\delta^{13}C$ ‰	dating material	68.2% probability	95.2% probability
14	70	tree-fall S90	072/Z46(1D)	UtC-9418	9060±70	-24.5	hazelnut	8420BC (4.4%) 8390BC 8380BC (1.7%) 8370BC 8360BC (1.5%) 8350BC 8340BC (60.6%) 8200BC	8550BC (95.4%) 7950BC
14	70	hazelnut-pit S89	075/Z46(3C)	UtC-9225	9270±60	-23.0	hazelnut	8610BC (57.3%) 8430BC 8420BC (1.7%) 8410BC 8400BC (2.7%) 8380BC 8370BC (6.6%) 8330BC	8690BC (1.7%) 8670BC 8640BC (93.7%) 8290BC

Table 47: Verrebroek "Dok 1"; calibrated radiocarbon dates from other features.

Another two pit-like features – a tree-fall feature S90 and a hazelnut-pit S89 – were absolutely dated, both on a sample of carbonized hazelnut shells (tab. 47). Both dates fit well with the bulk of surface-hearth dates (cf 3.2.3.1.). In the case of the tree-fall one may assume that, as with the hearth-pits and charcoal-pits, the nut shell is residual and thus does not date the uprooting of the tree. With respect to the hazelnut-pit, the obtained date falls perfectly within the range of both adjacent surface-hearths, one situated within artefact unit 70 (UtC-9223: 9080±60 BP), another one situated within artefact unit 71 (UtC-9419: 9070±70 BP).

3.2.3.5. | Discussion

Figure 140 represents the interquartile range and the 95% probability range of the dates from all analysed anthropogenic and (possible) natural features at Verrebroek "Dok 1". Outliers are not included, but have been represented separately. The calculation of the hearth-pit and charcoal-pit range is based only on the charcoal dates, since it is probable that the hazelnuts in these features are residual.

From figure 140 it can be deduced that there is no chronological relationship between the bulk of surface-hearth dates and the bulk of the hearth-pit dates. The former are clearly restricted in time to the latest quarter of the Pre-Boreal and the first half of the Boreal (Early Mesolithic), while the latter only occur in the second half of the Boreal and exceptionally towards the transition to the Atlantic, corresponding to the Middle Mesolithic in Belgium. Clearly the hearth-pits belong to a younger occupation phase of the site. Strangely, only a little material evidence of a Middle Mesolithic occupation has been identified so far within the lithic inventory of the site. Apart from a few dispersed microliths with flat retouch[5], only one lithic concentration might date to this

5 Some of these might also be of a younger (Late Mesolithic) date.

period. This is a partially excavated unit (unit 98) situated at the northern limit of trench 13. It yielded a series of small backed bladelets and points with flat retouch, part of which was made in Wommersom quartzite, typical of the Middle Mesolithic in Belgium (Crombé, Cauwe 2001)[6]. Remarkable, however, is the considerable distance (ca. 50m) between this unit and the cluster of hearth-pits, which is situated at the opposite end of trench 13. Furthermore, within the filling of the hearth-pits not one single artefact that can be related to the Middle Mesolithic has been found. In fact, the recovered artefacts do not differ in any respect from the lithics found in association with the surface-hearths belonging to the Early Mesolithic. Hence it must be questioned whether the artefacts from the hearth-pits are not, as with the carbonized hazelnut shells, residual.

The interquartile range of the Verrebroek "Dok 1" surface-hearths is also synchronous with other hazelnut dates obtained on surface-hearths at other Early Mesolithic sites in Sandy Flanders (tab. 48), e.g. at Verrebroek "Aven Ackers" (Dauchot-Dehon *et alii* 1986), Doel "Deurganckdok-sector J/L" (Bats *et alii* 2003) and Oostwinkel "Mostmolen" (Crombé 1998).

site	lab. reference	BP date	68.2% confidence	95.4% confidence
Verrebroek "Aven Ackers"	IRPA-633	8890±100	8240BC (64.6%) 7940BC 7930BC (1.6%) 7910BC 7900BC (1.0%) 7880BC 7860BC (1.1%) 7840BC	8300BC (95.4%) 7650BC
Doel "Deurganckdok-sector J/L"	KIA-20470	8830±45	8200 (0.21) 8110 BC 8090 (0.09) 8040 BC	8210 (0.33) 8040 BC 8020 (0.67) 7750 BC 8000 (0.69) 7800 BC
Oostwinkel "Mostmolen"	UtC-3438	9250±160	8436BC (1.00%) 8076BC	8919BC (1%) 8876BC 8856BC (3%) 8794BC 8703BC (0%) 8695BC 8668BC (96%) 7965BC

Table 48: calibrated radiocarbon dates on hazelnut shells from other Early Mesolithic sites in Sandy Flanders.

3.2.4. | Doel "Deurganckdok-sector B"

The radiocarbon analyses on the Doel-samples were performed a few years later than the ones on the Verrebroek-samples. In the mean time the precision of AMS dates had considerably improved. This explains the smaller standard deviation on the Doel-samples.

6 In the absence of carbonised hazelnut shells this unit could not be dated absolutely.

3.2.4.1. | Surface-hearths

A series of seven samples from presumed surface-hearths and/or hearth-dumps was measured at Doel "sector B". Three dates were obtained on charcoal, another three on carbonised seeds and pips and one on a charred hazelnut shell. Four samples originate from the north-western sector of the site, which was water-screened. Two samples were retrieved from the disturbed north-eastern area, while the hazelnut sample was taken from the south-eastern zone. Altogether at least five different presumed surface-hearths/dumps have been dated, one of which (SH2) by means of more than one sample.

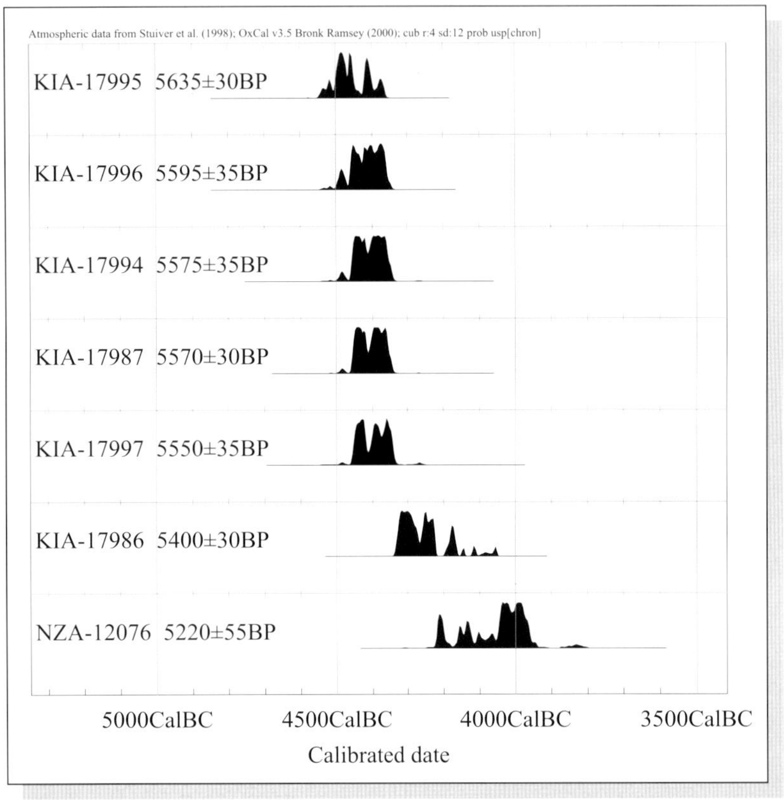

Figure 134: Doel "Deurganckdok-sector B"; calibrated radiocarbon dates from surface-hearths.

sector	feature reference	sample location	lab. reference	BP date	$\delta^{13}C$ ‰	dating material	68.2% probability	95.2% probability
NW	SH1	W16/Z7(1)	KIA-17995	5635±30	-25.3	*Sorbus* charcoal	4500BC (48.1%) 4440BC 4420BC (17.2%) 4400BC 4380BC (2.9%) 4370BC	4540BC (95.4%) 4360BC
NW	SH2	W19/Z1(4)h	KIA-17996	5595±35	-26.3	*Quercus* charcoal	4460BC (14.8%) 4435BC 4425BC (53.4%) 4360BC	4500BC (95.4%) 4350BC
NW	SH2	W20/Z2(1)	KIA-17994	5575±35	-28.4	*Cornus* charcoal	4455BC (28.9%) 4415BC 4405BC (39.3%) 4355BC	4490BC (1.0%) 4480BC 4460BC (94.4%) 4340BC
NE	SH3	W6/Z6(3)	KIA-17987	5570±30	-24.3	carbonised seeds	4450BC (30.4%) 4415BC 4405BC (37.8%) 4355BC	4460BC (95.4%) 4340BC
NW	SH2	W19/Z1(4)	KIA-17997	5550±35	-27.1	carbonised *Hedera helix* seeds	4450BC (29.1%) 4420BC 4400BC (39.1%) 4340BC	4460BC (95.4%) 4330BC
NE	SH4	W6/Z18(4)	KIA-17986	5400±30	-25.8	carbonised *Prunus spinosa* pip	4330BC (42.0%) 4270BC 4260BC (24.4%) 4220BC 4180BC (1.8%) 4170BC	4340BC (80.6%) 4220BC 4200BC (11.9%) 4160BC 4120BC (1.5%) 4110BC 4070BC (1.4%) 4050BC
SE	SH5	R069	NZA-12076	5220±55	-26.4	charred hazelnut shell	4220BC (8.1%) 4200BC 4160BC (1.7%) 4150BC 4140BC (4.2%) 4120BC 4050BC (54.2%) 3960BC	4230BC (11.9%) 4180BC 4170BC (83.5%) 3940BC

Table 49: Doel "Deurganckdok-sector B"; calibrated radiocarbon dates from surface-hearths.

The series contains no real outliers (tab. 49, fig. 134). The interquartile range is situated between 4440 and 4310 BC, the 95% probability range between 4510 and 3990 BC. The dates from the three northernmost surface-hearths (SH1, SH2 and SH3) are very similar, whereas those from SH4 and SH5, situated more southwards, seem to be slightly younger.

Furthermore the three dates from surface-hearth SH2 are very consistent, suggesting that this feature is most likely the result of one occupation phase. This possibility is further strengthened by the fact that these dates were obtained on different dating materials, i.e. charcoal and seeds/pip. Of course the hypothesis of different visits within a short time range can never be excluded. Radiocarbon is not precise enough if one wants to obtain results with a time resolution of one or a few decades.

3.2.4.2. | Hearth-pits

A series of six charcoal samples and two hazelnut samples from hearth-pits at Doel sector B were measured. The selection of the charcoal samples was restricted to the hearth-pits in which the charcoal had been determined as to species (cf 3.4.3.1.). All samples consisted of *Quercus* charcoal. One of the hazelnut samples was selected from a hearth-pit which was also dated on charcoal (S22). The other one was retrieved from a hearth-pit situated in a sector where relatively many patinated artefacts had been recovered (cf 3.1.3.2.). Altogether seven features were sampled for dating. They included five typical hearth-pits with double filling and one typical hearth-pit with one single fill (S24). Feature S79 is the only atypical hearth-pit so far dated.

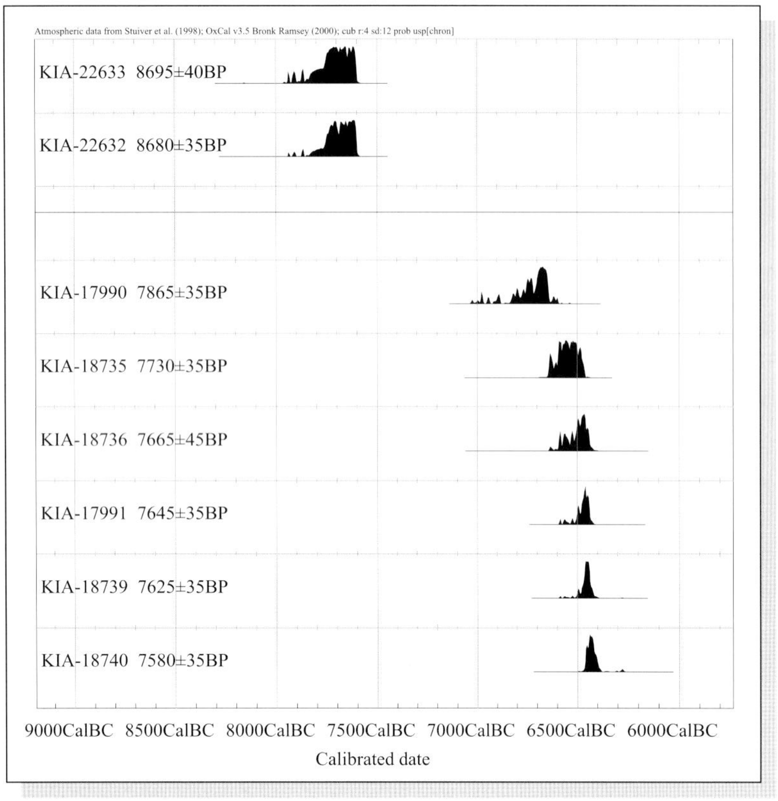

Figure 135: Doel "Deurganckdok-sector B"; calibrated radiocarbon dates from 7 hearth-pits. The upper 2 dates are on carbonised hazelnut shells, the remaining on charcoal samples.

feature reference	lab. reference	BP date	$\delta^{13}C$ ‰	dating material	68.2% probability	95.2% probability
S22	KIA-22633	8695±40	-24.6	charred hazelnut shell	7750BC (1.00%) 7600BC	7940BC (0.03%) 7900BC 7870BC (0.02%) 7860BC 7830BC (0.95%) 7590BC
S79	KIA-22632	8680±35	-25.8	charred hazelnut shell	7740BC (1.00%) 7600 BC	7820BC (1.00%) 7590BC
S47	KIA-17990	7865±35	-24.7	*Quercus* charcoal (2 rings)	6800BC (2.6%) 6790BC 6770BC (65.6%) 6640BC	7000BC (2.8%) 6960BC 6950BC (1.6%) 6930BC 6910BC (3.2%) 6880BC 6830BC (87.8%) 6590BC
S24	KIA-18735	7730±35	-26.2	*Quercus* charcoal (6 rings)	6600BC (68.2%) 6480BC	6640BC (95.4%) 6460BC
S34	KIA-18736	7665±45	-27.3	*Quercus* charcoal (4 frag.)	6590BC (1.9%) 6580BC 6570BC (5.0%) 6550BC 6530BC (4.0%) 6520BC 6510BC (57.3%) 6440BC	6600BC (95.4%) 6420BC
S105	KIA-17991	7645±35	-24.7	*Quercus* charcoal (15 rings)	6500BC (6.9%) 6490BC 6480BC (61.3%) 6435BC	6590BC (1.5%) 6580BC 6570BC (4.1%) 6540BC 6530BC (89.8%) 6420BC
S22	KIA-18739	7625±35	-25.8	*Quercus* charcoal (5 rings)	6471BC (68.2%) 6433BC	6510BC (95.4%) 6400BC
S96	KIA-18740	7580±35	-26.3	*Quercus* charcoal (9 rings)	6456BC (68.2%) 6416BC	6470BC (94.4%) 6380BC 6280BC (1.0%) 6270BC

Table 50: Doel "Deurganckdok-sector B"; calibrated radiocarbon dates from hearth-pits.

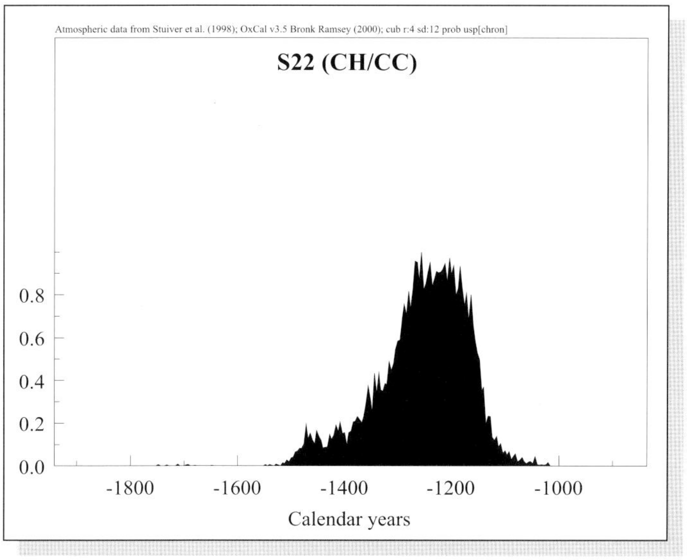

Figure 136: Doel "Deurganckdok-sector B"; difference in ^{14}C age between a charred hazelnut sample and a charcoal sample from hearth-pit S22.

The series of charcoal dates contains no outliers (interquartile range: 6570-6460 BC, 95% probability range 6810-6300 BC). The results are grouped in table 50 and figure 135. Both hazelnut dates, on the other hand, are clearly older than the charcoal dates (interquartile range: 7740-7650 BC, 95% probability range: 7900-7610 BC), showing no overlap, even not after 2 sigma calculation. In this respect the hearth-pits of Doel are very much comparable to those of Verrebroek "Dok 1" (cf 3.2.3.2.). Hence, it may also be assumed here that the dated hazelnuts are residual material, belonging to an older (Early Mesolithic) occupation phase. This assumption is further corroborated by the double dating of feature S22. Statistical analysis (fig. 136) points to an age difference between the hazelnut and charcoal sample from this feature in the range of 1 to 1.5 millennium (68.2% probability -1150/-1310; 95.4% probability -1480/-1110). Material evidence of an Early (Boreal) Mesolithic occupation of the site, as indicated by the hazelnuts, however, is very scarce. The only find in the immediate vicinity of both hazelnut-dated hearth-pits that can be linked with such an occupation is a triangle found in pit S61. This absence of large amounts of Early Mesolithic finds is most likely due to the fact that 1° the top of the Pleistocene sand in this particular sector (NE) had already been removed mechanically before the excavations started and 2° the deeper horizons of the Pleistocene substrate could not be excavated due to shortage of time. Further excavations in this sector would certainly have yielded much more Early Mesolithic material[7].

7 Numerous traces of Early Mesolithic occupations have been registered at two nearby sand dunes in the Deurganckdok, e.g. sector J/L (Bats *et alii* 2003) and sector M. These finds clearly demonstrate the intensity of land-use during the Pre-boreal and Boreal.

It is reasonable to suppose that both the above arguments can also be used to explain the lack of archaeological finds typologically related to the charcoal dates. Indeed, not a single artefact typical of the time-span of 6810-6300 BC – i.e. the late Middle and early Late Mesolithic (such as points with flat retouch, small backed bladelets, broad trapezes and Montbani blades) – has been found, neither in the hearth-pits, nor in the surrounding Pleistocene sand. On the contrary, some hearth-pits yielded finds, such as lithic artefacts and potsherds, associated with older – Final Palaeolithic – or younger – Final (Swifterbant) Mesolithic – periods. Unfortunately none of these hearth-pits has been dated directly so far, hence the existence of older or younger hearth-pits cannot be fully excluded. On the other hand, one may seriously doubt the existence of hearth-pits dating back to the Final Palaeolithic or to the Final Mesolithic at Doel. In the light of recent studies by H. Groenendijk (1987) and H.T. Waterbolk (1999) on the chronology of hearth-pits excavated in the Netherlands, it is sufficiently proved nowadays that hearth-pits are typically Mesolithic features which were constructed mainly between 9000 and 6000 BP. Apparently, hearth-pits only rarely occur on Final Palaeolithic, Early (Preboreal) Mesolithic and Final Mesolithic/Early Neolithic (Swifterbant culture) sites. Furthermore the fact that the dated features of Doel are distributed randomly over the excavated site-surface reinforces the likelihood of a certain contemporaneity among these features.

Considering that all hearth-pits excavated at Doel might date back to the transition from the Middle to the Late Mesolithic, we have to accept the presence of residual and intrusive archaeological objects within some features. Following this line of thought, the Final (patinated) Palaeolithic artefacts within some hearth-pits should be interpreted as residual material that once was lying at the sand surface but slipped into the pit during its digging, use or filling. This scanario is partly corroborated by the spatial distribution, showing a clear coincidence between the Final Palaeolithic artefacts from the "undisturbed" coversand deposits and those retrieved from the hearth-pits (cf 3.1.3.2.). The younger-dated material, such as the rectangular trapeze from S89, the small pottery fragments from S14 and S25 and possibly also some of the burnt bone fragments, on the other hand, most probably must be considered as intrusive. Most of this material might have ended up in the hearth-pits as a result of bioturbation. Indeed, pedological analysis has clearly proved that the top of the Pleistocene sand deposits in which the hearth-pits were dug is intensively bioturbated up to a depth of -40 to -45cm. Hence, it is not surprising that some minor archaeological fragments/debris from the abundant Swifterbant level have migrated downwards into these pits.

3.2.4.3. | Pottery dates

Three Swifterbant potsherds with black, charred organic (food) residue on the inner wall side were selected for radiocarbon dating. One sherd was selected from the north-western sieved sector of the site, the remaining two from the south-eastern sector.

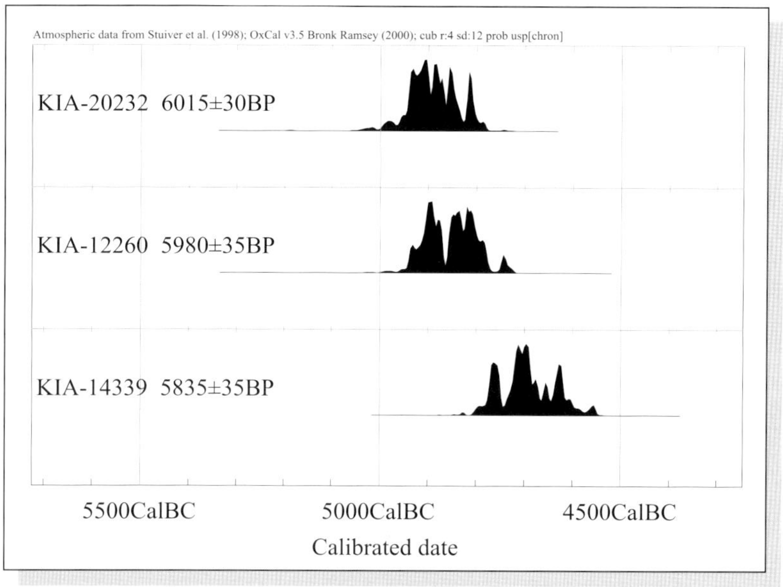

Figure 137: Doel "Deurganckdok-sector B"; calibrated radiocarbon dates of organic residue (foodcrust) on Swifterbant potsherds.

sector	find reference	lab. reference	BP date	$\delta^{3}C$ ‰	dating material	68.2% probability	95.2% probability
NW	W24/Z2(4)	KIA-20232	6015±30	-25.2	organic residue	4940BC (62.2%) 4840BC	4990BC (95.4%) 4780BC 4820BC (6.0%) 4810BC
SE	109	KIA-12260	5980±35	-28.0	organic residue	4910BC (27.9%) 4870BC 4860BC (40.3%) 4800BC	4950BC (93.3%) 4770BC 4750BC (2.1%) 4730BC
SE	742	KIA-14339	5835±35	-27.0	organic residue	4780BC (15.7%) 4750BC 4730BC (41.6%) 4670BC 4640BC (10.9%) 4610BC	4800BC (94.4%) 4590BC 4570BC (1.0%) 4550BC

Table 51: Doel "Deurganckdok-sector B"; calibrated radiocarbon dates of foodcrust on potsherds.

Tests performed on the organic residue found on one of these sherds (cf 3.2.1.5.) prove that the dates represent the original ^{14}C content of the organic deposit. A T-test reveals that the differences between the dates are significant enough to conclude that the three samples cannot date the same event. The results are represented at figure 137 and table 51. Although the number of dates is limited, the interquartile range was calculated (95% probability range: 4960-4620 BC; interquartile range: 4900-4740 BC).

The obtained dates are in perfect agreement with the typological dating of the pottery, which points to an Early Swifterbant age, with a possible extension into the first half of the Middle Swifterbant phase (Crombé *et alii* 2000, 2002). The Doel dates are consistent with the Swifterbant chronology of The Netherlands, in which the Early Swifterbant is situated between 4900 and 4600 BC (Raemaekers 1999, 108-109) and the first half of the Middle Swifterbant phase between 4600 and 4450 BC (Raemaekers 2001, 148). Apparently the three dated potsherds belong to the Early Swifterbant occupation phase(s). No sherds from the Middle Swifterbant phase were selected for dating[8].

3.2.4.4. | Discussion

Figure 140 represents the interquartile ranges from the surface-hearths, the hearth-pits (hazelnut dates excluded) and the potsherds from Doel "Deurganckdok-sector B". Although the number of dates per set is limited, the results can be considered valid because they form coherent series.

From figure 140 it can be concluded that the three series of dates must be considered as independent. As was the case at the Verrebroek "Dok 1" site, hearth-pits and surface-hearths at Doel are clearly diachronic in use. Contrary to Verrebroek, however, at Doel the hearth-pits are older than the surface-hearths. They are even older than the dated potsherds, as already suggested by the (nearly) total absence of ceramic finds in their filling. The chronological gap between the hearth-pits on the one hand and the surface-hearths and potsherds on the other hand clearly cannot be due to the so called old-wood effect of the dated charcoal samples. Therefore the gap is way too large. Assuming that all hearth-pits date to the same period, these features represent an occupation of the site of Doel during the final stage of the Middle Mesolithic and/or the early stage of the Late Mesolithic. Judging by the great number and density of hearth-pits, fairly intensive occupation can be assumed. It is therefore very surprising to find so few material remains which can be correlated with this occupation either in the hearth-pits or in the undisturbed sand deposits. Of course one should take into account that large parts of the original Pleistocene surface, in particular the A and Bw-horizons, were not investigated due to earlier perturbation and /or time pressure. Hence it is possible that the Middle/Late Mesolithic occupation area(s), as well as the Early Mesolithic one(s) (indicated by both charred hazelnut dates) have been missed during salvage excavations.

8 On the basis of pottery technology, morphology and decoration it seems quite difficult, generally even impossible, to make a clear distinction between pottery sherds from the Early and first half of the Middle Swifterbant phase (Raemaekers 2001). Furthermore, the selection is even more problematic when only those fragments with ^{14}C-datable organic residue have to be chosen. Indeed, organic residue mostly occurs on the least diagnostic pottery fragments, i.e. on the bottom parts.

Another problem concerns the clear age difference between the potsherds and the surface-hearths. Although both have been found in the same stratigraphic level – i.e. the A-horizon within the top of the coversand – the latter are dated between 4960-4620 BC, the former between 4510-3990 BC (at 95% confidence level). Explaining this disparity is not easy. A first possibility is that the carbonisation of the dated charcoal, seeds and pips from the selected surface-hearths wasn't humanly induced. Forest-fires may be responsible, although this is in contradiction to the conclusions of the palaeobotanical analyses (cf 3.4.3.)[9]. p203

Another way to explain the observed age difference is by assuming that it is an artificial effect of sampling. Indeed, as already mentioned earlier, the relative dating of the recovered pottery encompasses the Early and first half of the Middle Swifterbant, i.e. a period from ca. 4900 to 4450 BC. Although it is hard to believe, it is not entirely impossible that during sampling potsherds from the Middle phase and surface-hearths from the Early phase were accidentally not selected. The only way to verify this hypothesis is by extending the radiocarbon program of the site.

RESERVOIR EFFECT.

A better explanation is to be found in a possible reservoir effect due to the processing of fish in the ceramic vessels dated at Doel. Recently, a Danish group (Fischer, Heinemeier 2003) found a freshwater reservoir effect in the radiocarbon dates of food residue on Ertebølle pottery. In a preliminary study O.E. Craig (cf Appendix F), using both gas chromatography mass spectrometry (GCMS) and bulk stable (Carbon/Nitrogen) isotope analysis, also found indications of fish components in a series of potsherd residues from Doel, including the ^{14}C-dated sample KIA-20232[10]. The results also indicated that these residues are most probably a mixture of aquatic and terrestrial material. This makes it impossible to correct the samples for the reservoir effect. First of all because the reservoir age of the River Schelde in that period is not known, secondly because the ratio between the freshwater derived carbon and terrestrial carbon is unknown. However, considering that the pottery was dumped at the site during the same occupation event as the remaining settlement waste (charcoal, seeds, nut shells, flint artefacts, etc.), as indicated by the stratigraphic and spatial distribution, a reservoir effect of 200-300 BP-years approximately can be estimated, i.e. the time difference between the potsherd dates and the surface-hearth dates. Of course further research is necessary in order to verify and improve this estimation.

In conclusion, based on the presently available information it seems better not to consider the potsherd dates from Doel "Deurganckdok-sector B" when dating the Swifterbant occupation. Based on the surface-hearth samples, the latter should rather be situated between 4510-3990 BC. Combined with the dates obtained on the natural sequence at the nearby "sector A" (cf 2.4.), the Swifterbant occupation should be situated most likely within the phase of increased marine influence and the deposition of the so-called Calais-deposits (fig. 138).

[9] It was hoped that the burned bones recovered from these surface-hearths could be used to solve the problem, but unfortunately the burning degree of these bones is too low to allow dating by means of structural carbonate.

[10] At the Swifterbant site of Almere "Hoge Vaart-A27" in the Netherlands - a site very similar in type, settlement location (wetland environment) and economy (game and fish) - no discrepancy between the dates on food crusts and on charcoal from surface-hearths was found. Unfortunately, so far no $\delta^{15}N$ measurements were performed on the dated food crust samples.

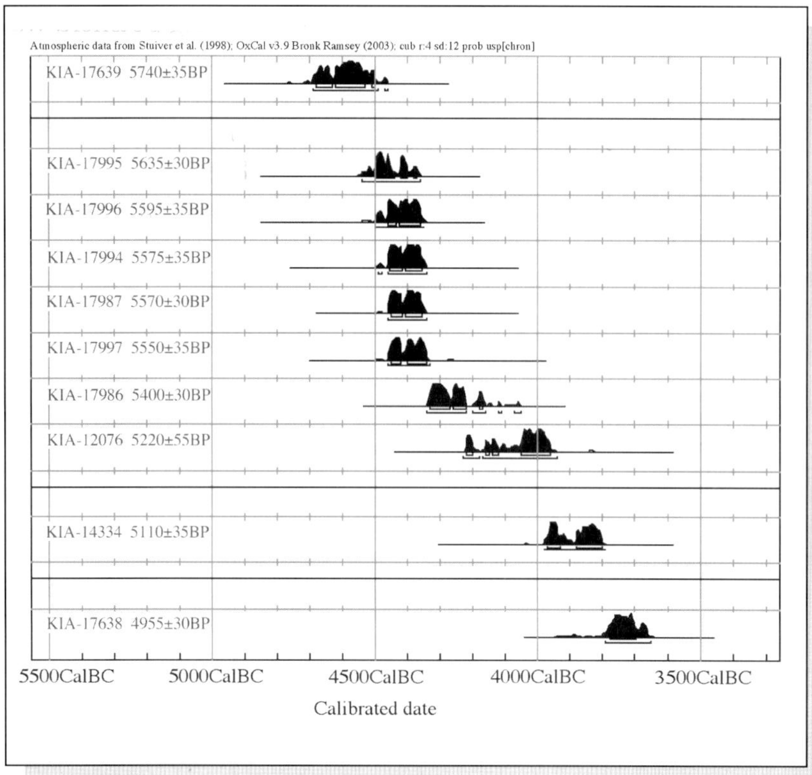

Figure 138: Doel "Deurganckdok-sector B/C"; diagram showing the distribution of the archaeological dates on surface-hearths (sector B) and potsherd residue (sector C) against the environmental dates (base and top of so called Calais-deposits).

3.2.5. | Doel "Deurganckdok-sector C"

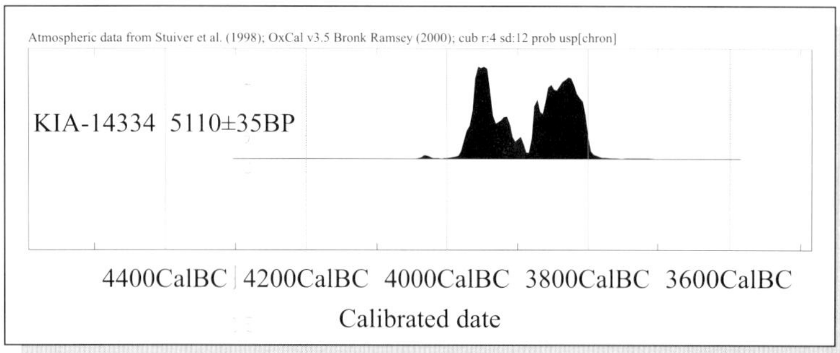

Figure 139: Doel "Deurganckdok-sector C"; calibrated radiocarbon date of organic residue (foodcrust) on a Michelsberg potsherd.

lab. reference	find reference	BP date	$\delta^{13}C$ ‰	dating material	68.2% probability	95.2% probability
KIA-14334	989	5110±35	-26.4	organic residue	3970BC (23.0%) 3930BC 3880BC (45.2%) 3800BC	3980BC (95.4%) 3790BC

Table 52: Doel "Deurganckdok-sector C"; calibrated radiocarbon date of foodcrust on a potsherd.

At Doel "sector C" the black charred food residue found on a Michelsberg potsherd was dated (tab. 52, fig. 139). The obtained result falls within the expected time range for the Michelsberg culture in Belgium, i.e. 5400/5300 BP till 4700/4600 BP (Cauwe *et alii* 2001). Although this sample was not analysed by GCMS or stable isotopes by O.E. Craig (cf Appendix F), other samples from the same site show $\delta^{13}C$ and $\delta^{15}N$ values that are typical for plants. Hence, it may be assumed that this date is not influenced by the reservoir effect and thus may be considered as valid.

3.2.6. | Conclusions

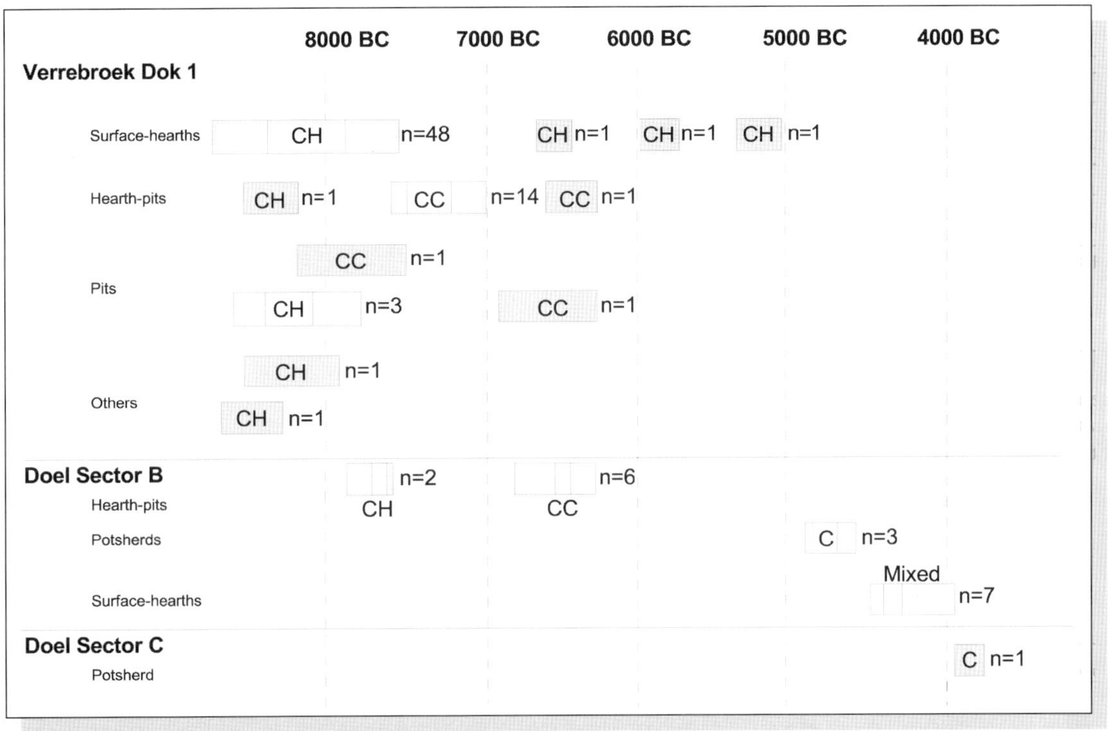

Figure 140: the interquartile range and the 95% probability range of the different groups (light grey) and the 95% probability range of each outlier (dark grey) at Verrebroek "Dok 1" and Doel "Deurganckdok-sector B and C". Key: CH=carbonised hazelnut shell sample; CC=charcoal sample; C=organic crust on potsherd.

All radiocarbon results, except the Final Palaeolithic date on burned bark from Verrebroek "Dok 2", are grouped in figure 140. Although the standard deviation on this sample was relatively small, after calibration at the 2 σ (or standard deviation) level a date range of about two millennia has to be allowed for. This is caused by the imprecision of the, until recently accepted, calibration curve in that period (cf 2.3.1.). If, however, we anticipate the publication of the new curve, a date range between 11900 and 11400 cal. BC is much more likely. This age fits better with the current view on the recolonisation of the sandy lowlands of Belgium (Crombé, Verbruggen 2002). So far no hard evidence has been found for human presence in the area earlier than the Creswello-Hamburgian and *Federmesser* culture. In southern Belgium and surrounding countries the former is generally dated in the Late Bølling/Dryas II phase (Otte *et alii* 1984), the latter in the Allerød phase or shortly before (De Bie, Vermeersch 1998; Baales 2000; Fagnart, Coudret 2000). The fact is that the date of Verrebroek "Dok 2" is so far the oldest radiocarbon evidence for human presence in Sandy Flanders during the Late Glacial.

From the radiocarbon projects at Verrebroek "Dok 1" and Doel "Deurganckdok-sector B" four important conclusions can be made:

1° On both sites the date series obtained on the hearth-pits and the surface-hearths are not compatible with each other. At Verrebroek "Dok 1" hearth-pits are younger than the bulk of surface-hearths (and older than the hazelnut outliers), while at Doel the chronology is the other way around. At Doel the time lag between the two features is even larger than at Verrebroek. Similar chronological discrepancies have been observed on other coversand sites in The Netherlands, such as Almere (Peeters *et alii* 2001) and Urk-E4 (Peters, Peeters 2001). At the former site the hearth-pits are dated between 6400 and 6050 BP, while the oldest surface-hearths date back to ca. 6000 BP. At Urk-E4 the hearth-pit samples (7900-6200 BP) seem to pre-date those on potsherds and cereal grains (ca. 5300/5200 BP).

2° The chronological range of the dated hearth-pits at Doel and Verrebroek together (N= 21) is limited to the second half of the Boreal according to the biozonation of Sandy Flanders (Verbruggen *et alii* 1996), i.e. roughly between 8500 and 7600 BP. This coincides surprisingly well with the temporal spread of the majority of the hearth-pits excavated in the northern and middle parts of The Netherlands. According to a recent analysis by H.T. Waterbolk (1999) the temporal distribution of Dutch hearth-pits (N=230) shows two peaks: the most important one occurs roughly between 8800/8400 BP and 7200 BP. A smaller second peak is situated between ca. 6600 BP and 6000 BP.

3° The dates obtained on burnt hazelnut fragments found in hearth-pits always deviate considerably from the charcoal dates. Cross dating of some hearth-pits both at Verrebroek and Doel has clearly demonstrated that hazelnut shells in these features should be considered residual (as probably should some of the archaeological finds as well). Hence, it is clear that hearth-pits cannot be associated with the roasting of hazelnuts.

4° At Verrebroek and Doel little direct evidence of human presence during the period of use of the hearth-pits has been found within the lithic and/or ceramic inventories. On the other hand, large amounts of settlement waste are associated with the surface-hearths at both sites. This discrepancy might to a large extend be the result of erosion (Verrebroek) or destruction (Doel) of the former occupation level. At Doel most hearth-pits were found in the north-eastern sector, which was already severely disturbed by the harbour works before the salvage excavations. At Verrebroek it should be noted that the hearth-pits are all situated in a small area that was severely truncated by Late Medieval ploughing and/or inundations. Yet it should also be emphasised that even in the better preserved parts of both sites, in areas adjacent to the hearth-pits, little or no material evidence of contemporaneous occupation phases have been reported. Hence it seems not unlikely that the use of hearth-pits should be linked with occupation phases which produced or left only a small amount of imperishable settlement waste. One might even consider a change in site function at the time the hearth-pits were dug and used.

3.3. | Pedological analyses

Geertrui Louwagie & Roger Langohr

3.3.1 | Verrebroek "Dok 1"

At Verrebroek both anthropogenic and natural structures were studied. Two important, most probably anthropogenic structure types could be distinguished: (i) hearth-pits and (ii) surface-hearths.

The hearth is diagnostically domestic and refers to a large fire, in which space (e.g. within a structure), fuel and temperature are controlled (McDonnell 2001).

The groundmass within these structures consists of dominant (> 50%) well to moderately sorted subrounded (polycrystalline) quartz and very few (< 2%) subrounded to rounded chert, feldspars, partly weathered glauconite and other mica minerals (tab. D.28).

3.3.1.1. | Anthropogenic features

Hearth-pits

These structures have in common that they are easily delineated because of their dark soil matrix colour (cf 3.1.2.1.). They are mostly half-spherical-shaped and measure about 0.5 to 0.7m in diameter and are on average 0.2 to 0.5m deep. They always contain charcoal fragments and charcoal powder. They are located at the higher parts of the coversand ridge.

Hearth-pits S100 and S101 (fig. 141, tab. 53, D.28, D.29, D.40, D.43)

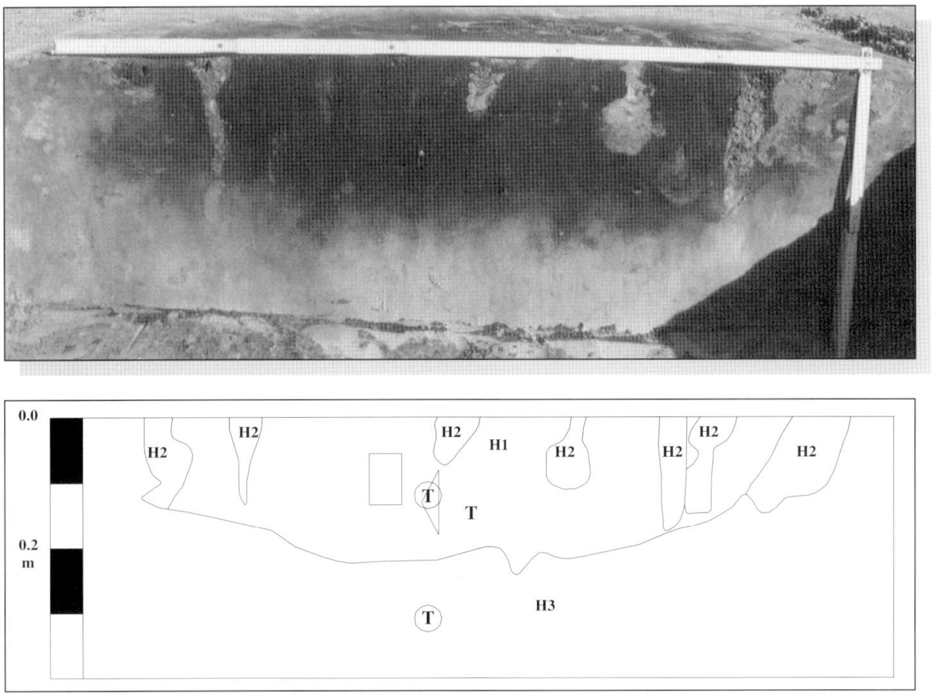

Figure 141: Verrebroek "Dok 1"; feature S100, photographic and schematic representation.

Horizon number	Symbol	Horizon depth (cm)	Profile description
H1	Aa?	(0)-11/24	Black matrix; sand; common charcoal fragments and charcoal powder
H2		(0)-7/15	Zonal; light grey matrix; bioturbated
H3	C?	11/24-(35)	Pale yellow matrix; sand

Table 53: Verrebroek "Dok 1"; hearth-pit S100, macromorphological observations.

These hearth-pit structures are situated in an area where the peat cover is absent today and where the Podzolised Soil B-horizons are eroded (fig. 33). Hence, chemical analyses must be interpreted with the necessary caution with respect to possible leaching from the nutrient-rich overlying flood sediments and present-day plough horizon.

On S100, micromorphological, particle size, chemical and tar analyses were performed. The undisturbed sample for micromorphology was taken within the darkest part of the infilling (Aa-horizon). Micromorphological observations show that the microstructure represents few (around 10%) voids having channels and chamber forms; these indicate that bioturbation processes were active here. Frequent (10 - 30%), up to 100µm large carbonised organic fine mass coatings and infillings (cf tab. D.47) and very few (< 2 - 5%), up to 80µm large carbonised organic particles and charcoal fragments up to 0.5mm large (cf tab. D.47) were observed within this horizon. This carbonised organic matter is the only direct evidence of fire activity. The presence of carbonised fine organic mass indicates a fair chance of *in situ*, incomplete burning of the organic material. Combined (i) presence of carbonised organic fine mass next to charcoal fragments, and (ii) absence of (morphological indicators for) ash, are indicative of intentional incomplete burning (Exaltus 2001, 19-35). Direct evidences of fire activity as a consequence of discrete changes in the oxides/hydroxides composition on the mineral soil material are absent here due to the nearly complete absence of free iron (Fe_d) in these soil horizons. Charcoal has an ignition temperature of 300°C in oxidising conditions (McDonnell 2001) and glauconite only starts to transform from crystalline to amorphous at 500°C (cf Appendix C. Methods). The presence of crystalline glauconite is this structure is another argument for termperatures that remained relatively low.

Particle size (tab. D.29, D.40) and chemical (tab. D.43) analyses were performed on samples of the Aa- and C-horizon; latter sample was taken below the hearth-pit structure. The pH in both the Aa- and C-horizon is slightly alkaline to neutral; the dominant cation is calcium (Ca). This is not surprising since the area where many hearth-pits are concentrated, is situated on the highest part of the coversand ridge and is directly overlain by the calcium carbonate containing flood sediments. For the cation exchange capacity (CEC) and the exchangeable cations, the following trends are observed; all figures are expressed in cmol(+) kg^{-1}. The BC-horizons in the reference profiles (P17, P18, referred to as Ref) at Verrebroek "Dok 1" have CEC-values of about 3. CEC in the Aa-horizon (CEC_Aa = 12.3) is higher than those in the reference BC-horizons. It should be mentioned here that P18 has evidences of fire activity throughout the profile, probably influencing cation contents even deeper down in the profile. Both exchangeable magnesium (Mg) and potassium (K) contents (Mg_Aa = 0.40,

K_Aa = 0.07; Mg_BC_Ref = 0.09 – 0.29, K_BC_Ref = 0.05) are respectively clearly to slightly higher as compared to the reference BC-horizons. This is also the case for the K-content in the underlying C-horizon (K_C = 0.06). Exchangeable magnesium and potassium percentages (EMP, EPP), indicating relative quantities of these cations with regard to the cation exchange capacity, give higher values in the C- than in the Aa-horizon. EMP- and EPP-values in the C-horizon are even considerably higher than in the reference BC-horizon without evidence of fire acitvity (EMP_C = 7%, EPP = 5%; EMP_BC_Ref = 3%, EPP_BC_Ref = 2%). This probably points towards leaching deeper into the soil of cations released on burning in the hearth-pit structure. The C/N-ratio in the Aa-horizon (C/N_Aa = 87) is higher than C/N-values around 60 commonly reported for wood ash. Phosphate in the Aa-horizon is mainly present as organic phosphates, whereas in the C-horizon the reverse is the case (Pi/Po_Aa = 0.6, Pi/Po_C = 3.3). The dominance of organic over inorganic phosphate in the hearth-pit as opposed to the surrounding soil sediment possibly indicates organic residues of dead plant and/or animal material, as e.g. lipids (Heron 2001).

Tar analysis (cf Appendix C. Methods), performed on the Aa- and C-horizon and on a charcoal fragment picked from the homogeneous dark infilling, didn't reveal any tar remnants. On S101, only tar analysis was performed, again on the Aa- and C-horizon and on a charcoal fragment picked from the Aa-horizon and additionally also on the transition from the Aa- to the C-horizon. Also here, remnants of tar are absent. It is questioned whether eventual original presence of tar was degraded since its formation, or that these structures are simply not related to cooking activities. With respect to the former, Heron (2001) reports bacterial reworking of the molecular residues of human activtiy as a possible mechanism for lipid degradation. Furthermore, Jansen and Peeters (2001, 45) mention that tar is only preserved in case the fire is covered in some way, so that tar doesn't burn on carbonising, but can escape and remain impregnated in the soil pore system.

Charcoal determination (cf 3.4.2.1.) revealed that most hearth-pits contain *Pinus* and some have *Pinus* and smaller amounts of *Quercus* and *Corylus*. According to Groenendijk (1987), *Pinus* produces charcoal that keeps glowing for a long time, preferentially indicating a slow, smouldering fire. Latter is anyhow a necessary prerequisite to preserve or obtain charcoal, with an ignition temperature of 300°C in oxidising conditions (McDonnell 2001). Radiocarbon dating on charcoal fragments from these hearth-pits gives dates between 8530±60 (UtC-9447) and 8190±70 (UtC-9451) BP (cf 3.2.3.2.).

Surface-hearths

As opposed to the hearth-pits, surface-hearths are not easy to delineate based on soil matrix colour. Concentrated presence of burnt artefact and ecofact material helps delineating them (cf 3.1.2.1.). The archaeologists defined possible surface-hearths based on the clustered presence of overheated lithic artefacts over one square metre. At Verrebroek "Dok 1", the surface-hearth always coincides with the highest quantities of burnt bone fragments, and frequently, but not always with the highest quantities of carbonised hazelnut shells (Sergant 2004).

Dating of the surface-hearths based on the hazelnut shells gives radiocarbon ages between 9490±60 (NZA-11022) and 8400±60 (UtC-9222) BP (Van Strydonck *et alii* 2001). These dates clearly indicate that the site has been occupied at several occasions (cf 3.2.3.1.).

Artefact unit 22 (fig. 142, tab. D.28, D.40, D.43)

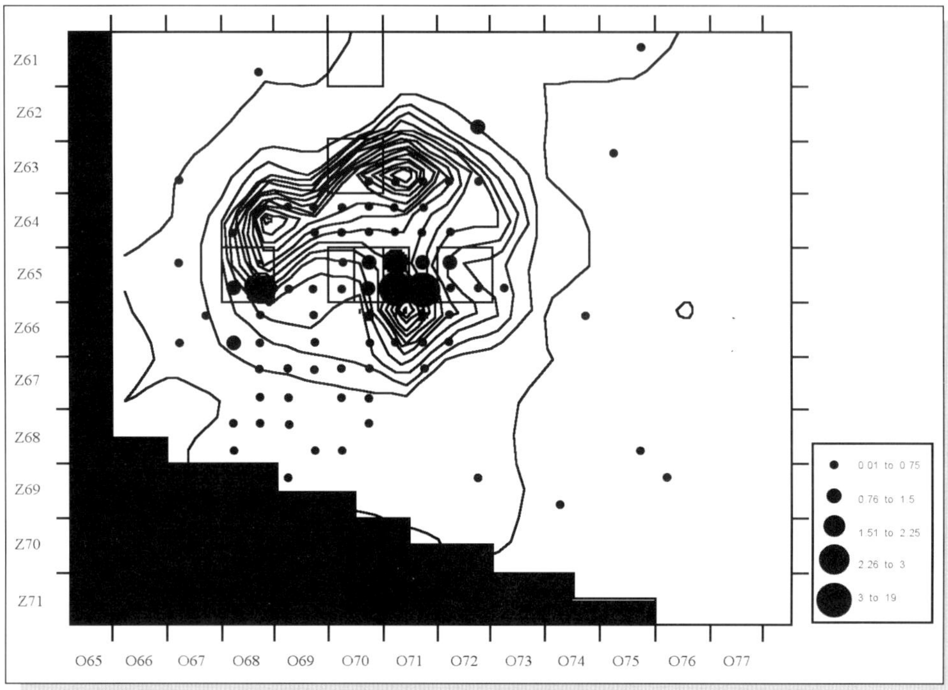

Figure 142: Verrebroek "Dok 1"; artefact concentration 22 marking the distribution of the charred hazelnut shells (contour lines; lowest limit 0.25gr., interval 5gr.), the burnt bone fragments (black dots) and the sampling units (quadrangles). The presumed position of the surface-hearth is indicated by a bold quadrangle.

Within artefact unit 22, the defined surface-hearth has 15.53g of burnt bone and 118.15g of carbonised hazelnut. To the west of the surface-hearth a second burnt bone cluster is observed. It also appears that the surface-hearth in this artefact unit does not coincide with the highest quantities of carbonised hazelnut shells; latter are located north of the hearth.

From a pedological point of view, this unit is situated in an area with Degraded Podzolised Soils (cf 2.1.4.8.), characterised by important organic matter decomposition due to microbiological activity and concomitant lower nutrient levels when compared to Podzolised Soils (cf 2.1.4.7.). Two orthogonal cross-sections (north-south and west-east) through what was in the field detected as the surface-hearth were investigated. Samples were taken in the centre of the analysed quadrangles. Micromorphology (tab. D.28) and phosphate content (tab. D.43) were analysed. Also phytolith analysis (cf 3.4.2.4.) was performed. Detailed archaeological study of the artefact and ecofact distribution over these sections however learnt that some of the sampled quadrangles, namely O70/Z57 and O70/Z59, belong to another artefact unit (unit 67) with its own surface-hearth(s). These quadrangles are not further commented in the discussion below.

Few evidences of fire were found: very few (< 2%) carbonised organic fine particles up to 150µm, very few (< 2%) charcoal, varying in size up to 1.5mm and very few (< 2%) carbonised cracked hazelnut fragments (cf tab. D.47), also varying in size up to 2mm. Only two samples relatively close to the surface-hearth (O70/Z63 and O72/Z65) had very few (< 2% and only at 1 spot) carbonised organic fine mass infillings (cf tab. D.47), probably indicating deliberate incomplete burning; latter were however absent in the quadrangle covering half of the surface-hearth (O70/Z65), further called the surface-hearth. The soil material didn't show any reddening either, possibly because the parent material contains too little iron.

The surface-hearth and the immediate perimeter have the highest total phosphate content. The samples on either side of the surface-hearth on the west-east section are the only samples that have dominantly inorganic phosphate (Pi/Po_O68/Z65 = 2.0, Pi/Po_O72/Z65 = 1.2). It is surprising that only the Pi/Po-ratios around the surface-hearth have a value over or close to 1, whereas the surface-hearth itself has only a Pi/Po-ratio of 0.4. Especially at or close to the surface-hearth, we would expect a high Pi/Po-ratio, as the lithic artefact morphology suggests conditions of complete burning.
High total phosphate content and dominantly inorganic phosphate coincides only in one of the samples with the earlier-mentioned, local presence of carbonised organic fine mass infillings (O72/Z65). The other spot where the carbonised organic fine mass infillings were observed has a Pi/Po-ratio close to 1 (Pi/Po_O70Z63 = 0.7) and is located north of the surface-hearth, coinciding with dominant presence of carbonised hazelnut shells (cf tab. D.47).

Phosphate input can be either organic, as e.g. dead plant material or organic waste, or inorganic, as e.g. bone would dominantly provide. Dominance of inorganic over organic phosphate might be due to (i) direct input of inorganic phosphate, or resulting from (ii) organic phosphate oxidation and mineralisation by contact with air; (iii) latter, relatively slow process can speed up on contact with fire (cf 2.1.3.4.). Furthermore, relatively high total phosphate content and a Pi/Po-ratio over 1 was also registered in a burnt hazelnut cluster (cf below). Although all the samples along the analysed sequence have comparable exposure to air and bioturbation, only a few samples close to the surface-hearth have higher inorganic than organic phosphate content. This relatively high Pi/Po-ratios can be the consequence of either (i) presence of bone, (ii) additional input of organic matter, (iii) increased mineralisation through fire activity, or (iv) a combination of these processes.

The distribution pattern of burnt bone and carbonised hazelnut shells might help clarifying the analysed phosphate quantities and presence of organic fine mass infillings. The quadrangles O68/Z65, O70/Z65 and O72/Z65 (with O, east and increasing numbers indicating eastern direction and Z, south and increasing numbers indicating southern direction) are located in or at the edge of burnt bone clusters and are linked with relatively high burnt bone weight (tab. D.43): 6.87, 4.26 and 1.55g respectively. Along the same west-east section, carbonised hazelnut weight decreases from 146.73g in O68/Z65 to 74.65g in O72/Z65. Along the analysed sections, quadrangle O70/Z63 has the highest carbonised hazelnut content of 159.55g and coincides with the hazelnut cluster situated to the north of the surface-hearth (tab. D.43).

Altogether, we find an increasing, albeit not continuous, trend of indicators, such as presence of carbonised organic fine mass, a Pi/Po-ratio close to or over 1 - with a surprising low at the surface-hearth: O70/Z65 -, and/or increasing total phosphate content towards the surface-hearth. Within the particular context of this surface-hearth, these indicators stand for anthropogenic input of bone (along the west-east axis: O68/Z65, O70/Z65 and O72/Z65) and/or hazelnut shells (O70/Z63, O68/Z65, O70/Z65 and O72/Z65) with subsequent burning and are thus possible indicators of *in situ* fire activity. As the samples were located at only 0.10m below the Early Mesolithic bare occupation surface, most of the carbonised organic fine mass is probably lost due to surface processes as wind activity or trampling.

Artefact unit 41 (fig. 143, tab. D.28, D.29, D.40, D.43)

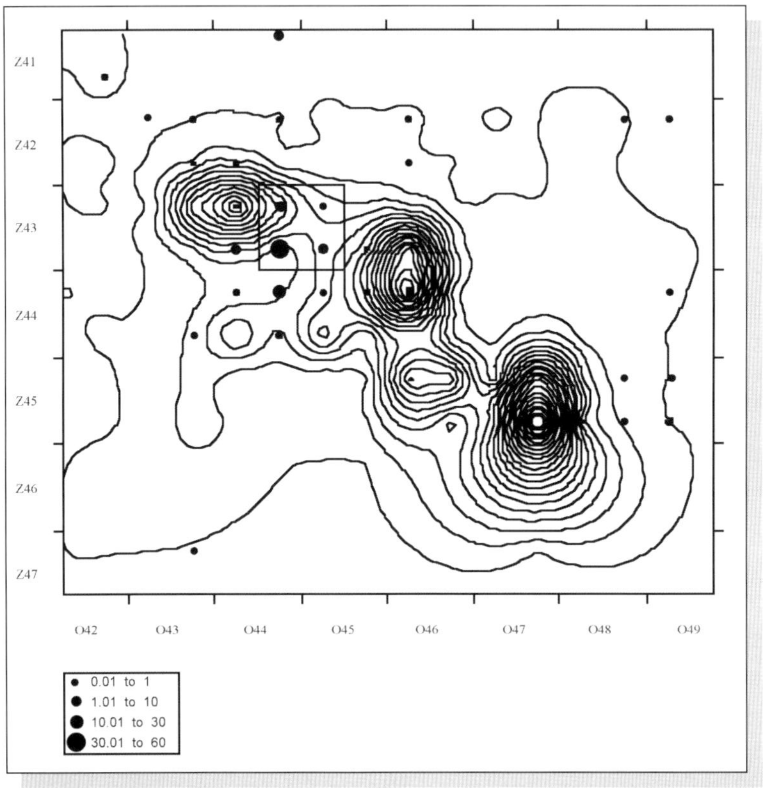

Figure 143: Verrebroek "Dok 1"; artefact concentration 41 marking the distribution of the charred hazelnut shells (contour lines; lowest limit 0.01 gr., interval 10 gr.), the burnt bone fragments (black dots) and the surface-hearth (quadrangle).

Within this unit too a possible surface-hearth (O44/Z43(1-2), O45/Z43(3-4)), coinciding with a cluster of overheated lithic artefacts and burnt bone, was detected by the archaeologists. In the immediate vicinity at least six clusters of carbonised hazelnut shells (O47/Z45, O46/Z45, O46/Z44, O46/Z33, O45/Z44, O43/Z45) could be observed (fig. 143). In one of these clusters (O43/Z45) in the field some charcoal was also noticed, further described as hazelnut/charcoal cluster. In this artefact unit, the different clusters were investigated separately.

This artefact unit is situated in an area where the peat cover is absent; the flood sediments are in direct contact with the Mesolithic occupation level. Hence, nutrient leaching from the overlying flood sediments and particularly from the present-day plough horizon might severely influence nutrient contents within the surface-hearth structures. Only one of these hazelnut concentrations has been analysed for micromorphological characteristics, particle size, soil chemical characteristics and tar contents.

Burnt hazelnut cluster (O46/Z44)

Micromorphological observations (tab. D.28) show that this concentration contains very few (2 - 5%), up to 5mm large carbonised cracked hazelnut fragments and very few (< 2%), up to 2mm large charcoal fragments, both delivering direct evidence of fire activity. However, not all the organic material present is carbonised. Very few (2 - 5%), up to 10µm thick dusty organic fine mass coatings and infillings, very few (< 2 - 5%), up to 80µm large amorphous brown organic fine mass particles and very few (< 2%) non-birefringent organ residue (root) and amorphous dark brown organic matter fragments up to 350µm large were observed too. The amount and type of organic fine mass coatings and infillings correspond to what is found in the eluvial E- and EB-horizons of the slightly podzolised, imperfectly drained reference profile P16.

The pH is slightly alkaline and probably reflects leaching from the overlying carbonate-containing flood sediments. Total phosphate content is higher than in the E- and EB-horizons or BC-horizons of the reference profile P16, not covered with peat either. Inorganic and organic phosphate is more or less in balance (Pi/Po = 1.3), as in the reference EB- and BC-horizons. Nitrogen content, cation exchange capacity and exchangeable magnesium and potassium contents are comparable to those of the reference EB-horizon. Also the exchangeable magnesium and potassium contents related to the cation exchange capacity have comparable values. It is questioned whether burning of hazelwood results in a particular release of phosphorus and doesn't especially affect other nutrient cycles or stores, as charcoal or carbonaceous particles (Macphail 1998; Macphail *et alii* 2000; Davidson *et alii*, submitted), appreciable amounts of phosphate on its surface and in its porous matrix. Nevertheless, leached phosphorus from the overlying flood sediments and particularly plough horizon could partly account for the relatively high phosphate content too. Alternatively, lateral migration of phosphate from the nearby burnt bone cluster can contribute to the relatively high phosphorus content here.

Tar analysis on samples of three burnt hazelnut clusters gave negative results.

Hazelnut was an important dietary constituent of Mesolithic populations in Europe (Rankine *et alii* 1960, in Williams 1975, 97; Simmons 1996, 25; Mithen *et alii* 2001). Possible reasons for hazelnut roasting were: (i) prolonged storage (Hamburg *et alii* 2001, 13) and facilitating (ii) processing, (iii) digestion, (iv) taste and (v) portability (Mithen *et alii* 2001).

Burnt bone cluster (O44/Z43)

As mentioned above, this cluster coincides with the surface-hearth (O44/Z43(1-2), O45/Z43(3-4)) of artefact unit 41. Micromorphological observations (tab. D.28) confirm the particular presence of a burnt bone fragment of 8mm and furthermore show very few (< 2%), up to 1mm large carbonised cracked hazelnut fragments and very few (< 2%) charcoal fragments up to 2mm in size (cf tab. D.47). This carbonised material is the only direct evidence of fire. Different forms of non-carbonised organic matter, such as few (< 5 - 10%), up to 150µm large amorphous brown organic fine mass particles and very few (< 2%) organ residue (root) and amorphous (dark) brown organic matter fragments up to 1.5mm large are present too (cf tab. D.45). Furthermore very few (2 - 5%), up to 10µm thick dusty organic fine mass coatings and infillings were observed. The latter type of organic matter pedofeature is observed as well in the eluvial E- and EB-horizons of the slightly podzolised, imperfectly drained reference profile P16. This concentration however has more amorphous organic fine mass particles.

As in the burnt hazelnut cluster, the pH is slightly alkaline, probably reflecting calcium carbonate leaching from the overlying flood sediments. Total phosphate (P_2O_5) content is high and mainly consists of inorganic phosphate (P_2O_5t = 2924mg kg^{-1}, Pi/Po = 6.0). The high inorganic phosphate content can be related to the presence of bone; burnt bone weight in this quadrangle is indeed high: 60.79g (tab. D.43). Presence of very few (< 2%) amorphous phosphate (cracked) infilling fragments observed in thin section (tab. D.28) call for bone weathering. When referring to the EB-horizon, cation exchange capacity (CEC), exchangeable magnesium (Mg) and potassium (K) values are relatively high. Exchangeable magnesium and potassium percentages (EMP, EPP) on the other hand are however comparable to those in the reference EB-horizon.

As opposed to the burnt hazelnut cluster, exchangeable calcium (Ca) content in the burnt bone concentration shows an appreciable increase as well when compared to the reference. Concomitant relative increase of Ca- and P_2O_5-contents confirms presence of Ca-phosphate related to bones.

No tar remnants could be detected within the burnt bone concentration.

Burnt hazelnut/charcoal cluster (O43/Z45)

Micromorphology (tab. D.28) shows that the only direct evidences of fire activity are the very few (2 - 5%), up to 1.5mm large charcoal fragments and very few (< 2%), up to 1.5mm large, carbonised, cracked, assumably hazelnut fragments. Furthermore very few (< 2 - 5%) amorphous brown organic fine mass particles up to 150µm and very few (< 2%) organ residue (root) and amorphous (dark) brown organic matter fragments up to 150µm are present. Very few (< 2%), up to 10µm thick dusty organic fine mass coatings and infillings again call for comparison with the EB-horizon of P16.

As in the burnt hazelnut and burnt bone cluster, the pH is slightly alkaline, probably reflecting calcium carbonate leaching from the overlying flood sediments. Particular attention should be paid to the relatively high exchangeable potassium (K) content and exchangeable potassium percentage (EPP) when compared to reference EB-horizon. Perlès (1977, 8) mentions traces of K in the residues of burnt wood; this could be a possibility here too. Nitrogen (N) and phosphate (P_2O_5) contents are comparable to those of the burnt hazelnut concentration. Also here inorganic phosphate is more or less in balance with organic phosphate (Pi/Po = 0.8). As in the burnt hazelnut cluster, leached phosphorus from the overlying flood sediments and particularly plough horizon could partly account for the relatively high phosphate content too. Alternatively, lateral migration of phosphate from the nearby burnt bone cluster (O44/Z43), with dominantly inorganic phosphate (Pi/Po_O44/Z43 = 6.0), could have significantly contributed to the relatively high phosphorus content here too. However, the slight dominance of organic over inorganic phophate (Pi/Po_O43/Z45 = 0.8) in the charcoal cluster rather fits with input of organic material, furthermore attested to by the presence of charcoal.

No tar remnants could be detected within the charcoal concentration.

Hazelnut-pits

Small and shallow pits characterised by a very high density of carbonised hazelnut shells were indicated as hazelnut-pits (cf 3.1.2.1.). S80, S89 and possibly S81were indicated as such, but were only discovered at 0.3 to 0.4m depth under the Early Mesolithic living floor. Based on the recognised A-Bw.bi-C horizon sequence within S89, latter feature was recognised as a tree-fall in pedological context (cf 3.3.1.2.).

Hazelnut-pits S80 and S81 in artefact unit 22 (fig. 77, tab. D.28)

S80 and S81 were only micromorphologically investigated. Very few (< 2 - 5%) carbonised organic particles up to 150µm, charcoal and carbonised cracked hazelnut fragments (cf tab. D.47) were observed. Charcoal and hazelnut fragments were slightly larger in S81 as compared to S80. As in S89 (cf 3.3.1.2.), no carbonised organic fine mass coatings or infillings were observed. As artefact and ecofact distribution is mainly limited to the upper 0.20m of the Early Mesolithic occupation level (cf 3.1.2.1., 3.3.1.3.), and the studied hazelnut-pit sections are located at 0.30 to 0.50m depth, it is less probable that the carbonised organic fine mass was prone to wind activity within these pits, unless they were open and possible subject to blow-out. Altogether, absence of carbonised organic fine mass within these hazelnut-pits rather points towards absence of *in situ* fire activity. The pedological data can at current not answer the question whether these features are man-made or are the result of natural processes, as for example faunal activity.

3.3.1.2 | Natural features

Bio-galleries

Roots are the most common evidences of bio-galleries at the Verrebroek site. Figure 26 shows the last generation of tree roots. In the centre, a well-preserved root is clearly visible in depth, but however fades at the level of the former surface horizon. Only the humiferous remains are still recognisable. This example provides evidence that the morphology of many former generations of root galleries can be erased due to subsequent soil processes.

Tree-fall features

In present-day forest sites, several types of tree-fall structures can be recognised (Langohr 1993). A simple tree-fall tilts the soil surface and a set of surface and subsurface horizons of the soil profile. Fragments of the exposed tilted horizons will gradually fill the tree-fall pit. Additional strokes of surface cover and horizon can be pulled out together with the tree and will be re-deposited in the tree-fall pit as well. Certain cases show that falling trees only pull the surface horizon over a large area. Although small-size tree-falls mix soils less deep, they may be much more numerous than the spectacular large tree-falls. In the soil, this disturbance will not be recognisable any more after a few centuries, except for the mixing down to a few decimetres of coarse fragments that were eventually present at the surface and down to this depth at the moment of the event.

At the Verrebroek "Dok 1" site, the original soil type, present during the Mesolithic occupation, was a Brunified Soil, with black surface A- and light greyish brown B-horizon. This horizonation is only preserved in some of the structures that can be attributed to a simple tree-fall whereby the tree rotated, leaving an open pit in front of the structure and wherein fragments of the upper soil horizons gradually accumulated. The lower boundaries of these horizons are rather gradual, illustrating relatively intensive disturbance processes. This preservation of the morphology of the original soils only occurs in those tree-fall structures that were deep enough and of a type (Langohr 1993) that enabled preservation. Mostly these events are related to areas where tree roots could grow deeper in the soil and where the deeper roots were still healthy when the tree fell. Hence, tree-fall disturbance is of utmost importance for understanding the impact of bioturbation processes on the artefact and ecofact migration on the Verrebroek "Dok 1" site. For a detailed description of the artefact and ecofact distribution pattern of these structures within a broader framework on bioturbation, we refer to Louwagie *et alii*, in press.

<u>Tree-fall S89</u> (fig. 144, tab. 54, D.30, D.31, D.40, D.43)

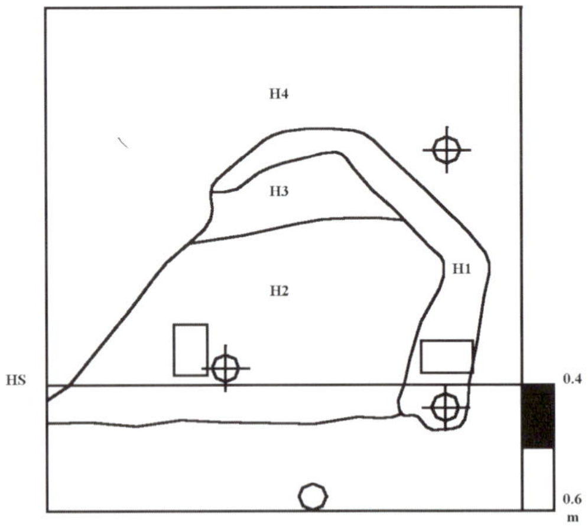

Figure 144: Verrebroek "Dok 1"; tree-fall S89, photographic and schematic representation, section at O74/Z45-Z46 (with O, east and Z, south), horizontal section at 0.4m below Early Mesolithic surface, vertical section from 0.4 to 0.6m.

Horizon number	Symbol	Horizon depth (cm)	Profile description
H1	Ab?	HS	Dark brown - very dark brown matrix; sand; charcoal fragments
H2	Bb.w.bi?	HS	Brown matrix: homogeneous; sand; carbonised hazelnut fragments; charcoal fragments; unburnt artefacts
H3	C	HS	Transition; light brown matrix; sand
H4	C?	HS	Very pale brown matrix; sand; parent material?

Table 54: Verrebroek "Dok 1"; tree-fall S89, macromorphological observations.

This area was first selected because of a concentration of numerous burnt hazelnut fragments (cf tab. D.47), seemingly indicating anthropogenic structures (cf 3.1.2.1.). Based on archaeological evidences, this feature is interpreted as a hazelnut-pit (cf 3.3.1.1.). This specific function was attributed as this feature contains a high amount of carbonised hazelnut, both in absolute quantity and when compared to surrounding quadrangles, and is located at the periphery of artefact unit 70. Careful examination showed however a sequence of tilted soil horizons that can be related to a tree-fall. Figure 144 shows a horizontal and a vertical section through the structure. The horizontal section is at 0.40m below the Mesolithic occupation surface. We observe a tilted surface A-horizon, followed in depth by a homogeneous brown, bioturbated B-horizon, possibly remnants of a Brunified Soil assumably present at the time of the Mesolithic occupation. The morphology informs that it corresponds to a tree that fell in southward direction. The size of the tree-fall at this depth is about 0.80 on 0.65m.

Texture analysis doesn't show appreciable differences between H1, H2 and H4 (SI ≥ 93).

Thin sections have been made in the A- and Bw.bi-horizons of the assumed buried Brunified Soil. Micromorphological observations (tab. D.30) show that both horizons are dominantly characterised by organic matter. In the Ab-horizon (H1), few (2 - 5%) organ residues (root, sclerotium), few (< 2 - 5%) amorphous (dark) brown plant fragments up to 1mm and few (2 - 5%) amorphous brown organic fine mass particles up to 150µm are observed. Additionally very few (< 2%) dusty organic fine mass coatings and locally, few (2 - 10%) dusty organic fine mass infillings, possibly excrement remnants, are present. Local presence of organic matter concentrations gives the Ab-horizon its somewhat patchy, heterogeneous aspect. In the homogeneous Bb.w.bi-horizon, the dusty organic fine mass coatings are somewhat thinner (< 5µm) and the organic fine mass infillings were not observed, even not locally. Nevertheless, the combined aspect of both pedofeatures provides H2 its homogeneous morphology. Very few (< 2% in H1, 2 - 5% in H2) charcoal fragments up to 1.5 (H1) or 3mm (H2) mark the minor presence of carbonised organic matter in both horizons. Absence of fine carbonised organic matter let us assume that burning didn't occur *in situ*, and that it probably doesn't concern a pit in which fire acivity, either naturally or intentionally, occurred on the spot. This would be a further argument for recovery of the discussed soil horizons in a tree-fall structure. Furthermore, phosphate contents in the Ab- and Bb.w.bi-horizons are roughly ranging from 430 to 470mg kg^{-1} (tab. D.43) and are slightly higher than in the C-horizon. Latter can possibly be linked to the presence of carbonised hazelnut shells (cf 3.3.1.1., burnt hazelnut cluster in artefact unit 41) in these horizons. As in the carbonised hazelnut concentration, phosphates are mainly inorganic. Furthermore, the Pi/Po-ratio in the Ab-horizon is relatively high as compared to that of the Bb.w.bi-horizon (Pi/Po_Ab = 3.5, Pi/Po_Bb.w.bi = 1.5). The C/Po-ratio in the Ab-horizon is around 250, relatively high and comparable to other buried surface horizons containing charcoal fragments at the Verrebroek site. As observed in other contexts, carbonised organic matter could also here account for an appreciable phosphate fixation.

The assumed A- and Bw.bi-horizons of the buried original Brunified Soil have no stronger chroma but have redder hue and higher clay content than the C-horizon, which are evidences of alteration. The Al_o+1/2Fe_o-content in the Bb.w.bi-horizon (Al_o+1/2Fe_o = 0.21%) is quite high when compared to other horizons with brunification characteristics, that commonly have values below 0.20% at the Verrebroek and Doel sites.
Both aluminium and iron, either amorphous or free, contribute to this relatively high value and reach peak levels when compared to other assumed Bw.bi-horizons.
The mentioned alteration characteristics in the assumed Bb.w.bi-horizon, such as redder hue and higher clay content and to some degree the still low Al_o+1/2Fe_o-content, are in favour of Brunified Soil remnants within this tree-fall structure. Moreover, absence of fine carbonised organic matter doesn't allow assuming *in situ* burning, which further supports this hypothesis.

Even in case the feature would finally be interpreted as anthropogenic, preservation of Brunified Soil horizons is still preferential within this particular pit environment, somewhat hidden or protected from processes occurring directly at or in direct relation with the surface, as e.g. erosion-sedimentation or bioturbation processes.

Pollen analysis was performed on 3 samples within this tree-fall structure (fig. 144) (cf 3.4.2.3.). Pollen was not detected within the A- (H1), B- (H2) and C- (H4) horizons of this structure. This could either indicate (i) dry and well-aerated conditions over a long period, or (ii) a rapid sedimentation rate. The latter hypothesis is in favour of the presumed tree-fall structure. The assumption of a dry and well-aerated environment could indicate that this structure predates the wetting phase linked to peat development on the Verrebroek site and eventually corresponds to the genesis of the homogenised, highly bioturbated Bw.bi-horizon. Concluding, these findings are at least not contradictory with the formulated hypothesis of Brunified Soil remnants within a tree-fall structure.

Tree-fall S90 (fig. 145, tab. 55, D.32, D.33, D.40, D.43)

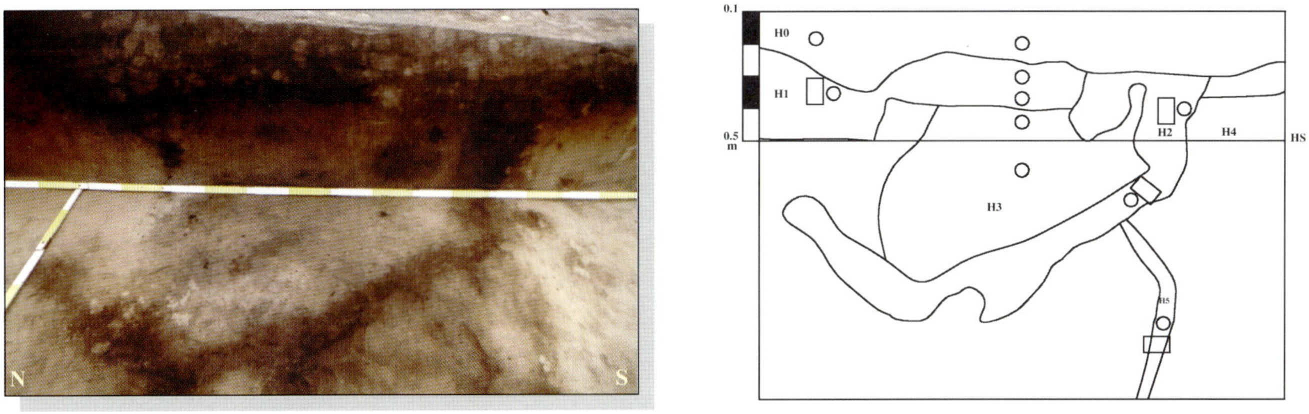

Figure 145: Verrebroek "Dok 1"; tree-fall S90, photographic and schematic representation, section at O72/Z44-Z45.5 (with O, east and Z, south), horizontal section at 0.5m below Early Mesolithic surface, vertical section from 0.1 to 0.5m.

Horizon number	Symbol	Horizon depth (cm)	Profile description
H0	A	(10)-21/35	Dark grey matrix; humiferous sand; A horizon of Podzolised Soil
H1	Bh	21/35-36/50	Brown - strong matrix; sand; B horizon of Podzolised Soil pale yellow mottles due to consumption of organic matter by microbiological activity
H2	Ab?	29/30-32/50 + HS	Very dark greyish brown-dark grey matrix; humiferous sand
H3	Bb.w.bi?	32/40-(50) + HS	Grey - light brownish grey matrix; sand; charcoal fragments
H4	C?	36/38-(50) + HS	Very pale brown matrix; sand
H5		HS	Grey matrix; sand; infilled periglacial ice fissure ("periglacial fissure pattern")?

Table 55: Verrebroek "Dok 1"; tree-fall S90, macromorphological observations.

The peat cover is absent here: about 0.10m Medieval cultivation horizon and 0.90m flood sediments buried the coversands. Figure 145 shows a horizontal and a vertical section through the structure. The horizontal section is at 0.50m depth where the tree-fall structure has a diameter of 0.80m. At the southern edge appears a surface A-horizon and an underlying, tilted, greyish B-horizon containing charcoal fragments. This morphology indicates that the tree fell in southward direction.

Particle size analysis (tab. D.33) shows that the buried assumed Brunified Soil horizons (H2, H3) have higher intra-similarity indices, mostly equal to or higher than 95, than the inter-similarity indices with the Podzolised Soil horizons (H0, H1) that are equal to or lower than 95. This confirms the presence of two different soil units or soil groupings.

Only the Bh-horizon (H1) was investigated micromorphologically. The morphological characteristics correspond to that of BC-horizons in the poorly drained Podzolised Soil profiles. The weak presence of organic matter (pedofeatures) might be attributed to the consumption of organic matter by microorganisms, indicated by the pale mottles. Latter process was discussed when indicating degradation mechanisms in Podzolised Soils (cf 2.1.4.8.). Organic matter content in H1 (Bh-horizon) is indeed somewhat lower than in most studied Bh(.s)-horizons, but is still above the required threshold value of 1.03% as defined for a spodic diagnostic horizon. It is questioned whether evidence of intenser microorganism activity should be related to some nutrient input, eventually related with neighbouring artefact or ecofact concentrations. The A-horizon (H0) indeed has a relatively high total phosphate content of about 415mg kg^{-1}, as compared to the Bh(.s)-horizon (H1) that contains around 200 to 230mg kg^{-1} total phosphate (tab. D.43). Phosphate in the A-horizon is mainly organic (Pi/Po_A = 0.5) which rather points towards input of organic phosphorus, eventually from the overlying peat horizon, if present at this spot, or from peat admixture within the overlying cultivation horizon. Unfortunately, no nitrogen or cation contents were determined within this structure. As in the buried Bw.bi-horizon, some charcoal (< 2%) is present in the Bh-horizon as well.

The crushed Bb.w.bi-horizon has redder hue, but no stronger chroma than the C(B)-horizon material that commonly has a matrix colour around 8.0 - 8.5YR 6.5 - 7.0 / 3.0 - 3.5.; its clay content is also markedly higher, from 1.1 to 1.7% as opposed to less than 0.5% commonly found in C(B)-horizon material. The $Al_o+{}^1/_2\,Fe_o$-content in the Bb.w.bi-horizon ($Al_o+{}^1/_2\,Fe_o$ = 0.10 - 0.12%) is markedly lower than in the Bh-horizons ($Al_o+{}^1/_2\,Fe_o$ = 0.18 - 0.23%) of Podzolised Soil profiles and fits well with what is found for other Bw.bi-horizons. Total phosphate content in the buried, assumed Brunified Soil Bw.bi-horizon is rather low when compared to other Brunified Soil or Podzolised Soil horizons and rather corresponds with values from eluviated E-horizons. As in S89, the Pi/Po- and the C/Po-ratio in the Ab-horizon are relatively high (Pi/Po_Ab = 2.5; C/Po_Ab = 409). The combined high Pi/Po- and C/Po-ratio possibly indicates appreciable inorganic phosphate fixation within this assumed relatively old, buried surface horizon of a Brunified Soil.

The mentioned alteration characteristics, such as redder hue, higher clay content, low Al_o + 1/2 Fe_o -content in the Bb.w.bi-horizon, together with the textural difference of the assumed Brunified Soil horizons with the Podzolised Soil horizons support the presence of Brunified Soil remnants within this tree-fall structure. However, beginning podzolisation with formation of an eluviated horizon is not excluded here (cf 2.1.3.8.).

Phytolith analysis (cf 3.4.2.4.) on the assumed Bb.w.bi-horizon only revealed very few (< 2%) phytoliths.

Tree-fall S93 (fig. 146, tab. 56)

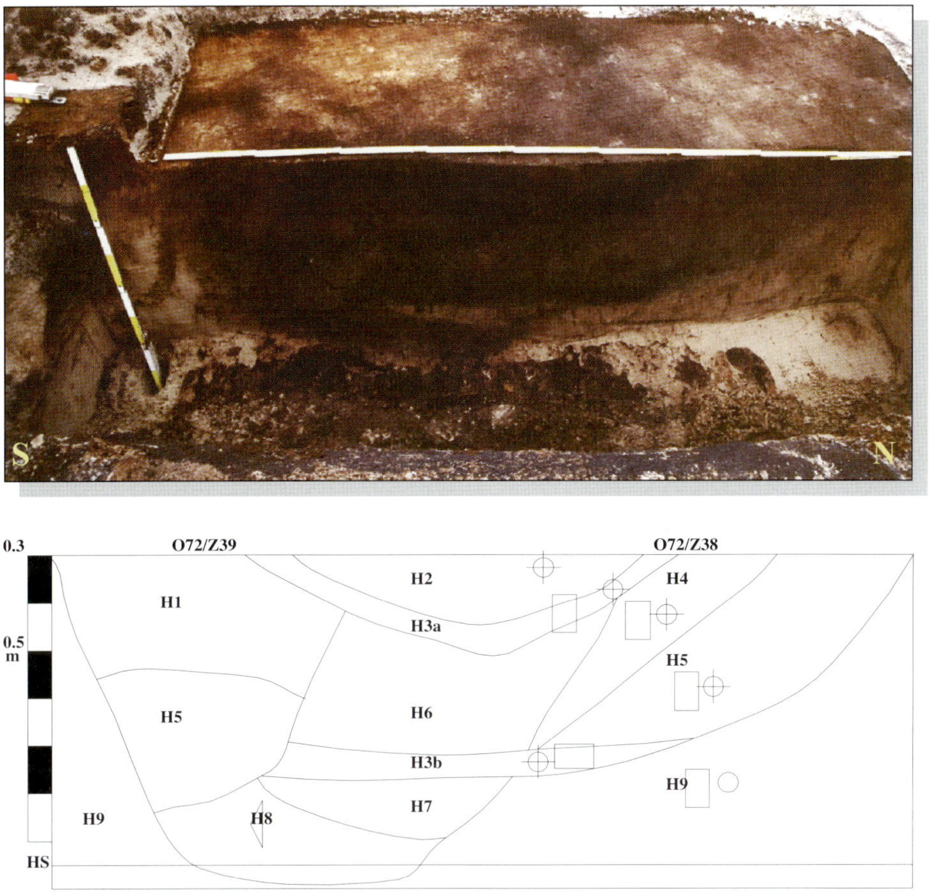

Figure 146: Verrebroek "Dok 1"; tree-fall S93, photographic and schematic representation, section at O72/Z37.5-39.5 (with O, east and Z, south), horizontal section at 0.3m below Early Mesolithic surface, vertical section from 0.3 to 0.9m

Horizon number	Symbol	Horizon depth (cm)	Profile description
H1	B?	(30)-42/60	Brown - dark brown matrix; sand; B horizon of Podzol Soil?
H2	EB	(30)-(30)/45	Dark greyish brown - dark grey matrix; sand
H3a	Ab3	(30)-(30)/51	Very dark brown matrix; humiferous sand; locally charcoal fragments
H3b	Ab2?	69/72-69/78	Very dark brown matrix; humiferous sand
H4	Bb.w.bi?	(30)-(30)/70	Brown - dark brown matrix; homogeneous; sand
H5	E?	54/60-72/83	Greyish brown - grey matrix; sand; light grey mottles due to consumption of organic matter by microorganisms
H6	E+B?	40/51-69/71	Dark grey + brown matrix; sand
H7		76/78-80/90	Pinkish grey matrix; sand
H8	Ab1?	76/90-(95) + HS	Very dark brown-grey matrix, humiferous sand
H9	C	(30)-(95)	Very pale brown - light grey matrix; sand; parent material

Table 56: Verrebroek "Dok 1"; tree-fall S93, macromorphological observations.

Figure 146 provides a horizontal and a vertical section through the structure. The horizontal section shows that the structure has a diameter of 1.80m at 0.30m depth. A surface A-horizon and an underlying, tilted and unfortunately hardly visible, brownish B-horizon are observed at the northern side of the structure. The tree-fall morphology once more illustrates that the tree fell in southward direction. Within the vertical section other fragments of surface A-horizon are visible. These might be peripheral strokes of surface A-horizon pulled off during the tree-fall and afterwards collapsing into and partly filling the tree-fall pit. A complex structure caused by several generations of tree-falls is not excluded either.

Pollen analysis was performed on 5 samples from 2 different horizon sequences within this structure (fig. 146) (cf 3.4.2.3.). Based on the field morphology, horizon sequences can be recognised:
i) H3b, H5 and H4, respectively corresponding to an A-, E- and Bw.bi-horizon;
ii) H3a and H2, respectively corresponding to an A- and EB- horizon.
The different horizon sequences indeed indicate different vegetation phases. Compared to sequence (ii), the spectra in the E- (H5) and Bw.bi- (H4) horizons of sequence (i) have relatively high *Betula*, *Corylus*, *Tilia* and *Ericaceae* values that point towards rather dry and acid conditions, within the context of the Verrebroek site favourable for podzolisation. Sequence (ii) with strongly dominant *Alnus* and *Quercus* values shows an overall wetter spectrum, definitely too wet to correspond with the environment of the Mesolithic hunter-gatherers and rather fitting with the onset of peat growth conditions. The A-horizon in sequence (i) (H3b) occupies an intermediate position. Concluding, the pollen analysis doesn't provide evidences for remnants of Brunified Soil horizons within this tree-fall structure. As mentioned already at several occasions, the Brunified Soil type probably corresponds to the environmental conditions prevailing during Mesolithic times.

Tree-sway features

On the Verrebroek "Dok 1" site we detected a puzzling type of disturbance (Langohr, Louwagie 2000). Figure 147 shows how this disturbance looks like in horizontal section; figure 148 gives a vertical cut. In horizontal sections, clear traces of water-transported sediment are visible within the existing soil profile. In vertical sections, the feature is composed of up to several centimetres thick layers, consisting of the laminated, bleached coarse fractions of the soil matrix, alternating locally with black humiferous laminae. Frequently, but not always, these traces occur parallel or slightly concave to the surface. Their position in the soil is always at the level of the subsurface horizons, mostly between 0.25 and 0.50m from the original occupation surface or living floor. In Podzolised Soils as is the case here, these structures commonly occur at the level of the Bh- and/or Bs-horizons.

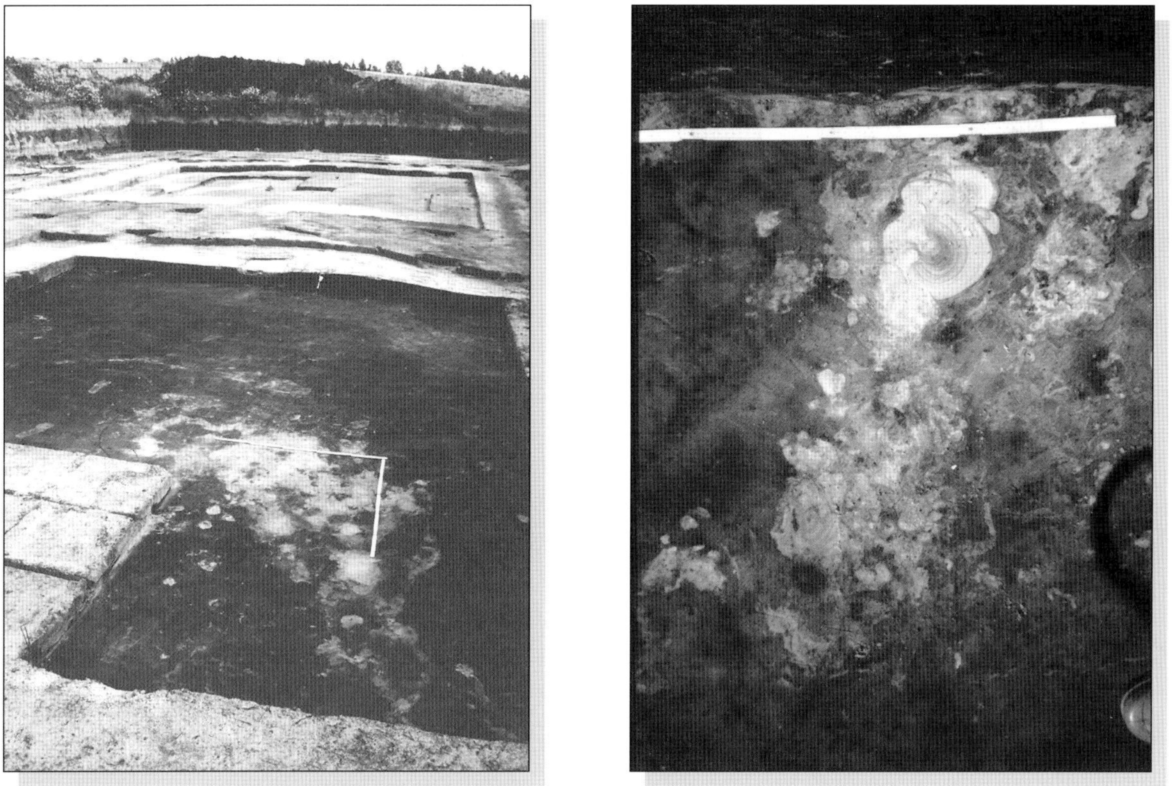

Figure 147: Verrebroek "Dok 1"; horizontal view of soil structures that are attributed to the tree-sway process; the structure is situated at the level of the Bh-horizon.

Figure 148: Verrebroek "Dok 1"; vertical section through a tree-sway structure with irregular, but mostly horizontal units of bleached stratified sand.

The feature has a somewhat similar morphology to what is observed on an irregular bare soil surface of humiferous sands after a strong rainfall. In such environmental conditions, a sorting process of lighter organic and heavier mineral soil fractions occurs as result of rain drop impact, splash erosion and consequent surface runoff. As the running water collects in the microdepressions, they become filled with stratified sediments composed of alternating strata or laminae of black organic colloids and bleached quartz sand. The position of the structures in Podzolised Soil Bh(.s)-horizons, does not allow to envisage that the stratified layers with abrupt upper and lower boundaries, once occurred on a bare soil surface. In such a case it would be unlikely that the burial process by new soil material and all the pedogenic processes active before and during the podzolisation processes, would not have disturbed the nicely stratified surface lain sediments.

The described morphology calls for a type of disturbance that occurs in stormy periods when soil moisture content is at or nearby saturation conditions. Trees swaying in stormy conditions exert huge alternating pushing and pulling forces on the root system. If the latter is mainly situated in the surface horizons, without the presence of a well-developed tap root system, it may move the upper part of the soil upwards, creating a cracking system that is more or less parallel to the soil surface. The suction forces created by this movement when all the soil pores are water-filled, will aspire free water and suspended earth coming from the crack and pore walls towards the newly created cracking system. When moving downwards, the uplifted soil mass will push the water and sediment backwards through the cracking system that originated during the uplifting phase. As a consequence, a new sedimentary structure is created in the middle of an existing soil profile. The pulling and pushing movement will have a tendency to sort the transported soil material and results in stratified, sorted sediment filling the newly created cracking system within the pre-existing soil profile. Possibly, this sorted sedimentation process is accompanied by a geyser-type eruption of mud originating from the soil, with concomitant deposition of sediments at the soil surface as can be observed in present-day forest sites (fig. 149).

Figure 149: Bleached mud deposits at the soil surface near a tree that was swaying during rainy, windy periods, with earth expulsed as small mud geysers from small holes near the tree trunk.

At the Verrebroek "Dok 1" site several of such structures were observed in the Bh-horizon of a Podzolised Soil that developed after the recurrent Mesolithic occupations. Later, as result of a gradual groundwater rise, this Podzolised Soil became buried under peat that gradually invaded larger part of the landscape. The horizontal sections show very few traces of roots that cross the stratified sediments of the tree-sway structure. These data permit to propose that the tree-sway structure developed in a phase when the Podzolised Soil development was already non-active, but originated before the burial by the Peat Soil. Most probably these structures are associated to the period when the deeper tree roots started to die as a result of the raising, permanent groundwater table. The gradual rise of the groundwater table could even have accelerated this type of bioturbation, as the deeper root system might have become absent in such conditions. Hence, trees during this period were probably of smaller size. This may explain why the tree-sway structures, although less common than the tree-fall structures, are not so uncommon on this site.

For a detailed description of these structures, we refer to Langohr and Louwagie (2000).

Spek *et alii* (2001a, 20) mention comparable morphological features in the "Hoge Vaart" sections (Flevoland, The Netherlands). They observed a particular microlamination with alternating thin black humiferous and very thin white sand layers, in and below the A-horizon. The A-horizon was originally considered to be the surface horizon of a Podzolised Soil. From micromorphological observations, they however concluded that the surface horizon was disturbed during posterior subaqeuous soil conditions, when the groundwater table rose above field level and subsequently enabled peat growth.

3.3.1.3 | Artefact and ecofact distribution

Method

At the level of the Early Mesolithic occupation, quadrangles of 0.50 by 0.50m large and 0.10m thick were wet sieved for artefacts and ecofacts. Chips (< 1cm) and artefacts (> 1cm) were separately counted. Ecofacts, such as carbonised hazelnut shells and burnt bone, were weighed in air-dry condition.

In a preliminary phase, only a limited number of sections were studied to detect possible artefact and ecofact migration patterns (see also Louwagie *et alii*, in press). For a more extended and in depth analysis of the artefact and ecofact distribution pattern at Verrebroek "Dok 1", we refer to Sergant (2004).

The sections in the preliminary study were chosen because of their particular soil morphology and distinguished horizon sequence (A-Bw.bi-C) that allowed interpreting them as tree-fall structures. Sections in north-south direction were made through the structures: S89, S90 and S93. Artefact and ecofact counts through tree-fall structures might tell us how tree-fall processes influenced artefact and ecofact migration. Additionally, orthogonal sections (in north-south and west-east direction) were made through artefact unit 15, an artefact unit visibly not disturbed by tree-fall processes.

The diagrams (Figures D.1 - D.11) give counts for chips (< 1cm) and artefacts (> 1cm) and masses (g) for carbonised hazelnut shells. These were recorded from the top of the Early Mesolithic occupation surface (= 0.0m) down to a depth where almost no chips or artefacts were found. The blocks represent the mentioned quadrangles of 0.50 by 0.50m large and 0.10m deep. Those in black indicate the cut through the tree-fall.

Chip and artefact distribution pattern

Surface-hearth in artefact unit 15 (fig. D.1, D.2)

This is a structure where the artefact concentration is not disturbed by biological processes. Figures D.1 and D.2 give an overview of the chip and artefact distribution pattern of artefact unit 15. Figure D.1 shows the north-south section, Figure D.2 the west-east section.
In north-south direction, the chip (< 1cm) and artefact (> 1cm) concentrations occur at relatively shallow depth and continue occasionally till 0.40m depth. Major part of the migration is however limited to about 0.20m. Along the west-east section, a similar pattern is observed.
This example shows that there must have been some, but however limited migration of chips and artefacts from the Mesolithic occupation surface towards deeper soil horizons. This limited migration can be attributed to at least three environmental factors.

i. After the Mesolithic occupation, soils evolved gradually towards a Podzolised Soil type with well-developed and slightly cemented accumulation Bh-horizon. Except for tree-fall processes, bioturbation by burrowing animals and roots in such very acid, nitrogen-poor soils becomes largely restricted to the A-, and to some degree to the E-horizon. Locally a root mat may develop on top of the cemented Bh-horizon and as such contribute to the accumulation of organic matter at this level.
ii. After the development of the Podzolised Soil, a Peat Soil gradually buried and sealed the whole site. Later on flood sediments further buried this peat cover. Over large areas the thickness of the peat and flood sediments has an amplitude of more than 0.50m, sufficiently thick to protect the large areas of Mesolithic remains from later human disturbances such as ploughing and house and field road construction.
iii. The presence of a groundwater table that, at least from the moment on that the Peat Soil developed, kept the artefact concentrations under the water table, and thus free from root penetration and burrowing by animals such as earthworms, ants and moles.

Tree-fall S89 (fig. 144, D.3, D.4, D.5)
This area was first selected because of a concentration of numerous burnt hazelnut fragments, seemingly indicating anthropogenic structures (Crombé *et alii* 1998). Careful examination showed however a sequence of tilted soil horizons that can be related to a tree-fall. Figure 144 shows a horizontal and a vertical section through the structure. The horizontal section is at 0.40m below the Mesolithic occupation surface. We observe a tilted surface A-horizon, followed in depth by a homogeneous brown, bioturbated B-horizon. The morphology informs that it corresponds to a tree that fell in southward direction. The diameter of the tree-fall at this depth is about 0.50m.
The artefact and ecofact distribution through S89 was followed from the Mesolithic surface down to 0.60m. Figure D.3 shows that the distribution pattern, averaged over this depth for chips, artefacts and carbonised hazelnut shells, is comparable. Figure

D.4 gives a detailed view of the changes in the chip distribution pattern. Relative to the quadrangles surrounding the tree-fall structure, we observe a decrease in chip concentration close to the Mesolithic occupation surface and an increase from 0.20m below this surface on. The highest chip counts appear at 0.30 to 0.40m depth, corresponding with the depth where the tree-fall morphology started to be visible. The highest chip concentration occurs at the side of the original surface A-horizon, appearing at the southern side of the tree-fall. The artefact distribution pattern (fig. D.5) clearly shows comparable results. However, as the tree-fall structure diameter of S89 hardly exceeds the quadrangle size, the overlap of the highest chip and artefact quantities with the appearance of the buried surface A-horizon, can be coincidental.
Nevertheless, the distribution pattern of S89 argues that a disturbance that occurred after the Mesolithic occupation but before the Podzolised Soil development is at least preserved in depth. It possibly shows that tree-fall processes at least locally influence the distribution of Mesolithic artefacts.

<u>Tree-fall S90</u> (fig. 145, D.6, D.7, D.8)
Figure 145 shows a horizontal and a vertical section through the structure. The horizontal section is at 0.50m depth where the tree-fall structure has a diameter of 0.80m. At the southern edge appear a surface A-horizon and an underlying, tilted, greyish B-horizon containing charcoal fragments. The morphology indicates that the tree fell in southward direction.
Figure D.6 illustrates that the distribution patterns averaged over 0.00 to 0.70m for chips, artefacts and carbonised hazelnut shells are comparable. Figure D.7 gives a detailed view of the changes in the chip distribution pattern with depth. Also here an increase in chip concentration with depth in the tree-fall structure, relative to the quadrangles surrounding the structure, is observed. The highest chip concentration occurs at the southern side of the surface A-horizon. However this side of the section also coincides with the appearance of a surface-hearth in southward direction, which explains the increasing chip concentration in the upper horizons. As for S89, the artefact distribution pattern in S90 (fig. D.8) shows results comparable with those of the chip distribution. Also here, the distribution patterns argue that a disturbance that occurred after the Mesolithic occupation but before the Podzolised Soil development is at least preserved in depth. It also shows that tree-fall processes at least locally influence the distribution of Mesolithic artefacts.

<u>Tree-fall S93</u> (fig. 146, D.9, D.10, D.11)
Figure 146 provides a horizontal and a vertical section through the structure. The horizontal section shows that the structure has a diameter of 1.80m at 0.30m depth. A surface A-horizon and an underlying, tilted and unfortunately hardly visible, brownish B-horizon are observed at the northern side of the structure. The tree-fall morphology once more illustrates that the tree fell in southward direction. Within the vertical section other fragments of surface A-horizon are visible. These might be peripheral strokes of surface A-horizon pulled off during the tree-fall and afterwards collapsing into and partly filling the tree-fall pit. A complex structure caused by several generations of tree-falls is not excluded either.

Figure D.9 shows the distribution patterns averaged over 0.00 to 0.50m for chips, artefacts and carbonised hazelnut shells. As compared to S89 and S90, this tree-fall structure gives a less clear pattern of the artefact and ecofact distribution. Figures D.10 and

D.11 give a detailed view of the changes in the chip and artefact distribution pattern with depth. The maximum chip and artefact concentration is situated at the side of the surface A-horizon (southern side). The highest artefact (> 1cm) counts appear at 0.30 to 0.40m depth, corresponding with the depth where the tree-fall morphology was visible in the horizontal section. This is however not true for the chip distribution pattern at that depth. Here the size of the artefacts might play a role in the migration pattern.

Conclusions

Over the different structures studied, we see that Mesolithic artefact and ecofact migration is limited, dominantly not more than 0.20m. It can be attributed to at least three environmental factors:

i) after the Mesolithic occupation, a relatively fast development of an acid, nitrogen-poor Podzolised Soil type with slightly cemented B-horizon of only a few decimetres depth and associated with poor faunal burrowing activities;
ii) the burial of the site by a Peat Soil and later by flood sediments;
iii) the gradual rising of the groundwater table creating an environment that was strongly restricting the bioturbation by fauna, such as worms, ants and moles.

Bio-galleries, tree-fall and tree-sway structures are the main disturbances that affected the remnants of the Mesolithic occupation level.

The analysed tree-fall structures nearly all date from the period after the Mesolithic occupation, but before the development of the Podzolised Soils. This may reflect an environment largely devoid of large trees during the period of Podzolised Soils and till shortly before the Peat Soil development.

In the tree-fall structures, detected on basis of the soil morphology, the spatial distribution pattern in depth for chips, artefacts and carbonised hazelnut is comparable. There is a relative increase in depth of the chip and artefact concentration. The highest concentration of chips and artefacts occurs at the side of the surface A-horizon of the buried soil. This side corresponds with the direction of the tree-fall. However, care should be taken when the diameter of the structure is only about the size of a quadrangle, as is the case with S89. Peaks in the artefact and ecofact distribution could then simply reflect those quadrangles where the intersection between the grid system and the detected tree-fall structure is the largest.

Nevertheless, the study of the artefact and ecofact distribution shows that, spite of the relatively small tree-fall structures (0.50 to 80m diameter in depth) observed, the quadrangle excavation method used at the Verrebroek "Dok 1" site, with 0.50 by 0.50m large and 0.10m thick units, was appropriate to detect and interpret the disturbances among these elements, at least when the structure diameter exceeded the grid size.

3.3.2 | Doel "Deurganckdok-sector B"

The type of archaeological structures discovered here are so-called hearth-pits. The archaeological characterisation of these structures is restricted to those horizons that are dark-coloured and contain visible amounts of charcoal and charcoal powder (cf 3.1.3.1.); they are further described as the hearth-pit structure *sensu stricto*. These structures were dug out deeper for pedological purposes in order to observe eventual associated soil characteristics; S55 and S58 are good examples of these and moreover show a comparable morphology and horizon sequence.

Hearth-pits S55 (fig. 150, tab. 57, D.34, D.35, D.41, D.44)
Hearth-pit S58 (fig. 151, tab. 58, D.36, D.37, D.41, D.44)

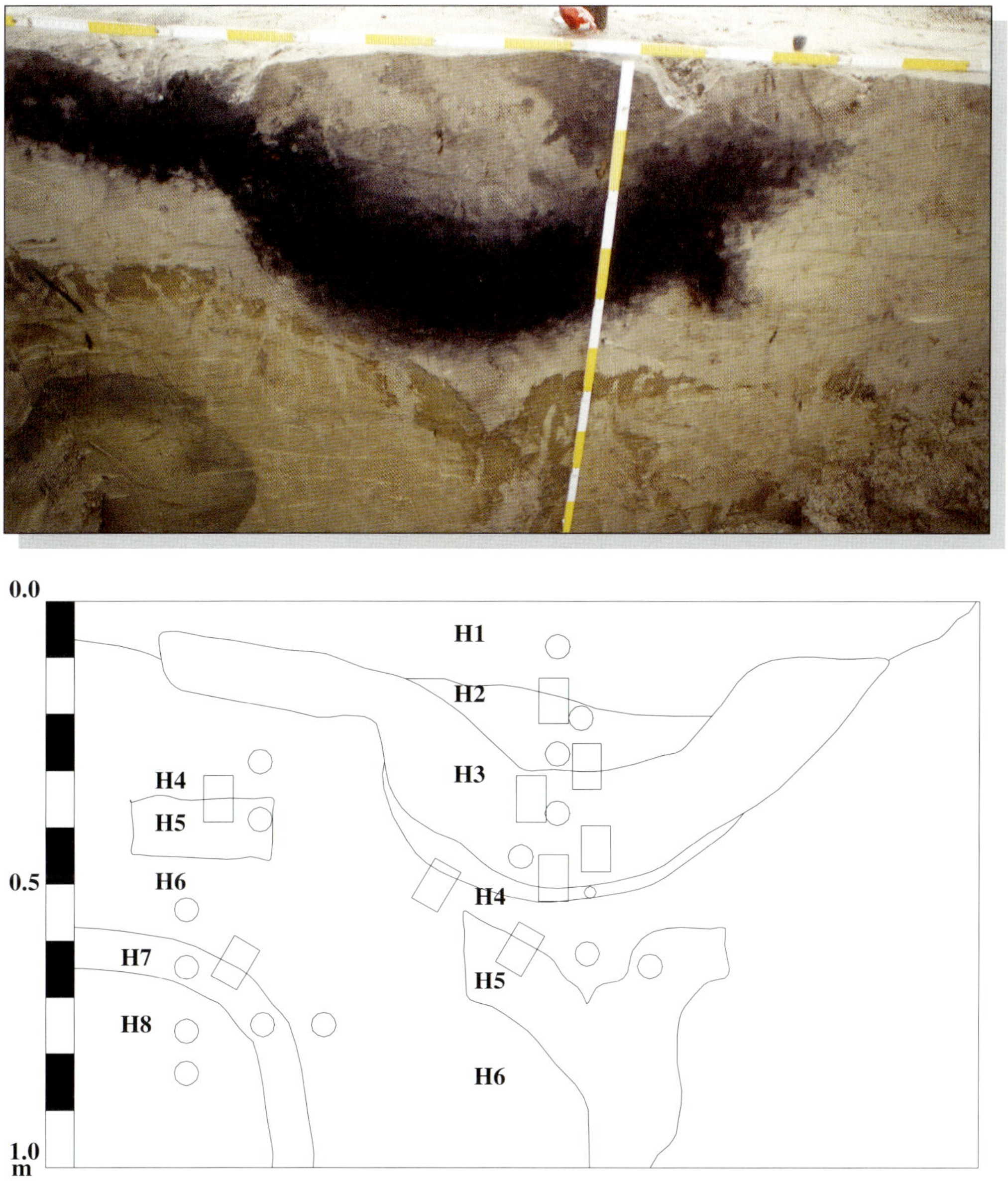

Figure 150: Doel "Deurganckdok", hearth-pit S55, photographic and schematic representation.

Horizon number	Symbol	Horizon depth (cm)	Profile description
H1		(0)-0/21	Light grey matrix; zones with grey - dark grey matrix; sand; bioturbated; few roots
H2		13/21-13/30	Grey matrix; sand; bioturbated; few roots
H3	Aa?	5/30-11/51	Black matrix; sand; common charcoal and charcoal-powder; clear-gradual upper and lower boundary
H4	1C?	0/52-35/71	Light grey matrix; sand; parent material?; few roots
H5	Bh?+Bt	35/71-45/(100)	Light olive brown matrix; loamy sand?; discontinuous clay accumulation horizon; lamina of very dark grey matrix at top, possibly due to organic matter accumulation
H6	2C?	45/(100)-57/(100)	Light brownish grey - light yellowish brown matrix; glauconiferous sand?; parent material?
H7	Bh?+Bt	57/(100)-65/(100)	Local; greenish grey matrix; loamy sand?; clay accumulation horizon; lamina of very dark grey matrix at top; possibly due to organic matter accumulation
H8		65/(100)-(100)	Local; olive grey matrix; sand

Table 57: Doel "Deurganckdok-sector B"; hearth-pit S55, macromorphological observations.

Figure 151: Doel "Deurganckdok", hearth-pit S55, photographic and schematic representation. The bluish colour is a reproduction artefact.

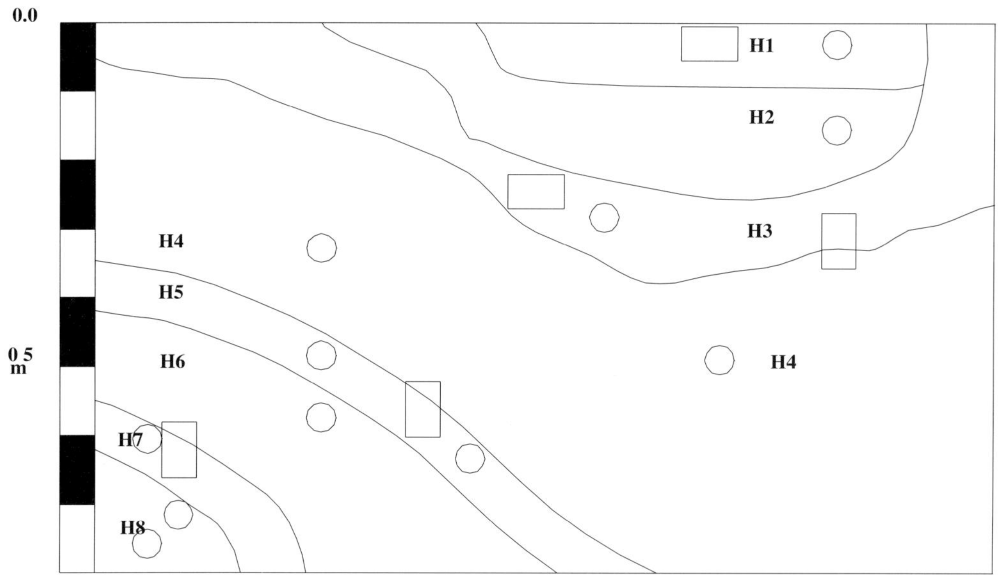

Figure 151: Doel "Deurganckdok", hearth-pit S58, photographic and schematic representation.

Horizon number	Symbol	Horizon depth (cm)	Profile description
H1		(0)/8/10	Grey - dark grey matrix; sand; bioturbated; few roots
H2		(0)/10-(0)/27	Dark grey - very dark grey matrix; sand; bioturbated; few roots; cf clear-gradual boundary of H3 in S55?
H3	Aa?	(0)-(0)/38	Black matrix; sand; common charcoal and charcoal powder; clear upper and lower boundary
H4	1C?	5/38-35/(80)	Light grey matrix; sand; parent material?; few roots
H5	Bh? + Bt	35/(80)-42/(80)	Light olive brown matrix; loamy sand?; discontinuous clay accumulation horizon; lamina of very dark grey matrix at the top, possibly due to organic matter accumulation
H6	2C?	42/(80)-55/(80)	Light brownish grey - light yellowish brown matrix; glauconiferous sand?; parent material?
H7	Bh? + Bt	55/(80)-65/(80)	Local; greenish grey matrix; loamy sand?; clay accumulation horizon; lamina of very dark grey matrix at the top, possibly due to organic matter accumulation
H8		65/(80)-(80)	Local; olive grey matrix sand

Table 58: Doel "Deurganckdok-sector B"; hearth-pit S58, macromorphological observations.

As mentioned above, both structures show a comparable morphology and horizon sequence. The hearth-pit structure *sensu stricto* consists of H1 to H3. It is strongly bioturbated, with resulting gradual boundaries. The clay accumulation horizon, H5, is not parallel to the present-day surface but rather follows the hearth-pit morphology. Hence, it is considered as an associated pedogenic horizon to the hearth-pit. H5 is strongly disturbed due to biological, probably mainly root and faunal activity, however leaving sharp horizon boundaries. Surrounding horizons H4 and H6 are considered to be the original soil or eventually parent material in which the hearth-pit structure and associated pedogenic horizons were formed. Both horizons show traces of biological activity, mainly by roots in H4 and only features of faunal activity in H6. Observations below the hearth-pit structure *sensu stricto* revealed that H7 and H8 symmetrically follow the structure outline (fig. 151). H7 represents another clay accumulation horizon. As opposed to the overlying horizons, H7 and H8 still show root activity, but no other forms of bioturbation. As a consequence, the strongly bioturbated zone has a somewhat pocket-like form. The top of H7 possibly represents the groundwater table level.

The faunal bioturbation is characterised by, from a few mm up to a few cm wide, mainly vertically oriented channels (fig. 150, 151). Their size could point towards channels created by mound-building ants, varying from 0.16 to 2.5cm in diameter (Green *et alii* 1998). With regard to this, it is questioned whether the structure could correspond to an abandoned and collapsed ant mound, rather than to an anthropogenic structure. Effect of mound-building ants on soil characteristics is discussed in Appendix B. Glossary.

The groundmass in and around the hearth-pits consists of dominant (> 50%) subrounded (polycrystalline) quartz, few (5 - 10%) in H4 to H7 or very few (2 - 5%) in H1 to H3 rounded glauconite that is partly, mainly chemically weathered and very few (< 2%) subrounded to rounded chert, feldspars and mica. The standard deviation of the particle size distribution (SD_S55 = 74 - 90µm, SD_S58 = 70 - 116µm) (tab. D.41) points toward a very poorly sorted sediment. Referring to the sorting charts used in micromorphology (Bullock *et alii* 1985, 27), sorting is moderately to poor. Micromorphological observations reveal as well that packing is dense in H7 and H5 and is loose in the other horizons, which could explain why clay accumulated preferentially in these horizons.

The particle size analysis (tab. D.35, D.37) confirms that H1 to H3, or the hearth-pit structure *sensu stricto*, have a high textural similarity (SI ≥ 95). It should be mentioned here that treatment of the samples with 30%-H_2O_2 prior to particle size analysis doesn't destroy carbonised organic matter, the characteristic component in H1 to H3. Hence, common presence of carbonised organic fine mass and particles contributes to the high similarity. H5, recognised both at macro-and microscale as a clay accumulation horizon, shows an important difference with its surrounding horizons H4 and H6 (SI around 80); H5 indeed contains considerably more clay and fine to coarse silt (tab. D.35, D.37). Texture of H4 and H6 has furthermore a high similarity above 95. Hence, the textural difference of H5 with its surrounding horizons is clearly due to pedogenesis. Particular characterisistics of this clay accumulation horizon will be discussed below.

H7 is horizontally to slightly obliquely laminated; the lamination is possibly original and thus independent of the hearth-pit structure that formed within the sediment. This horizon has two types of clay coatings; (i) very few (< 2%) thin (< 5µm) slightly dusty light green (PPL), probably glauconitic clay coatings are found next to (ii) very few (< 2 - 5%), up to 100µm thick, very dusty greyish brown clay coatings and infillings; here clay is mixed with contrasting black organic micro-particles. The glauconitic clay coatings were regularly observed in the Upper Pleniglacial coversands of the Verrebroek site and are hence probably characteristic of the parent material here in Doel as well. These coatings rather point towards mechanical glauconite breakdown (Van Ranst, De Coninck 1983). The dusty clay pedofeatures show a strong continuous orientation, parallel to the void surface and have colours of 2nd order birefringence (tab. D.34, D.36, cf tab. D.49). The parallel orientation calls for illuviation, rather than neoformation. Furthermore, based on the optical characteristics of the clay minerals, it is known that clay coatings with high order birefringence are in favour of dominant 2/1-clay type components, such as e.g. micas. Parallel, this horizon also has a high cation adsorption capacity, reflected in relative peaks of CEC, Ca, Na, Mg and K.

In S58, additional very dusty slightly oriented fragmented clay infillings are present. They possibly represent former surface crust fragments formed at the surface at the moment of sand deposition and with time incorporated in the groundmass.
The transition of H7 to H6 is less than 0.5cm thick and consists of alternating laminae of fine- and coarse-grained quartz. Clay coatings of the same type and amount as in H7 are present here as well. Additionally, very few (< 2%) coatings composed of carbonised organic punctuations (cf tab. D.47) are visible; they are deposited on clay coatings and their genesis is thus posterior to (mica) clay migration.

H6 has very few (< 2%), thin (< 5µm) slightly dusty light green (PPL), probably glauconitic, clay coatings. The chitonic coarse/fine related distribution, corresponding to a pellicular grain microstructure, indicates that these coatings are present throughout the horizon. They might account for the more brownish matrix colour when compared to H4.

The clay accumulation horizon, H5, has, next to very few (< 2%) thin, probably glauconitic clay coatings, few (5 - 10%), up to 100µm thick, very dusty greyish brown clay coatings and infillings; the clay has contrasting black organic micro-particles. The clay pedofeatures show a strong continuous orientation, parallel to the void surface and have colours of 2nd order birefringence (tab. D.34, D.36, cf tab. D.49). The parallel orientation calls for illuviation, rather than neoformation. Clay coatings of the same type were observed in H7 as well, however in smaller quantitiy. Also here, the illuviated clay might be of a mica type; micas, amongst others glauconite, are indeed observed in the groundmass. Since a peat cover sealed the coversands, it is unlikely that the observed 2/1-type clay coatings originated from the overlying flood clay. Seen the associated morphology of H5 and H7 to the hearth-pit *sensu stricto*, we assume that the (mica) clay migration is related to the fire event. The different clay illuviation horizons (H5 and H7 in S55, S58; H3 and H9 in S75) possibly correspond to different clay migration events or might be synchronous and simply reflect different zones within the soil profile where clay accumulation was favoured, e.g. a consequence of dense packing. Clay migration is favoured by a cation status on both the organic and mineral colloids that is neither dominated by Ca^{2+} (reflected in a high pH), nor dominated by Al^{3+} (reflected by a low or acid pH). A strong Na^{+}-saturation may also activate the colloid migration process. Increased sodium (Na) content may result from either (i) fire activity, or (ii) influx of Na through incoming brackish or salty water during posterior flood events. Parallel, those horizons also have a high cation adsorption capacity, reflected in relative peaks of CEC, Ca, Na, Mg and K. Very few (< 2 - 5%) coatings and link cappings composed of carbonised organic punctuations (cf tab. D.47) are present as well, mainly in the top of H5; these pedofeatures are eventually formed on the former deposited clay coatings. Carbonised organic matter from the hearth-pit evidently moved here as organic punctuations of about 1µm large through the soil.

The transition of H5 to H4 is about 0.5cm thick and contains very few (2 - 5%) clay coatings and infillings of the same type as described in H5. It has very few (2 - 5%) in

S55, to few (5 - 10%) in S58, carbonised organic punctuations coatings and infillings (cf tab. D.47), in S58 obviously a higher quantity than observed in H5.

The horizon underlying the hearth-pit structure *sensu stricto*, H4, contains very few (2 - 5%) carbonised organic punctuations coatings and infillings, very few (< 2%) carbonised up to 50µm large organic particles and very few (< 2%) up to 1mm large charcoal fragments (cf tab. D.47).

In the field no stratification was detected on the transition from H4 to H3. In thin section however, lamination was observable in S55 and S58, however less clear in S58 due to diagenetic bioturbation processes. The boundary between H4 and H3 is about 0.5cm thick and shows in thin section a lamination sequence of alternating groundmass with gefuric and monic coarse/fine related distribution, corresponding respectively to bridged grain and single grain microstructure (tab. D.34, D.36). The bridged grain microstructure refers to the carbonised organic fine mass coatings and infillings as observed in H3. The single grain microstructure is observed along voids. It is questioned whether the single grain microstructure corresponds to sliding down of sand grains during eventual emptying of the hearth-pit for subsequent re-use. Latter multiple use of hearth-pits was suggested by Exaltus 2001 (34-35) for the hearth-pits at the "Hoge Vaart" (Flevoland, The Netherlands). Other possibilities are the diagenetic sedimentary fill of galleries originated by decayed roots or drilled by animals. However, the alternating laminae possibly show an eluviation - illuviation pattern of carbonised organic fine mass. Latter would explain the lamination as the result of a migration, rather than of a sedimentation process. The morphology of the coatings and infillings is however markedly different from the carbonised organic matter coatings consisting of 1µm-size carbonised organic punctuations (cf tab. D.47), mainly observed in H5 and the transition to H4.

In H3 characteristics of the hearth-pit *sensu stricto* are most pronounced. It has a black matrix and in its centre it contains frequent (20 - 30%) carbonised organic fine mass coatings and infillings (cf tab. D.47), very few (2 - 5%) up to 80 µm large carbonised organic particles and very few (2 - 5%) up to 1cm large charcoal fragments (tab. D.34, D.36). The C/N-ratio in this Aa-horizon is around 60, which fits with common values for wood ash. Phytoliths were not observed in thin section. Latter characteristics could indicate use of wood, rather than dried grass e.g., as fuel. Charcoal determination indeed revealed oak (*Quercus*) (cf 3.4.3.1.). Both the upper and lower boundaries of H3 are clear to gradual and have bioturbated morphology.

H1 and H2 have a light to dark grey matrix colour and show clear traces of bioturbation as well. They contain very few (< 2 - 5%) carbonised organic fine mass coatings and infillings, very few (< 2 - 5%) in H1 to few (5 - 10%) in the base of H2, up to 80µm large carbonised organic particles and very few (< 2%) up to 0.5mm large charcoal fragments (tab. D.34, D.36) (cf tab. D.47). H1 to H3 all contain very few (< 2 - 5%) non- to slightly birefringent organ residue (root, sclerotium) and very few (< 2%) amorphous dark brown plant fragments (tab. D.34, D.36).

Organic matter content determined by loss on ignition is highest in H3 and shows a decreasing gradient over H2 and H1. In H3 values are from 2.6% to 4.8%; H2 still has an organic matter content of 1.0 to 1.2%. In H3, the organic matter content determined by loss on ignition appears definitely higher than the non-carbonised organic matter recognised in thin section. It is questioned whether the present carbonised organic matter in the soil, with an ignition temperature of 300°C (McDonnell 2001), still looses weight by heating at a higher temperature of 500°C in the laboratory.

Nitrogen (N) and phosphate (P_2O_5) contents show a parallel decrease from H3 to H1. Nitrogen (N) content varies between 0.03 to 0.04% in H3 and 0.01 to 0.02% in H1 and is not considerably higher than in the reference profile P1. This is not in favour of the ant mound hypothesis, since Green *et alii* (1998) report up to ten times higher N-contents in ant mound horizons, as compared to adjacent soil horizons. Total phosphate contents range from 869 to 1038mg kg^{-1} in H3 to 379 to 103mg kg^{-1} in H1. Phosphate is dominantly inorganic with Pi/Po-ratios moving between 1 and 11. Even cation exchange capacity (CEC) and the exchangeable cations calcium (Ca), sodium (Na), magnesium (Mg) and potassium (K) show the same decreasing gradient from H3 to H1. The cation exchange capacity of 10 to 18cmol(+) kg^{-1}, as well as the cation (Ca, Na, Mg, K) contents are relatively high in H3 when compared to the reference profile P1. High nutrient release must be linked to *in situ* burning and hence mineralisation of wood, also proven by common presence of charcoal fragments, carbonised particles and organic fine mass. Especially K, Ca, Mn, Mg and P are typical elemental components of wood ash (Jansen, Peeters 2001, 42-43). Phytholiths were not observed in thin section, which indicates that input of grass or other non-arboreal vegetative material as fuel must have been restricted, as these produce relatively high amounts of phytoliths upon burning. Absence of direct evidences of fire impact on the soil in the hearth-pit structures might point towards anaerobic burning conditions, as is e.g. the case in charcoal production where the charcoal heap is covered with mud or in particular types of Polynesian earth-ovens where food is rather vaporised than cooked. However, as in Verrebroek "Dok 1" (cf 3.3.1.1.), the parent material possibly contains too little iron to result in reddening on fire. Combined (i) presence of carbonised organic fine mass next to charcoal fragments, and (ii) absence of (morphological indicators for) ash as is the case in the hearth-pits at Doel "Deurganckdok", are also indicative of intentional incomplete burning (Exaltus 2001, 19-35). In H1 and H2 of S55, exchangeable sodium percentages (ESP) exceed the threshold value of 15; sodium adsorption ratios (SAR), taking the neutralising effect of calcium and magnesium into account as well, are however below the threshold value of 13 for soils with sodic properties.

Considering the complete pedo-lithostratigraphy of the hearth-pits *sensu largo*, cation exchange capacity and cation contents show relative peaks in H3, H5 and H7, seemingly horizons with high adsorption capacity. Either clay or (carbonised) particulate organic matter can account for the latter property. In S55, absolute peaks for these attributes, except K, occur in H3; K has its peak in H5. In S58, absolute peaks for Na and K occur deeper down in the profile; Na moved further (into H7) than K (into H5) and are indicative of a leaching process active from the soil surface. As in H1 to H3, high ESP-values, exceeding the threshold value of 15, are also determined in the horizons underlying the hearth-pit *sensu stricto*. The highest values appear in H8 and H7, to a minor degree in H6, showing a decreasing gradient from bottom to top. H7 and H6 contain well-distributed glauconitic clay coatings. SAR-values in H8 and H7 are above or approaching the threshold value of 13; the highest values appear in H7. As mentioned above, H8 and H7 also follow the hearth-pit morphology, but are not bioturbated to the same degree as the overlying horizons, including the hearth-pit *sensu stricto*. As at Verrebroek "Dok 1" (cf 2.1.4.3.,

2.1.4.7.), either (i) input/discharge of salts from the underlying marine substratum (Tertiary Kattendijk and Kallo sands) through a fluctuating groundwater table, or (ii) flooding by salt-containing, brackish or salty water can explain this phenomenon. Discharge of salts via the groundwater table is most probable in the deepest soil horizons, especially those below the assumed groundwater table (top of H7); moving from H8 to H6, ESP-values show a decreasing trend. Incoming of brackish or salty water, concomitantly with flood events, seems more probable for the top horizons (H1 and H2) in S55. As sodium moves more rapidly than the other cations, the highest sodium quantity in those horizons that easily retain cations would reflect the leaching front in latter case.

In S55, a decreasing gradient from bottom to top in exchangeable calcium (Ca), sodium (Na) and potassium (K) can be followed over the horizons H8, H6 and H4. For magnesium (Mg), this decrease is only observed when moving from H6 to H4. Free iron (Fe_d) contents follow the same pattern. H8, H6 and H4 don't show any particular feature directly or indirectly associated with the hearth-pit structure and are considered to represent the original soil wherein the structure developed. The decreasing cation and free iron pattern possibly points towards chemical weathering of mica (K), amongst others glauconite (Mg, Fe) and feldspars (Ca, Na) minerals, both present in the groundmass. Relative peaks of these cations and free iron in both the clay accumulation horizons, H5 and H7, support the hypothesis that the clay coatings (partly) consist of mica clay types. Furthermore, high Na-values enable clay dispersion and hence migration and are partly responsible for moderately alkaline pH values, both discussed earlier. Parallel, all these relatively high amounts of soluble cations observed here reflect that these soil horizons have never been leached strongly by the percolating rainwater, probably because of the presence of a permanent groundwater table nearby. Latter was probably situated at the level of H7, as evidenced by the absence of faunal bioturbation from this level onwards.

The cation distribution over the different horizons probably reflects the combined or superposed influence of (i) fire activity, (ii) mineral weathering in the groundmass, (iii) discharge from the underlying Tertiary sands via the groundwater table, and (iv) incoming of brackish or salty water with flood events. Incoming of brackish or salty water from the surface with flood events at Doel "Deurganckdok-sector B" must have been posterior to the Swifterbant occupation. Based on archaeozoological and botanical findings, a fresh water environment is proposed for the Doel Swifterbant occupation level (cf 3.5.4.). Dating on charcoal from the hearth-pits *sensu stricto* gives dates between 7865±35 and 7580±35 BP; datings on charcoal, carbonised hazelnut and pottery situate the Swifterbant occupation at Doel "Deurganckdok-sector B" between 6015±30 (W24/Z2(4)) and 5220±55 BP (NZA-12076) (cf 3.2.4.). The Swifterbant occupation is thus posterior to the development of the hearth-pits. As the site was not yet flooded by brackish or salty water at the time of the Swifterbant settlement, it is unlikely that the sea level would have had any impact on the hydrology of the site a few thousand years earlier.

Hearth-pit S75 (fig. 152, tab. 59, D.38)

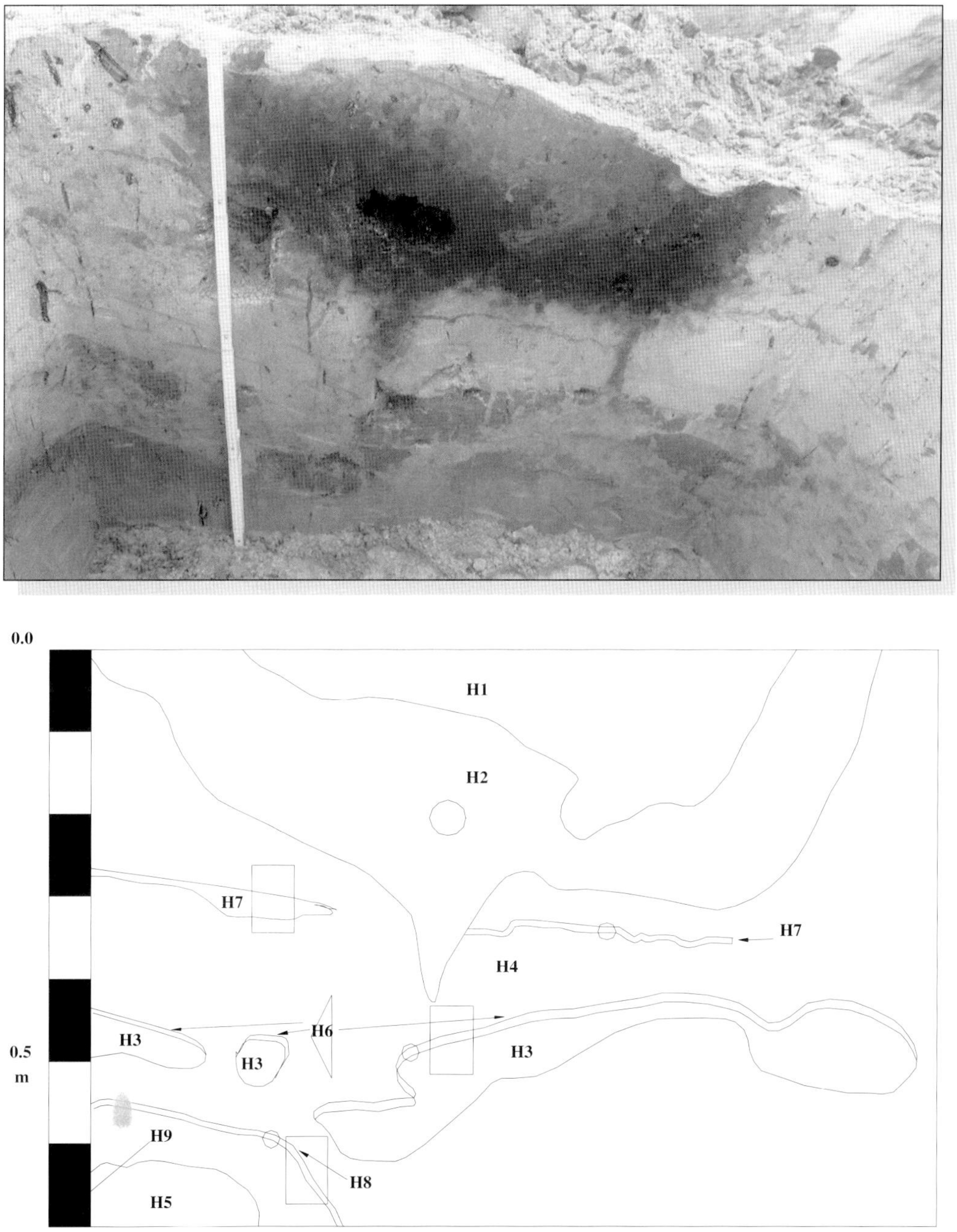

Figure 152: Doel "Deurganckdok", hearth-pit S75, photographic and schematic representation.

Horizon number	Symbol	Horizon depth (cm)	Profile description
H1		(0)-(0)/23	Light grey - grey matrix; sand; bioturbated; few roots
H2		(0)/23-(0)/43	Very dark grey - black matrix; sand; common charcoal and charcoal powder; clear upper and lower boundary
H3	Bt1?	27/57-28/62	Greyish olive matrix; loamy sand?; discontinuous clay accumulation horizon?
H4	C?	(0)/43-42/55	Light grey matrix; sand; parent material?; few roots; locally horizontal laminae of brown matrix (H7)
H5	Bt3?	62/(70)-(70)	Local; light greenish grey matrix; (loamy?) sand; reduced?
H6	Bh1?	41/55-42-56	Local; horizontal laminae (3-5 mm) of very dark grey matrix; organic matter accumulation?
H7		32-35	Local; horizontal laminae (2 mm) of brown matrix; loamy sand?
H8	Bh2?	53/(70)-55/(70)	Local; horizontal laminae (3-5 mm) of very dark grey matrix; organic matter accumulation?
H9	Bt2?	55/(70)-62/(70)	Local; light bluish grey matrix; loamy sand?; clay accumulation horizon?

Table 59: Doel "Deurganckdok-sector B"; hearth-pit S75, macromorphological observations.

The macromorphology of this hearth-pit structure is comparable to those of S55 and S58. Horizon number designations are however not corresponding. Only morphological analysis, neither particle size nor chemical analysis has been performed on this structure.

Micromorphologcial characteristics of the different horizons are comparable to those of the corresponding horizons in S55 and S58. S75 additionally has a few mm thick clay accumulation band (H7) that is cut by the hearth-pit structure *sensu stricto*. It is characterised by few (5 - 10%), up to 100µm thick, very dusty greyish brown clay coatings and infillings; also here the clay is dotted with contrasting black microparticles and has a strong continous orientation with birefringence colours of 2nd order (cf tab. D.49). The top of H7 in S75 furthermore has few (5 - 10%) carbonised organic punctuations and very few (< 2%) carbonised organic punctuations coatings, deposited on clay coatings (cf tab. D.47). This type of clay coating corresponds to those observed in the above-described clay illuviation horizons H7 and H5 in S55 and S58. However, as H7 is cut by the hearth-pit structure *sensu stricto*, it might correspond to the discontinuous clay accumulation bands (H8) as observed in the reference profile P1. The macromorphology of the discontinuous clay accumulation bands (H8) in P1 and H7 in S75 is moreover comparable. Duchaufour (1982, 270-305) considers clay migration as a secondary process superposed on the brunification process. The originally present clay accumulation band H7 thus represents a favourable zone for additional clay accumulation and easily withholds clay and organic matter migrating as a result of fire activity. As a consequence, H7 on the one hand and H3 and H5 on the other hand possibly correspond to different clay migration events. All three of them represent zones within the soil profile where clay accumulation was favoured.

3.3.3 | Conclusions

3.3.3.1 | Verrebroek "Dok 1"

At Verrebroek "Dok 1" both anthropogenic and natural structures were studied. The groundmass within these structures consists of dominant well to moderately sorted, subrounded quartz and polycrystalline quartz and very few subrounded to rounded chert, feldspars, partly weathered glauconite and other mica minerals.

Anthropogenic features

Two different anthropogenic structure types could be distinguished: (i) hearth-pits and (ii) surface-hearths.

The hearth-pits have in common that they are easily delineated based on their dark soil matrix colour. They are half-spherical-shaped and measure commonly 0.5 to 0.7m in diameter and are on average 0.2 to 0.5m deep. They commonly contain charcoal fragments and charcoal powder. Carbonised organic matter is the only direct evidence of fire activity. Presence of carbonised fine organic mass indicates a fair chance of *in situ*, incomplete burning. Direct evidences of fire activity on the soil mineral material, as e.g. reddened soil, are absent, probably as a consequence of very low iron content in the parent material. There are indications of cations released on burning in the hearth-pit structure that leached deeper into the soil. Tar remnants are absent in and below the hearth-pit structures, which leaves the question open whether these structures were related to cooking activities. However, the dominance of organic over inorganic phosphate in the hearth-pit as opposed to the surrounding soil sediment possibly indicates organic residues of dead plant and/or animal material.

As opposed to the hearth-pits, surface-hearths are not easy to delineate based on soil matrix colour. The archaeologists defined possible surface-hearths based on the clustered presence of overheated lithic artefacts over one square metre. At Verrebroek "Dok 1", the surface-hearth always coincides with the highest quantities of burnt bone fragments, and frequently, but not always with the highest quantities of carbonised hazelnut shells.
An extended cross-section through a surface-hearth indicates an increasing, albeit not continuous, trend of indicators, such as presence of carbonised organic fine mass, a Pi/Po-ratio close to or over 1, and increasing total phosphate content towards the surface-hearth. Within the particular context of the studied surface-hearth, they indicate anthropogenic input of bone and/or hazelnut with subsequent burning and are thus possible indicators of *in situ* fire activity.
Different types of ecofact clusters were investigated separately. The hazelnut cluster has dominance of inorganic over organic phosphate, but no exceptionally high cation content. It is questioned whether burning of hazelwood results in a particular release of phosphorus, without affecting other nutrient cycles, or stores, as charcoal, appreciable amounts of phosphate. The burnt bone cluster has a high, mainly inorganic phosphate content that is related to the presence of bone. Amorphous phosphate

infilling fragments call for bone alteration. The hazelnut/charcoal cluster has a slight dominance of organic over inorganic phosphate that fits with input of organic material, also attested to by the presence of charcoal.
All investigated anthropogenic disturbances gave negative results on tar analysis.

Natural features

Bio-galleries, tree-fall structures and tree-sway structures are the main disturbances recognised at the site.

The original soil type, present during the Mesolithic occupation, was a Brunified Soil, with a black surface A- and light greyish brown B-horizon. Preservation of this horizonation at Verrebroek "Dok 1" only occurred in those tree-fall structures that were deep enough and of a type that enabled preservation. In a particular case, absence of fine carbonised organic matter excluding *in situ* burning within the structure further supports the tree-fall hypothesis. (i) Alteration characteristics in the assumed Bw.bi-horizon, such as redder hue and higher clay content and to some degree the relatively low $Al_o + 1/2\ Fe_o$ content, and (ii) textural difference between the assumed Brunified Soil horizons and the adjacent Podzolised Soil horizons, are in favour of Brunified Soil remnants within those tree-fall structures.

Tree-sway structures show clear traces of water-transported sediment in horizontal sections. In vertical sections, the feature is composed of up to several centimetres thick layers, consisting of the laminated, bleached coarse fractions of the soil matrix, alternating locally with black humiferous laminae. The morphology calls for a type of disturbance that occurs in stormy periods when soil moisture content is at or nearby saturation conditions. These structures probably developed in a period when the Podzolised Soil development was already non-active, but originated before the burial by the Peat Soil and are then associated to the raising, permanent groundwater table.

Artefact and ecofact distribution

Mesolithic artefact and ecofact migration is limited, dominantly not more than 0.20m. It can be attributed to at least three environmental factors:
i) after the Mesolithic occupation, a relatively fast development of an acid, nitrogen-poor Podzolised Soil type with slightly cemented B-horizon of only a few decimetres depth and associated poor faunal burrowing activities;
ii) the burial of the site by a Peat Soil and later by flood sediments;
iii) the gradual rising of the groundwater table creating an environment that was strongly restricting the bioturbation by fauna, such as worms, ants and moles.

Bio-galleries, tree-fall and tree-sway structures are the main disturbances that affected the remnants of the Mesolithic occupation level. The analysed tree-fall structures nearly all date from the period after the Mesolithic occupation, but before the development of the Podzolised Soils. This may reflect an environment largely devoid of large trees during the period of Podzolised Soils and till shortly before the Peat Soil development. In the tree-fall structures, detected based on the A-, E- and Bw.bi-horizon sequence, the spatial distribution pattern in depth for chips (< 1cm), artefacts (>

1cm) and carbonised hazelnut is comparable. There is a relative increase in depth of the chip and artefact concentration. The highest concentration of chips and artefacts occurs at the side of the surface A-horizon of the original Brunified Soil. This side corresponds with the direction of the tree-fall. However, care should be taken when the diameter of the structure is only about the size of a quadrangle, as is the case with S89. Peaks in the artefact and ecofact distribution could then simply reflect those quadrangles where the intersection between the grid system and the detected tree-fall structure is the largest.

Nevertheless, the study of the artefact and ecofact distribution shows that, spite of the relatively small tree-fall structures (0.50 to 80m diameter in depth) observed, the quadrangle excavation method used at the Verrebroek "Dok 1" site, with 0.50 by 0.50m large and 0.10m thick units, was appropriate to detect and interpret the disturbances among these elements, at least when the structure diameter exceeded the grid size.

3.3.3.2 | Doel "Deurganckdok-sector B"

The so-called hearth-pit structures at Doel "Deurganckdok" are dark-coloured and contain visible amounts of charcoal and charcoal powder; they are further described as the hearth-pit structure *sensu stricto*; the darkest part is indicated as an anthropogenic Aa-horizon. These structures were dug out deeper for pedological purposes in order to observe eventual associated soil characteristics. The groundmass consists of dominant moderately to poorly sorted, subrounded quartz and polycrystalline quartz, few to very few rounded glauconite that is partly, mainly chemically weathered and very few subrounded to rounded chert, feldspars and mica. Based on the different characteristics observed in and around the hearth-pit structures, the following time sequence is proposed.

1. Decreasing gradients from bottom to top for cations and free iron provide evidence for chemical weathering of mica and feldspars minerals. Both weathered and unweathered glauconite was observed, which favours weathering prior to or independent of fire events.

2. Some clay accumulation horizons are cut by the hearth-pit *sensu stricto* and are comparable to the discontinuous clay accumulation bands (H8) observed in the reference profile P1. This clay migration phase probably represents a secondary process superposed on the brunification process as evidenced in the reference profile P1.

3. Hearth-pits have been installed between 7865±35 and 7580±35 BP. The only direct evidences of fire present within the hearth-pit structure *sensu stricto* are carbonised organic matter present in different forms: fine mass, particles up to 80µm large and fragments up to 1cm large. They are however indicative of incomplete burning. Phytoliths are absent, probably indicative for the use of wood as fuel. Indeed, C/N-ratios in the Aa-horizon are around 60, common for wood ash, and charcoal determination revealed mainly oak (*Quercus*).

4. In and around the hearth-pit structures, the phosphate is mainly inorganic. Inorganic phosphate, nitrogen and cation contents are especially high in the Aa-horizon where accumulation of carbonised organic matter is most pronounced. High nutrient release must be linked to *in situ* burning and hence mineralisation. The low iron content of these soils probably favoured the immobility of the phosphorus compounds which otherwise would have migrated when the groundwater table rose afterwards.

5. Absence of direct evidences of fire impact on the soil in the hearth-pit structures might point towards anaerobic burning conditions, as is e.g. the case in charcoal production where the charcoal heap is covered with mud or in particular types of Polynesian earth-ovens where food is rather vaporised than cooked. Absence of direct evidences asks for indirect evidences, associated characteristics that are possibly linked to fire events. However, as in Verrebroek "Dok 1", the parent material possibly contains too little iron to result in reddening on fire.

6. As opposed to the strongly to moderately acid reference profile, pH values in and around the hearth-pits are neutral to moderately alkaline. Fire events possibly raised the pH of the soil complex from acid to alkaline. At some moment during this rapid increase, physico-chemical conditions must have been favourable for clay migration (pH 5 - 7). High sodium values moreover enable clay dispersion and hence migration and are partly responsible for moderately alkaline pH values.

7. The clay accumulation horizons below the hearth-pit structures *sensu stricto*, evidenced by an important increase in the clay and silt fraction and by the presence of dusty clay coatings, follow the hearth-pit morphology. Their formation thus seems to be associated to the fire event and possibly even belongs to the hearth-pit structure *sensu largo*. The parallel continuous orientation of the clay to the grain surface or void wall calls for clay illuviation. The 2nd order birefringence of the illuviated clay calls for a 2/1-clay type, as e.g. mica. Since a peat cover sealed the coversands, it is unlikely that the observed 2/1 clay coatings originated from the overlying flood clay. Seen the associated morphology of these clay accumulation bands to the hearth-pit *sensu stricto*, we consider the (mica) clay migration as a consequence of the fire event. The different clay illuviation horizons possibly correspond to different clay migration events or might be synchronous and simply reflect different zones where clay accumulation was favoured. Clay migration was initiated by an increasing pH and concomitant clay dispersion on increasing sodium content resulting from either (i) fire activity, or (ii) influx of sodium through incoming brackish or salty water during posterior flood events. Parallel, those horizons also have a high cation adsorption capacity, reflected in relative peaks of cation exchange capacity and cations.

8. In addition to the dusty clay coatings in the just-mentioned clay accumulation horizons, coatings and link cappings composed of carbonised organic punctuations are present, mainly in the transition towards the overlying horizons; these pedofeatures are commonly formed on the (mica) clay coatings. Their deposition is thus posterior to the (mica) clay migration. Carbonised organic matter from the hearth-pit evidently moved as organic punctuations of about 1μm large through the soil.

9. Morphological observations showed that faunal activity disturbed the clay accumulation horizons linked to fire activity. This means that faunal activity had an important impact after the hearth-pit formation *sensu stricto*.

10. The (i) lamination at the bottom of the hearth-pit structure *sensu stricto*, (ii) particular morphology of the clay accumulation horizons following the hearth-pit morphology at distance and (iii) evidences of carbonised organic punctuations migration and accumulation at the top of these horizons, raise the question whether the dark-coloured hearth-pit structures could not have been originally markedly larger than suggested up to now. It is recommended to excavate deeper and larger when so-called hearth-pit structures are detected.

The same applies when comparing the hearth-pit structures at Doel "Deurganckdok" and Verrebroek "Dok 1"; the latter were only observed over the limited depth of the hearth-pit *sensu stricto*.

3.3.3.3 | Inter-site comparison

Hearth-pits were also found in the coversands of the "Veenkoloniën" (Province of Groningen, The Netherlands) (Groenendijk 1987). As in Verrebroek "Dok 1", their presence was restricted to and concentrated at the top and slopes of the largest sand dunes. Parallel, they were recognised up to 0.5m beneath the top of the coversands and generally measured 0.4 to 0.6m across, exceptionally reaching a width of 0.8m. They are interpreted as an innovation in hearth-type at the beginning of the Mesolithic. However, no detailed soil study has been undertaken at this site.

Both surface-hearths and hearth-pits were found at the Mesolithic site of the "Hoge Vaart" close to Almere (Province of Flevoland, The Netherlands) (Hogestijn, Peeters 1996; Hogestijn, Peeters 2001).

The shallow pits (< 0.15m) at the "Hoge Vaart", comparable to the surface-hearths at Verrebroek "Dok", are located at the higher part of the coversand ridge and consist of very humiferous or peaty sand, with large quantities of charcoal, often important quantities of burnt bone and minor presence of burnt and unburnt silex artefacts, pottery and hazelnut shells. As opposed to the surface-hearths at Verrebroek "Dok 1" that lack any soil morphological delineation, those at the "Hoge Vaart" are delineated in depth by a zone (1 - 10cm) of light orange to orange-brown mottles in the E-horizon, probably resulting from iron oxidation on heating of the subsoil following fire activity (Hamburg *et alii* 2001, 10). The strongly variable burning or heating degree of the bone material indicates that the surface-hearths were used for bone processing (Exaltus 2001, 43). Most of the surface-hearths are attributed an Early Neolithic age (Spek *et alii* 2001a, 17).

The deep pits (> 0.15 - 0.70m) at the "Hoge Vaart" are comparable to the so-called hearth-pits at Verrebroek "Dok 1" and Doel "Deurganckdok". They were only clearly visible in depth, at he transition from the E- tot the B-horizon, and mostly lack clear horizonation or layering of the carbonised organic material; anthropogenic emptying and refilling of these hearth-pits for multiple use is indicated as the reason for the latter

(Exaltus 2001, 34-35; Hamburg *et alii* 2001, 13). They contain variable quantities of charcoal and charcoal powder (cf carbonised organic fine mass in Verrebroek and Doel study), but seldom other ecofacts or artefacts. Combined (i) presence of carbonised organic fine mass next to charcoal fragments, and (ii) absence of ash, are indicative of intentional incomplete burning within these hearth-pits (Exaltus 2001, 19-35). Geochemical analysis on soil material and charcoal from the hearth-pits indicate that especially P, Ca, Sr, Mn, Mg, Cu and Zn are fixed at the void surfaces of the charcoal; the observed element spectrum would be in favour of flesh, bone and/or dried excrements, latter eventually used as fuel; wood ash would rather produce peaks in K and Mg (Jansen, Peeters 2001, 34-36). Locally, tar pellets were observed, indicating that at least those hearth-pits must have been covered while in use (Jansen, Peeters 2001, 45).

Based on pedological and palaeo-ecological arguments, Spek *et alii* (2001b, 42) situate the transition of deep to shallow pits or hearths within the gradual wetting of the "Hoge Vaart" environment.

3.4. | Palaeobotanical analyses

Jan Bastiaens, Koen Deforce, Bart Klinck, Lieselotte Meersschaert, Cyriel Verbruggen & Luc Vrydaghs

3.4.1. | Verrebroek "Dok 2": charcoal analysis

Bart Klinck & Koen Deforce

3.4.1.1. | Sampling

The analysed charcoal fragments were recovered from a large, irregular charcoal patch next to a small knapping area (cf 3.1.1.). The vertical distribution of the charcoal and flint artefacts was limited to the upper 2-3cm of a slightly humiferous soil, dated by radiocarbon and pollen to the Bølling and/or Dryas II (cf 2.3.2./2.2.2.).
The excavation method (cf 1.3.1.) produced bulk samples representing square units of 50 by 50cm, with a thickness of ca. 5cm.

3.4.1.2. | Methods and material

The bulk samples were all wet-sieved through two mm meshes. The residue was air-dried and the charcoal picked out. Thirty-two units produced charcoal, but only twenty-three contained enough good-quality charcoal for analysis. For identification, the method described by Vernet (1973) was used. It is a rapid way of proceeding, suitable for archaeological and physical contexts with a lot of charred wood. Each charcoal fragment was broken manually in the transverse, tangential and radial planes. On these sections it was possible to observe the anatomical features with a reflected light (darkfield) microscope with dark field illumination. The observed features were compared to slides from the existing reference collection of the Laboratory for Palaeoecology and Landscape Genesis (Krijgslaan 281 S8/A3, 9000 Ghent, Belgium) or by means of reference studies (Schweingruber 1990a, b; Grösser 1977; Greguss 1959). For the discussion on the distinction between *Salix* sp. (willow) and *Populus* sp. (poplar) we are greatly indebted to Dr. Ir. H. Beeckman from the Laboratory of Wood Biology of the Royal Museum for Central Africa.

3.4.1.3. | Results

The results of the charcoal analysis are presented in table 60. A total of 442 charcoal fragments were identified. Most of them (96.1%) are recognisable as *Populus* sp.. Only very small amounts of *Betula* sp. (birch)(2%), *Pinus sylvestris* (Scots pine)(0.7%), *Alnus* sp. (alder)(0.7%) and *Ulmus* sp. (elm)(0.5%) were found.

sample code	*Pinus sylvestris* Scots pine	*Populus* sp. poplar	*Betula* sp. birch	*Ulmus* sp. elm	*Alnus* sp. alder	fragments (n)
N4 W26(3)		14	1			15
N4 W27(4)						0
N5 W28(1)	1					1
N6		14			1	15
N6 W26(1)		7				7
N6 W26(3)						0
N6 W26(4)		20				20
N6 W27(1)		58		2		60
N6 W27(2)					1	1
N6 W27(4)		11	2			13
N6 W28(1)		12				12
N6 W28(2)	2	9	1			12
N6 W28(4)		36				36
N6 W30(2)						0
N7 W26(2)		9				9
N7 W26(3)		84	4			88
N7 W26(4)		6				6
N7 W27(1)		10				10
N7 W27(2)		3	1			4
N7 W27(3)		48				48
N7 W27(4)		10				10
N7 W28(1)		19			1	20
N7 W28(2)		40				40
N7 W28(3)		5				5
N7 W28(4)						0
N7 W29(1)						0
N7 W29(2)						0
N7 W29(4)						0
N7 W30(2)						0
N8 W27(1)		5				5
N8 W27(2)		5				5
N8 W29(3)						0
n fragments	3	425	9	2	3	442
n units	2	21	5	1	3	

Table 60: Verrebroek "Dok 2"; list of determined charcoal fragments.

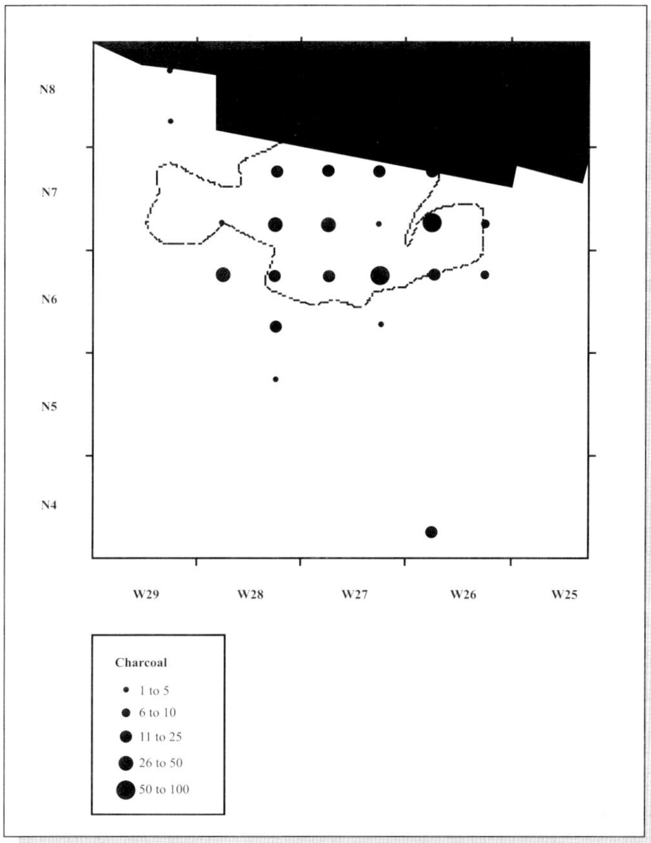

Fig. 153: Verrebroek "Dok 2"; spatial distribution of the charcoal fragments.

Figure 153 shows the distribution of the analysed charcoal particles. In general no specific spatial pattern can be observed. The analysed square units situated outside the patch contain only a few or no charcoal fragments, with one exception (N6/W28(4) with 36 pieces, bordering the patch). In the samples within the patch the number of charcoal fragments is significantly larger, in one case (square unit N7/W26(3) with 84 *Populus* sp. and 4 *Betula* sp. fragments) spectacularly so. One square unit, N7/W29, has no charcoal content.

3.4.1.4. | Discussion

As is known from former palaeoecological research (Verbruggen *et alii* 1996; Hoek 1997), elm did not occur in this region before the IIb-Allerød-biozone, while alder is only present from the Late Boreal onwards. As the stratigraphical and pedological situation of the layer excludes any bioturbation influence, the, albeit restricted, presence of these species poses an unsolved problem so far.

What is remarkable for the opposite reason is the dominance of *Populus* sp. in the charcoal assemblage of Verrebroek "Dok 2". It is very well known that pollen grains of *Populus* are rarely found in Late Glacial deposits. For the Netherlands and Northern Belgium we only know of pollen diagrams mentioned by Hoek (1997) and Bos (1998). Nevertheless the presence of *Populus* sp. in the Late Glacial is generally accepted. Taking into account the small anatomical differences between *Salix* sp. and *Populus* sp., the identification of *Populus* sp. is well-considered and based on conscientious study, comparison of the material with wood slides and consultation with experts (cf above).

Based on the presence of the analysed charcoal it can be said that the dimensions of the charcoal patch were pretty well estimated, although N7/W29 (1) and (2) has to be excluded and N6/W28 (3) and (4) included. The high number of charcoal fragments in N7/W26(3) corresponds with a flint concentration, perhaps an indication of human origin.

3.4.2. | Verrebroek "Dok 1"

3.4.2.1. | Charcoal analysis

Bart Klinck & Koen Deforce

Sampling

All charcoal fragments were sampled from the prehistoric features containing charred material, i.e. hearth-pits, charcoal-pits and surface-hearths. The last of these are weakly marked in the soil compared to the pit-features and also contain far fewer charcoal fragments.

As in Verrebroek "Dok 2" the surface of this site was excavated in a grid of units of 50 by 50cm. All units containing a presumed surface-hearth were bulk-sampled by the field-archaeologists. Due to the vast number of collected samples it was recommended to restrict the analysis to a selection of twenty-four samples, representing twelve different surface-hearths. All analysed surface-hearths were found within the limits of artefact concentrations. The latter usually contain one single hearth, but occasionally multiple hearths were observed within the same artefact unit (cf 3.1.2.1.). After analysis, eleven surface-hearths (representing eleven different artefact concentrations) yielded charcoal fragments.

The analysed hearth-pits (n=18) and charcoal-pits (n=20) were excavated, not as grid units but as whole structures. The volume of these samples varies between 6.5 and 112 litres. Some of the structures (9 hearth-pits and 1 charcoal-pit) were excavated and sampled in artificial layers. Charcoal-pit S97 was divided into two halves, one being sampled in horizontal layers of 5cm, the other as one bulk sample. Complete sampling in layers of 5cm was done for hearth-pits S108, S110 to S112 and S115. The filling of S100 to S103 was divided into two samples: one containing the upper 5cm and one with the underlying remaining part of the pits.

Methods and material

The sieving was started through meshes of 0.25, 0.5, 1 and 2mm. As too large a part of the sediment remained on the 0.25mm sieve, it was decided to restrict the sieving to the three largest meshes. However, in order to have a control on smaller particles a volume of 3 litres of some features was sieved through a 0.25mm mesh.
For the analysis of the charcoal, restricted to fragments larger than 2mm, a difference between the pits and the surface-hearths was made. As the surface-hearths had poor charcoal content, all the charcoal was analysed. For the two types of pits this way of working was not suitable and a selection suggested itself, making use of the methods for relevant random tests in large numbers of wood fragments developed by Jacomet & Kreuz (1999), Vernet (1973) and Chabal (1988). Identification of the charcoal was performed in the same way as at Verrebroek "Dok 2" (cf 3.4.1.)

Results

Hearth-pits (n= 18)

As stated above, nine pits were sampled in artificial layers. The results of these analyses are shown in tables 61-70. Except for S100 and S115, in which the ratio oak/pine varies considerably in depth, the vertical distribution of the infilling of the remaining pits (n=7) shows no significant differences. Therefore the numbers of the fragments have been added and treated as a total number for the pit.

The results of the hearth-pits show a clear dominance of *Pinus sylvestris* (Scots pine) in all pits. In six pits *Pinus sylvestris* is the only occurring species. For five pits the ratio between *Pinus sylvestris* and *Quercus* sp. (oak) is about 2 to 1. In four pits the percentage of *Quercus* sp. is lower than 10%. Two pits have a small amount of *Populus* sp. and one pit contains a considerable percentage (20%) of *Corylus avellana* (hazel).

Concerning the qualities of the charcoal from the hearth-pits, it can be said that a wide range of varieties in the state of preservation is present. Some pits contain more disintegrated material than others, but most pits combine a number of well preserved pieces with a number of small and weathered fragments. Some of the pits sampled in artificial layers show a dominance of small and weathered pieces in the top-layer (S115 and S101), but generally no trend in the distribution can be detected.

No pit shows clear evidence of a selection of only fresh or only dead wood. An assemblage of charcoal showing cracks, marks of fungal hyphae and insects represent a combination of fresh and dead wood as fuel for the fires. A small number of vitrified fragments were found as well. The cause(s) of the process of vitrification is the subject of an international discussion and has been dealt with by many researchers (Théry-Parisot 2001). In this study its presence is merely noted.

Special attention has to be given to an observation on the microscopic level: many charcoal pieces of *Pinus sylvestris* show an orientation of their rays not perpendicular to the ground tissue. Deviations up to 20° are no exception.

feature code	*Pinus sylvestris*	*Quercus* sp.	*Corylus avellana*	*Populus.* sp.	volume (l)	fragments (n)
	Scots pine	oak	hazel	poplar		
S66	126	12			47	138
S67	174	84			53	258
S68	155	12			20	167
S70	218	9			29.5	227
S73	107	53			60	160
S88	130				23.5	130
S91	67	10	23		21	100
S92	150				20	150
S100	256	107			112.5	363
S101	256	25			34	281
S102	186				15	186
S103	242	7		3	37	252
S108	93	54			?	147
S110	138				?	138
S111	203			4	13	207
S112	435				?	435
S114	50				6.5	50
S115	346	222			?	568
n fragments	3332	595	23	7		3957
n units	18	11	1	2		

Table 61: Verrebroek "Dok 1"; list of determined charcoal fragments from hearth-pits.

	volume (l)	*Pinus sylvestris*	*Quercus* sp.
		Scots pine	oak
top 5 cm	12	16	84
lower fill	100.5	240	23
sum	112.5	256	107

Table 62: Verrebroek " Dok 1"; hearth-pit S100.

	volume (l)	*Pinus sylvestris*	*Quercus* sp.
		Scots pine	oak
top 5 cm	5	71	13
lower fill	29	185	12
sum	34	256	25

Table 63: Verrebroek " Dok 1"; hearth-pit S101.

	volume (l)	*Pinus sylvestris*
		Scots pine
top 5 cm	8.5	92
lower fill	6.5	94
sum	15	186

Table 64: Verrebroek " Dok 1"; hearth-pit S102.

	volume (l)	*Pinus sylvestris*	*Quercus* sp.	*Populus* sp.
		Scots pine	oak	poplar
top 5 cm	8.5	97	0	3
lower fill	28.5	145	7	0
sum	37	242	7	3

Table 65: Verrebroek " Dok 1"; hearth-pit S103.

depth (cm)	volume (l)	*Pinus sylvestris*	*Quercus* sp.
		Scots pine	oak
0-5	no sample		
5-10	14.25	60	40
10-15	6.25	33	14
sum	20.5	93	54

Table 66: Verrebroek " Dok 1"; hearth-pit S108.

depth (cm)	volume (l)	*Pinus sylvestris*
		Scots pine
0-5	13.5	97
5-10	3.25	41
sum	16.75	138

Table 67: Verrebroek " Dok 1"; hearth-pit S110.

depth (cm)	volume (l)	*Pinus sylvestris*	*Populus* sp.
		Scots pine	poplar
0-5	7.5	96	3
5-10	11.5	107	1
sum	19	203	4

Table 68: Verrebroek " Dok 1"; hearth-pit S111.

depth (cm)	volume (l)	*Pinus sylvestris*
		Scots pine
0-5	4	130
5-10	5.5	150
10-15	4.5	150
15-20	2.5	15
sum	16.5	445

Table 69: Verrebroek " Dok 1"; hearth-pit S112.

depth (cm)	volume (l)	*Pinus sylvestris*	*Quercus* sp.
		Scots pine	oak
0-5	15.7	34	16
5-10	24.5	83	67
10-15	13.5	95	104
15-20	8.3	134	35
sum	62	346	222

Table 70: Verrebroek " Dok 1"; hearth-pit S115.

Charcoal-pits (n= 20)

The results of the charcoal-pits hardly differ from these of the hearth-pits (tab. 71). The dominance of *Pinus sylvestris* is even more pronounced: 16 pits contain *Pinus sylvestris* exclusively. The occurrence of *Quercus* sp. is restricted to four structures, among which S78 has it as the only wood species. In the three other pits (S60, S71 & S72) the *Quercus* sp. occurrence is restricted to maximum 5%. Twenty-one fragments of *Corylus avellana* charcoal (20%) are found in pit S71. One pit, S51, is different from all the other charcoal-pits and hearth-pits because its charcoal only consists of *Alnus* sp.

Generally the charcoal from these features shows the same qualitative characteristics as those found for the hearth-pits, although there seems to be a trend to more weathered and smaller material. Another subtle difference is that marks of attacks (fungal or from insects) are abundant, while clear marks of cracks are absent. Possibly, only dead wood was used as fuel for these fires. The presence of vitrified charcoal was observed as well. The microscopic deviation mentioned for the hearth-pits was also found in several fragments of *Pinus sylvestris*.

feature code	*Pinus sylvestris*	*Quercus* sp.	*Corylus avellana*	*Alnus* sp.	volume (l)	fragments (n)
	Scots pine	oak	hazel	alder		
S51				93	3.5	93
S59	100				23	100
S60	47	1			7	48
S61	20				16	20
S62	100				16	100
S63a	45				4.25	45
S63b	85				11	85
S64	94				7	94
S65	280				26.5	280
S69	32				15	32
S71	80	4	21		36	105
S72	126	4			28	130
S74	104				31	104
S78		20			3.25	20
S83	100				12	100
S85	100				15	100
S86	100				16	100
S95	100				11	100
S96	122				44	122
S97	124				8.75	124
n fragments	1759	29	21	93		1902
n units	18	4	1	1		

Table 71: Verrebroek " Dok 1"; list of determined charcoal fragments from charcoal-pits.

	volume (l)	*Pinus sylvestris*
		Scots pine
S97/1	4.5	60
S97/2/1	0.5	16
S97/2/2	0.5	19
S97/2/3	1	11
S97/2/4	1	2
S97/2/5	0.75	13
S97/2/6	0.5	3
sum	8.75	124

Table 72: Verrebroek " Dok 1"; charcoal-pit S97 (according to different layers).

Surface-hearths (n=12)

In the field an obvious difference in charcoal content between the two kinds of pit and the surface-hearths was observed. This is confirmed by the results of the charcoal analysis (tab. 73). Only a few and small pieces of charcoal were found in these features. Some of the investigated hearths do not contain any charcoal at all, others one or two pieces. The dominant species again is *Pinus sylvestris*. *Quercus* sp. and *Alnus* sp. each occur in two features.

As already noted in the description of the features (cf 3.1.2.1.), the charcoal in these features is strongly weathered and has small dimensions. Most fragments were too small for further analysis. No other qualities were observed for these features.

artefact unit	surface-hearth code	sample location	*Pinus sylvestris*	*Quercus* sp.	*Alnus* sp.	volume (l)	fragments (n)
			Scots pine	oak	alder		
16	SH1	087/Z26 (2C)				5	0
26	SH1	060/Z28(3A)	7	9		4.25	16
26	SH1	060/Z29(3A)	6	4		4	10
36	SH2	B15(1B)	5			3.75	5
37	SH1	E20(4C)	2			3.75	2
37	SH1	E20(2B) 99-12	1			3.5	1
38	SH1	I14(2A)				2.75	0
38	SH1	I15(4A)				2.75	0
38	SH1	I15(4B)	1			2.75	1
38	SH1	I14(2B)				3	0
44	SH2	045/Z36(2A)	6			3	6
44	SH2	047/Z36(3C)	8			3,5	8
44	SH2	047/Z36(3B)	6		9	3.5	15
49	SH1	040/Z15(2B)				3	0
49	SH1	040/Z15(2A)	2		2	3	4
58	SH1	0111/Z62(3A)	6			7.25	6
67	SH1	072/Z58(3B)	7			4	7
70	SH1	073/Z48(4B)	2			4.5	2
70	SH1	073/Z48(4A)	4			4	4
70	SH1	073/Z47(4A)				4.5	0
92	SH1	050/Z36(1A)	4			7	4
92	SH1	053/Z38(3A)	7			4	7
92	SH1	053/Z38(3B)	9			3.5	9
99	SH1	063/Z12(4B)	3			4.25	3
n total			86	13	11		110
n units			18	2	2		

Table 73: Verrebroek " Dok 1"; list of determined charcoal fragments from surface-hearths.

Discussion

The recorded wood species and the ratios between them are very similar for the three types of feature. The higher density of charcoal fragments clearly occurs in the hearth-pits and the charcoal-pits, while charcoal is hardly present within the surface-hearths. This difference might be related to taphonomic factors. Indeed, one may assume that the preservation conditions for charcoal are much better in pit-like features compared to hearths constructed at the former surface. In the latter the charcoal originally produced can easily be blown away while lying unsealed at the surface.

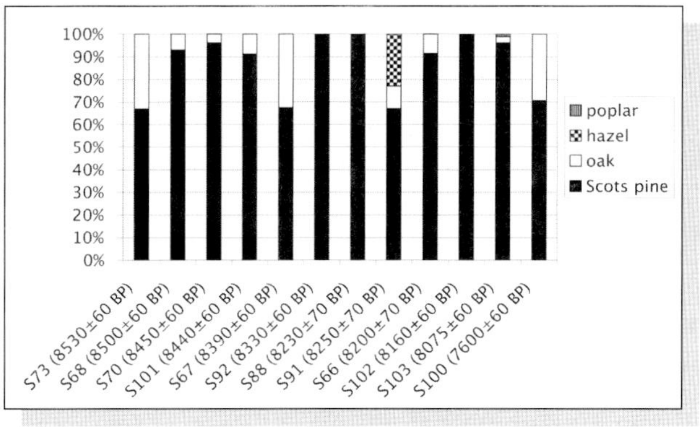

Figure 154: Verrebroek "Dok 1"; comparative chart of the charcoal frequencies found within the hearth-pits in relation to the radiocarbon dates.

Radiocarbon dating of the hearth-pits indicates a use-time situated between (NZA-11940) 8075±60 BP and (UtC-9447) 8530±60 BP, with the exception of S100, which dates back to (NZA-11941) 7600±60 BP (cf 3.2.3.2.). We know from extensive palynological investigation in Sandy Flanders (Verbruggen 1971,1979; Verbruggen *et alii* 1996) that at that time – i.e. the second half of the Boreal - *Pinus* (Pine) and *Corylus* (hazel) were the dominant trees. Apparently the wood of *Pinus sylvestris* was more frequently used as fuel than *Corylus avellana*, as the latter was found only in one hearth-pit (S91). The percentages of *Quercus* sp. wood correspond fairly well with those observed in contemporaneous pollen spectra. However, the generally accepted gradual expansion of *Quercus* sp. during the Boreal, is not reflected in the hearth-pits: the oldest pits contain as much *Quercus* sp. as the younger ones or even more (fig. 154).

3.4.2.2. | Botanical macroremains: seeds and fruits

Lieselotte Meersschaert

Sampling

The description of the samples taken for the purpose of the seeds and fruits analysis and the sampling methods are the same as those outlined in 3.4.2.1.

Methods and material

In the beginning of the study, all samples were processed using sieves with 2mm, 1mm, 0.5mm en 0.25mm meshes. Soon it became clear that this method was very time consuming and it was decided to make use of the three largest fractions. However, for the purposes of control, a volume of three litres was sometimes sieved through a 0.25mm mesh as well.

Most of the residues of the 2, 1 and 0.5mm processing were completely analysed. A few of them and the residues of the 0.25mm processing were analysed until the results became representative of the total volume of the samples. The air-dried samples were investigated by means of an Olympus stereomicroscope (magnification between 3.5 and 144x). Determination of the macroremains were carried out with the help of a reference collection and relevant literature from the Laboratory of Palaeoecology and Landscape Genesis from Ghent University.

Code	S66	S67	S68	S70	S88	S91	S100	S101	S101	S102	S103	S108	S108	S110	S112	S112	S112	S115	S115	S11
								top 5 cm				5-10 cm	10-15 cm	5-10 cm	0-5 cm	5-10 cm	15-20 cm	0-5 cm	5-10 cm	10-1 cm
Volume (l)	47	53	20	29.5	23.5	21	112.5	34	34	15	37	14.25	6.25	3.25	4	5.5	2.5	15.7	24.5	13.
Smallest fraction (mm)	0.25	0.25	0.25	0.25	0.25	0.5	0.25	0.5	0.5	0.5	0.5	0.5	0.5	0.5	0.5	0.5	0.5	0.5	0.5	0.5
Corylus avellana (hazel)	85	15	4	6	1	6	.	.	.	.	.	.	.	.	.	.	.	.	.	.
Stellaria cf. *media* (common chickweed)	.	.	.	.	.	.	.	.	1	.	.	.	.	.	.	.	.	.	.	.
Chenopodiaceae (goosefoot family)	.	.	.	.	.	.	.	1	.	.	.	1	1	.	.	.	1	.	.	3
Chenopodium album (fat-hen)	1	.	.	.	.	.	2	.	4	.	.	.	1	.	.	.	.	.	.	.
Fallopia convolvulus (black-bindweed)	.	.	.	.	.	.	.	.	1	.	.	.	.	.	.	.	.	.	.	.
Phragmites australis (common reed)	.	.	.	.	.	.	.	.	.	.	.	.	.	.	1	.	.	.	.	.
Pinus sp. (Scots Pine)	.	.	.	.	2	.	.	.	.	.	.	.	.	.	.	1	.	.	.	.
Polygonum cf. *minus* (small water-pepper)	.	.	.	.	.	.	2	1	.	.	.	.	.	.	.	.	.	.	.	.
Potentilla sp. (cinquefoil)	.	.	.	.	.	.	.	.	.	1	.	.	.	.	.	.	.	.	.	.
Indeterminata	.	.	.	.	.	.	.	.	.	.	1	.	.	1	.	.	.	3	5	.

Table 74: Verrebroek "Dok 1"; list of carbonised macroremains from hearth-pits.

Results

Hearth-pits (n=14)

Fourteen pits were sampled, some of them in artificial layers. The results of the analysis are shown in table 74.

The samples from Verrebroek "Dok 1" contain only charred seeds. The number of species found is very limited, as well as the number of seeds. Eight different species were attested. Macroremains of *Corylus avellana* (hazel) and Chenopodiaceae (goosefoot family)(including *Chenopodium album* (fat-hen)) were found in several pits, while the other six species were found in only one or two samples.

Charcoal-pits (n=20)

All charcoal-pits described in 3.4.2.1.were screened, but only five pits contained remains of seeds and fruits (tab. 75). One species was attested, namely *Corylus avellana,* in the charred fragments of nutshells. The quantity of recovered nutshells is very low.

	S69	S72	S74	S85	S86
volume (l)	15	28	31	15	16
smallest fraction (mm)	0.5	0.5	0.5	0.5	0.5
Corylus avellana (hazel)	1	2	1	2	2

Table 75: Verrebroek " Dok 1"; list of carbonised macroremains from charcoal-pits.

Surface-hearths (n=22)

Typical of the surface-hearths is the poor accumulation of charcoal (cf 3.4.2.1.). Nevertheless, these structures contain far more charred macroremains than the analysed charcoal-pits (tab. 76). *Corylus avellana* is again the only identified species.

artefact unit	surface-hearth code	sample location	volume (l)	smallest fraction (mm)	*Corylus avellana*
					hazel
26	SH1	060/Z28(3A)	4.25	0.5	33
26	SH1	060/Z29(3A)	4	0.5	18
36	SH2	B15(1B)	3.75	0.5	6
37	SH1	E20(2B) 99-12	3.5	0.5	9
37	SH1	E20(4C)	3.75	0.5	20
38	SH1	I14(2A)	2.75	0.5	11
38	SH1	I14(2B)	3	0.5	12
38	SH1	I15/4A)	2.75	0.5	34
38	SH1	I15(4B)	2.75	0.5	53
44	SH1	045/Z36(2A)	3	0.5	21
44	SH2	047/Z36(3B) 99-13	3.5	0.5	83
44	SH2	047/Z36(3C)	3.5	0.5	10
49	SH1	040/Z15(2A)	3	0.5	5
49	SH1	040/Z15(2B)	3	0.5	1
67	SH1	072/Z58(3B)	4	0.5	16
70	SH1	073/Z47(4A)	4.5	0.5	40
70	SH1	073/Z48(4A)	4	0.5	35
70	SH1	073/Z48(4B)	4.5	0.5	37
92	SH1	050/Z36(1A)	7	0.5	12
92	SH1	053/Z38(3A)	4	0.5	13
92	SH1	053/Z38(3B)	3.5	0.5	21
99	SH1	063/Z12(4B)	4.25	0.5	35

Table 76: Verrebroek " Dok 1"; list of carbonised macroremains from surface-hearths.

Discussion

Conservation and taphonomy

Only carbonised seeds and fragmented nutshells are preserved at the site of Verrebroek "Dok 1". The preservation of these remains is rather good, even of the more fragile seeds (e.g. the seeds of *Chenopodium album*). Uncarbonised seeds are absent as a result of the fairly dry conditions on top of the sandy dune during the occupation period; under such conditions uncarbonised material disintegrates.

The small numbers of attested macroremains may be the result of a certain selection, either by man or by nature, resulting in the recovery of the seeds of only a few species. The original number of plant species at or in the vicinity of the settlement will certainly have been greater. Because so few seeds were attested, it is difficult and hazardous to draw any further conclusions concerning taphonomic processes.

Useful plants

In the samples of the surface-hearths and charcoal-pits, only fragments of hazelnut shells could be recovered. *Corylus avellana* is also the most common species attested in the hearth-pits. However, compared to the surface-hearths the number of hazelnut shells in the hearth-pits and charcoal-pits is much lower. Furthermore, in the latter two features hazelnut shells were found only in a limited number of pits, while they occur in all analysed surface-hearths. This observation confirms the earlier hypothesis (cf 3.1.2.1.), that it was surface-hearths rather than pit features which were probably used to roast hazelnuts. Because of their high percentage of fat, hazelnuts were very popular food sources for Stone Age people (Van Zeist, Palfenier-Vegter 1981, 149; Zvelebil 1994). The nuts became tasty by drying them over a long period or by roasting them, which could be an explanation why the fragments found were all charred. However, it is also possible that, the shells were thrown into the fire as waste during eating of the dried nuts.

Several hearth-pits contained seeds of *Chenopodium album*. In early times this species was also gathered as a source of food; for instance, Van Zeist and Palfenier-Vegter (1981, 149)[1] refer to an Iron Age settlement in Denmark where convincing evidence for harvesting its seeds was found. Evidence of its earlier use has not yet been found, as far as we know. The seeds were crushed into meal and the leaves were used like spinach today (Weeda *et alii* 1985, 163). This should not surprise as *Chenopodium album* is an important source of vitamin C and several amino acids essential for the human body. It wasn't possible to detect whether this plant was also used for food on the Mesolithic site of Verrebroek "Dok 1". It should however be emphasised that *Chenopodium album* has been reported on many Mesolithic sites in NW-Europe, such as Ageröd 5 and Mount Sandel (Zvelebil 1994). The Relational Archaeobotanical Database for Advanced Research (RADAR) has ten prehistoric sites for the Netherlands dated before 5000 BC, among which are some Final Mesolithic and Early Neolithic (LBK) sites, where vegetable remains of *Chenopodium album* were recovered (tab. 77) (van Haaster, Brinkkemper 1995). Its frequent occurrence in the analysed samples at Verrebroek indicates that this species almost certain grew locally.

1 They refer to: Helbaek (1951).

site	occupation period
Arnhem-Schuytgraaf	6450-4200 BC
Hardinxveld-Giessendam Polderweg (Final Mesolithic)	5500-5000 BC
Hardinxveld-Giessendam De Bruin (Final Mesolithic)	5500-5100 BC
Elsloo Sanderboutlaan	5500-5000 BC
Geleen Haesselderveld	5500-5000 BC
Geleen Urmonderbaan	5500-5000 BC
Beek Molensteeg	5500-5000 BC
Geleen Kermisplein	5500-5000 BC
Sittard Landweringstraat	5500-5000 BC
Beek Kerkeveld	5500-5000 BC

Table 77: list of Mesolithic and Neolithic sites in The Netherlands, dated before 5000 BC and containing macroremains of *Chenopodium album* (RADAR) (van Haaster, Brinkkemper 1995).

Natural environment

It is not possible to reconstruct the natural environment of the Mesolithic settlement using only the rather poor results of the analyses of fruits and seeds. Most of the seeds that could be identified to species level derive from pioneer vegetations of dry to moist, moderate-to-high nutrient available soils (Runhaar *et alii* 1987, 277-359).

3.4.2.3. | Pollen analysis

Cyriel Verbruggen

Two archaeological features were analysed palynologically. The first feature (S89) is a tree-fall/hazelnut-pit (cf 3.1.2.1./3.3.1.2.). The samples from layers H4, H2 and H1 are completely pollenless, pointing either to dry and well aerated soil conditions before the deposition of the peat or to a rapid sedimentation rate. Taking into account the Upper-Pleniglacial age of the coversands, only the first explanation is feasible.

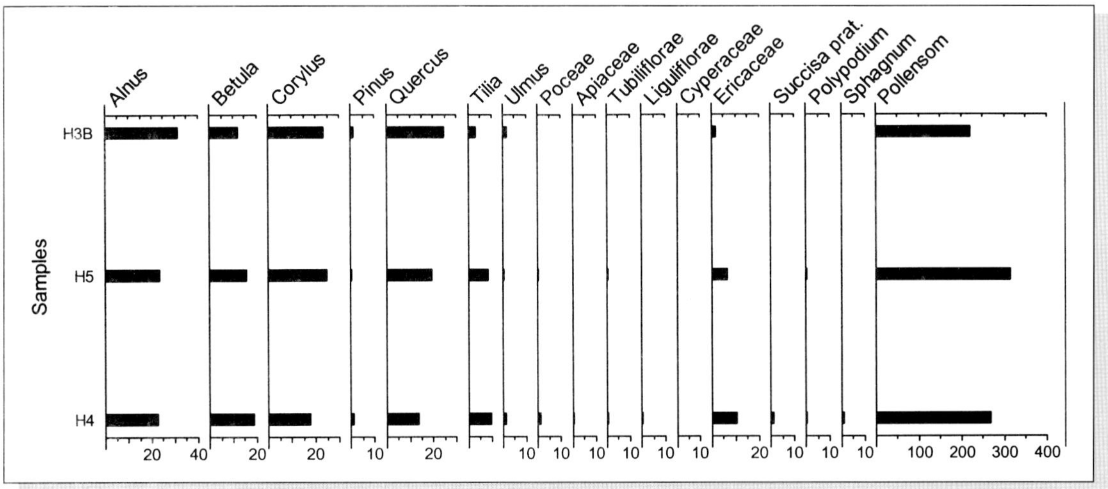

Figure 155: Verrebroek "Dok 1"; pollen diagram of tree-fall feature S93, levels H4, H5 and H3b.

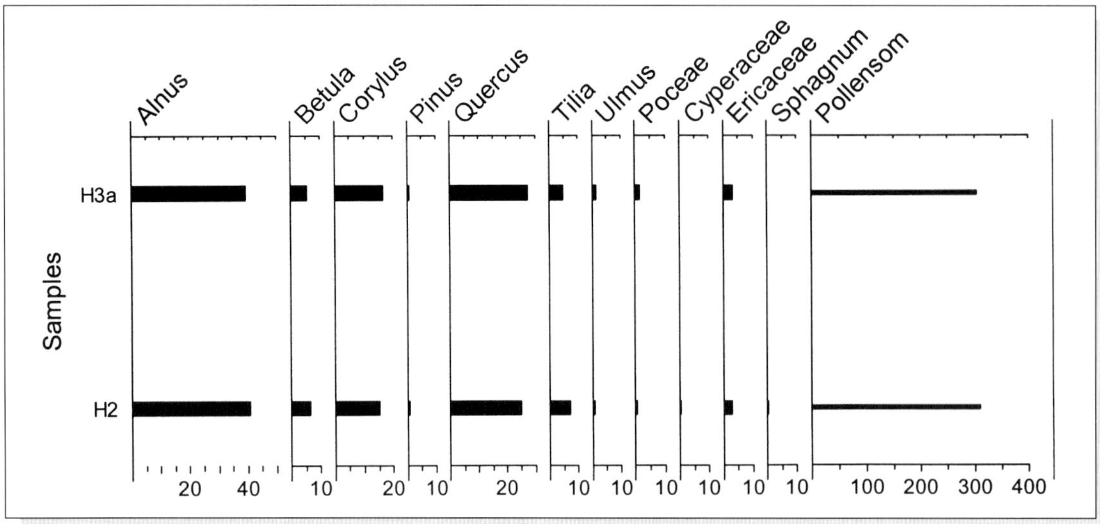

Figure 156: Verrebroek "Dok 1"; pollen diagram of tree-fall feature S93, levels H2 and H3a.

The second feature analysed (S93) has been identified as a tree-fall structure (cf 3.1.2.1./3.3.1.2.). The five analysed samples of this feature (fig. 155, 156) all show an early Subboreal forest spectrum. This dating is based on the absence of *Fagus* (beech) or herbs of human influence and on the low *Ulmus* (elm) values. There are smaller differences within the AP composition, pointing to probable alternating wetter and dryer conditions. The samples H2 and H3a (fig. 156) have strongly dominant *Alnus* (alder) and *Quercus* (oak) values and low *Betula* (birch) and Ericaceae (heather family) values, reflecting wet conditions, while the samples H4 and H5 (fig. 155) have much higher *Betula, Corylus* (hazel) , *Tilia* (lime) and Ericaceae values, reflecting dryer conditions. Sample H3b is an intermediate between the former wet and latter dry group of samples.

3.4.2.4. | Phytolith analysis

Luc Vrydaghs

In total five features – two hearth-pits, two surface-hearths and one tree-fall – were analysed. From hearth-pits S67 and S88 samples were taken from the basal filling (i.e. the charcoal-richest layer). In absolute quantities, very few (< 2%) phytoliths were observed. In S67, 5 individuals of the same morphotype were listed, whereas in S88, six individuals, mainly of the elongate morphotype, were counted.

Phytolith analysis was also conducted on two samples taken from a surface-hearth within artefact unit 22 (cf 3.3.1.1.). The first sample was retrieved from the upper 0.10m of the coversand, the second between 0.10 to 0.20m below the top of the coversand. In absolute quantities, very few (< 2%) phytoliths were observed. One sample contained 14 individuals, distributed over three morphotypes, with the dumbbell morphotype as dominant. Phytolith identification revealed a coherent spectrum of panicoid grasses. Doubt exists about the presence of the saddle morphotype, which would suggest chloridoid grasses. The other sample contained 19 individuals, distributed over five morphotypes; the elongate, trapezoid and one unknown morphotype are dominant. One dumbbell and one saddle morphotype were listed too. Phytolith identification revealed a spectrum of panicoid grasses and chloridoid grasses. The absolute presence of phytoliths is evaluated as very weak. When taking the phytolith identification of the observed grasses into account, an absolute presence of 5 to 10% phytoliths within the suspension preparation is expected.

Another analysed surface-hearth is situated within unit 21. Phytolith analysis reveals very little material (< 2%), represented by 18 individuals distributed over five morphotypes. The spectrum is dominated by the dumbbell morphotype. One saddle was also observed. The spectrum is coherent and allows us to identify panicoid and chloridoid grasses. However, when taking the phytolith identification of the observed grasses into account, an absolute presence of 5 to 10% phytoliths within the suspension preparation is expected.

The tree-fall feature S90 was analysed on a sample coming from the assumed Bb.w. bi-horizon (cf 3.3.1.2.). The analysis revealed very few (< 2%) phytoliths. In total, 14 individuals distributed over four morphotypes were listed. The trapezoid morphotype dominates the spectrum. Phytolith identification mainly revealed panicoid grasses but the dominance of the trapezoid morphotype within a coherent spectrum of Poaceae (grass family) might point towards dominant presence of one single organ.

3.4.3. | Doel "Deurganckdok-sector B"

3.4.3.1. | Charcoal analysis

Bart Klinck

Sampling

The charcoal analysed from this site was recovered from hearth-pits and surface-hearths, the latter situated in the upper humiferous layer (A-horizon). Due to a shortage of time, the investigation was restricted to a selection of these features.

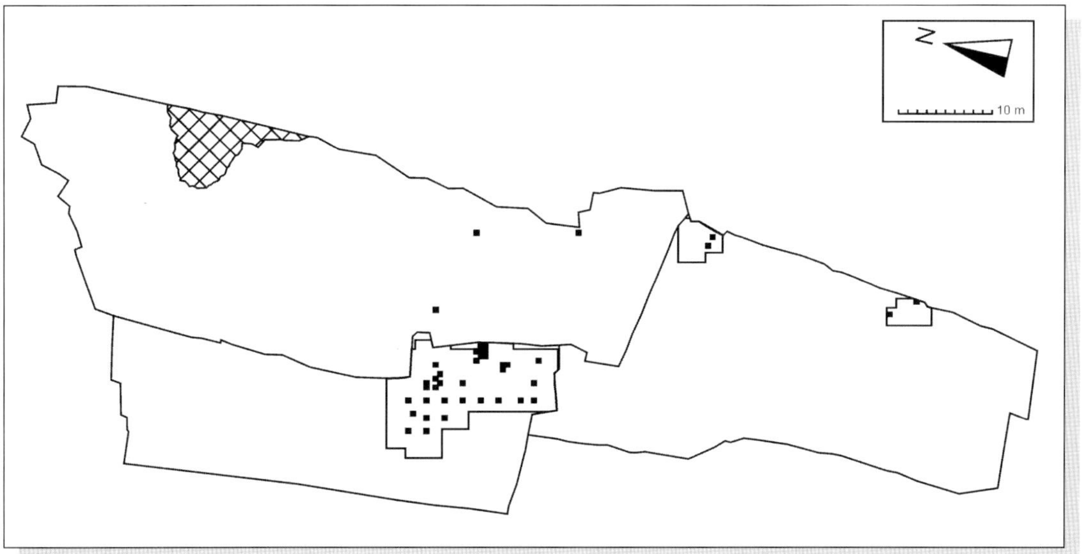

Figure 157: Doel "Deurganckdok-sector B"; distribution map of the analysed charcoal samples.

The sampling of the hearth-pits was done in the same way as described for these features in chapter 3.4.2.1. Because of the vast number of these structures, only 18 of the 102 pits were analysed. For the surface-hearths the same remark has to be made: only 42 quadrates, the majority situated in the sieved north-western sub-zone, were analysed (fig. 157). Initially 19 quadrates, which had already been selected for archaeozoological research (cf 3.5.4.), were analysed. Later, another 23 randomly selected samples were included in the study.

Methods and material

The sieving procedure was restricted from the beginning to 0.5, 1 and 2mm, because of the similarity of the sediment on this site with that at Verrebroek "Dok 1". All other steps in the treatment were the same as those described in 3.4.2.1.

Results

Hearth-pits (n=18)

A total of 1780 charcoal fragments were identified (tab. 78), most of them (97.5%) belong to *Quercus* sp. (oak). Only small amounts of *Pinus sylvestris* (Scots pine), *Populus* sp. (poplar), *Betula* sp. (birch), *Corylus avellana* (hazel) and *Fraxinus excelsior* (ash) were found. In all examined pits *Quercus* sp. is predominant; in eight cases it is actually the only occurring species.

Only a few structures (S86, S55) contain a considerable number of large pieces. Most pits, however, contain strongly crumbled charcoal. Marks of attacks by mould or insects are observed on many fragments and are an indication of dead wood. At the same time, fragments with cracks, caused by escaping gases during burning, indicating young, living wood, were attested.

feature code	*Betula* sp. birch	*Corylus avellana* hazel	*Fraxinus excelsior* ash	*Pinus sylvestris* Scots pine	*Populus* sp. poplar	*Quercus* sp. oak	Indeterm.	volume (l)	fragments (n)
S22						50		?	50
S24						50		35	50
S34	3					146		65	149
S47						150		32	150
S55						48	2	25	48
S58						100		22	100
S75						99		41	99
S84						100		30	100
S86						147	3	62	147
S90						110		16	110
S93			2		11	36		11	49
S95						97	3	30	97
S96						100		31	100
S103						97	2	21	97
S104		3				95	2	43	98
S105				1		22	3	7.5	23
S106	2			1		197	3	52	200
S109	4				3	91	3	30	98
n fragments	9	3	2	2	14	1735	21		1765
n units	3	10	1	2	2	18	8		

Table 78: Doel "Deurganckdok-sector B"; list of determined charcoal fragments from hearth-pits.

Surface-hearths (n=42)

Table 79, presenting the results from the charcoal analysis on the samples of the presumed surface-hearths and/or hearth-dumps, shows a much wider range of species. The 1036 examined fragments belong to 11 different species. Dominant are *Quercus* sp. and *Alnus* sp.(alder), with 456 and 366 pieces respectively. *Quercus* sp. occurs in every sample and *Alnus* sp. is absent only in two samples. They are followed by *Sorbus* sp. (whitebeam)(78 pieces) and *Fraxinus excelsior* (46 pieces). Considerable amounts of *Cornus* sp. (dogwood), *Corylus avellana, Prunus* sp. (cherry) and *Ulmus* sp. (elm) were found as well. *Salix* sp. (willow), *Tilia* sp. (lime) and *Viburnum opulus* (Guelder-rose) are represented by one, four and three pieces respectively.

sample code	*Alnus* sp.	*Cornus sanguinea/mas*	*Corylus avellana*	*Fraxinus excelsior*	*Prunus* sp.	*Quercus* sp.	*Salix* sp.	*Sorbus* sp.	*Ulmus* sp.	*Viburnum opulus*	*Tilia cordata/platyphyllos*	Indeterm. (n)	fragments (n)
	alder	dogwood	hazel	ash	cherry	oak	willow	whitebeam	elm	guelder rose sp.	lime		
W6/Z6(3)	4					5		1					10
W6/Z18(4)		7				13							20
W6/Z18(1or4)	3		1			5		10		1			20
W7/Z32(2)	5	4	5			11							25
W8/Z32(1)	4			11	1	4							20
W11/Z2(4)	14					3		1		2			20
W14/Z55(1)	8					11				1			20
W15/Z7(3)	7		1			2		1					11
W15/Z7(4)	11					4		2		1		6	24
W15/Z52(4)	3			4		39		3					49
W16/Z6(2)	11			5		1		3					20
W16/Z7(1)	13			3		6		5		1		2	30
W16/Z7(2)	13	1				14		6		3	1		38
W16/Z7(3)	17			5		13		10			1		46
W16/Z7(4)	12			5		4		3		1			25
W17/Z2(4)	8					5					1		14
W17/Z6(2)	8					7		3		2			20
W17/Z9(3)	8					8		3		1			20
W17/Z10(4)	12					7		1					20
W17/Z13(2)	8					5		3		3		1	20
W18/Z2(3)	20			2		16		10				1	49
W18/Z9(2)	12			1		15		2					30
W19/Z1(4)	21					7							28
W19/Z2(1)	26	1				4							31
W19/Z2(3)	22			4		12				2			40
W19/Z5(4)	6					10		3		1			20
W19/Z13(4)	2					5		1					8
W20/Z1(1)	12					6		1				1	20
W20/Z2(1)	6	1				9		2		2			20
W21/N2(4)	5			4		15		1					25
W21/Z1(4)	8					12							20
W21/Z3(4)	4					29				1			34
W21/Z5(4)	3				8	39							50
W21/Z7(4)	1				10	19							30
W21/Z9(4)	1	1	6	1		6		1	1				17
W21/Z11(3)	4	2				17		1				1	25
W21/Z13(4)	4					15	1						20
W23/N2(2)	28					2							30
W23/Z1(4)	6					13		1					20
W23/Z3(4)	2			1		14							17
W25/N2(1)	4				1	14				1			20
W25/Z1(1)						10							10
n fragments	366	17	13	46	20	456	1	78	1	23	3	12	1036
n units	40	7	4	12	4	42	1	25	1	15	3	6	

Table 79: Doel "Deurganckdok-sector B"; list of determined charcoal fragments from presumed surface-hearths/hearth-dumps.

Just as for the hearth-pits, the fragments are rather small, originating from larger crumbled pieces or small branches. The abovementioned indicators for both dead and living wood are found. A few glazed pieces were found as well.

Discussion

According to the radiocarbon dating (cf 3.2.4.2.) the hearth-pits at Doel were formed and used within a restricted time-span, i.e. between 7865±35 BP (KIA-17990) and 7580±35 BP (KIA-18740). Extensive palynological investigation in different parts of Sandy Flanders has shown that in this period (the first part of the Atlantic) *Quercus* sp. had become the dominant tree in the forest, followed by *Alnus* sp., *Tilia* sp., *Corylus* sp. and *Ulmus* sp. (Verbruggen *et alii* 1996). Therefore it may be assumed that the absolute dominance of oak wood in the hearth-pits of Doel reflects a human selection rather than a natural situation.

Concerning the higher positioned surface-hearths, the dominance of *Quercus* sp. and *Alnus* sp. corresponds much better with the contemporaneous environment as reconstructed on the basis of pollen grains. *Corylus avellana, Fraxinus excelsior, Tilia* sp., *Ulmus* sp. and *Viburnum opulus* are subdominant and expected species in that forest, whereas the presence of *Cornus* sp., *Sorbus* sp. and *Prunus* sp. may be regarded as remarkable. An extensive discussion of this striking assemblage is included in 3.4.4.

3.4.3.2. | Botanical macroremains: seeds and fruits

Jan Bastiaens & Lieselotte Meersschaert

Sampling

The samples for the seeds and fruits analysis were recovered from the upper humiferous layer (A-horizon), which contained large numbers of Swifterbant settlement waste (fig. 158). The majority of samples originate from presumed surface-hearths and/or hearth-dumps situated in the sieved north-western sub-zone (cf 3.1.3.3.).

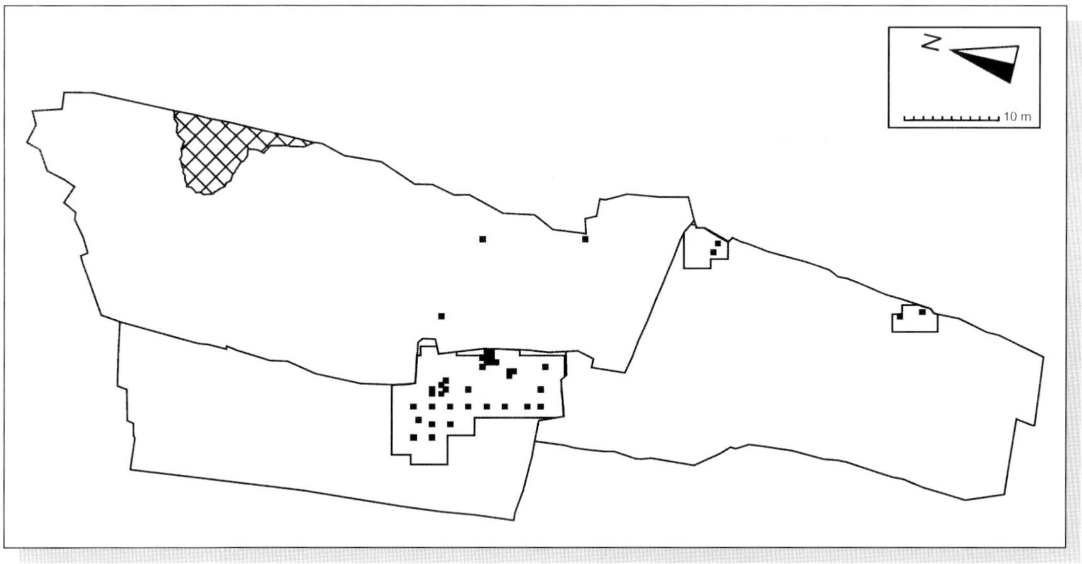

Figure 158: Doel "Deurganckdok-sector B"; distribution map of the analysed seeds and fruits samples.

Because of the vast number of samples, only a selection of 42 quadrates were analysed. In a first stage, 21 quadrates were selected, based on the examined samples in the archaeo-zoological research (cf. 3.5.4.). Afterwards, this selection was increased with other quadrates to obtain a better representation of the total area (21 quadrates, tab. 1). Finally, 13 samples were selected for more detailed analysis.

Methods and material

All samples were sieved in the field with a 2mm mesh. Ca. 3 litres per quadrant were separated for more detailed analysis. The samples of the 13 quadrants selected for detailed analysis were processed using sieves with 2mm, 1mm and 0.5mm meshes. All other steps in the treatment were the same as those described in 3.4.2.

Results

			Waterlogged					Charred														
			Wild plants					Cereal	Wild plants													
Quadrant	**Volume (l)**	**Smallest fraction (mm)**	*Ajuga reptans* (bugle)	*Alnus* sp. (alder)	*Cornus sanguinea* (red dogwood)	*Menyanthes trifoliata* (bogbean)	*Sonchus asper* (prickly sow-thistle)	*Triticum aestivum* (bread wheat)	*Cladium mariscus* (great fen-sedge)	*Cornus sanguinea* (red dogwood)	*Corylus avellana* (hazel)	*Crataegus monogyna* (hawthorn)	*Galium aparine* (cleavers)	*Hedera helix* (ivy)	Lamiaceae (dead-nettle family)	*Lathyrus* sp./*Vicia* sp. (pea/vetch)	*Malus sylvestris* (crab appel)	*Prunus spinosa* (blackthorn)	*Quercus* sp. (oak)	Rosaceae (rose family)	*Sparganium* sp. (bur-reed)
W6/Z6 (3)	?	2	.	.	.	.	.	.	.	1	.	1	.	.	.	.	.	.	.	.	.
W6/Z18 (4)	?	2	.	.	.	.	.	.	.	1	3	.	.	.	.	.	.	1	.	.	.
W6/Z18 (1 of 4)	?	2	.	.	.	.	.	.	.	1	.	.	.	.	.	.	.	1	.	.	.
W7/Z32 (2)	?	2	.	1	.	.	.	.	.	.	7	.	.	.	.	.	.	.	.	.	.
W8/Z32 (1)	?	2	.	.	.	.	.	.	.	.	2	.	.	.	.	.	.	.	.	.	.
W8/Z32 (1)	1.7	0.5	.	.	.	.	.	1	.	.	.	.	.	.	.	.	.	.	.	.	.
W11/Z2 (4)	?	2	.	.	.	.	.	.	.	.	.	.	.	.	.	.	.	.	.	.	.
W14/Z55 (4)	?	2	1	.	.	6	.	.	.	.	74	.	.	.	.	.	.	.	.	.	.
W15/Z7 (3)	?	2	.	.	.	.	.	.	.	.	3	.	.	4	.	.	.	.	.	.	.
W15/Z7 (3)	3.1	0.5	.	.	.	.	.	.	.	.	6	.	.	5	.	.	.	.	.	.	.
W15/Z7 (3)bis	?	2	.	.	.	.	.	.	.	.	2	.	.	.	.	.	.	.	.	.	.
W15/Z7 (4)	?	2	.	.	.	.	.	.	.	.	12	.	.	11	.	.	.	.	.	.	.
W15/Z52 (2)	?	2	.	.	.	.	.	.	.	.	61	.	.	.	.	.	.	.	.	.	.
W16/Z6 (2)	?	2	.	.	.	.	.	.	.	1	4	.	3	6	.	.	.	.	.	.	.
W16/Z7 (1)	?	2	.	.	.	.	.	.	.	1	23	1	9	9	.	3	.	.	2	.	.
W16/Z7 (1) A	?	2	.	.	.	.	.	.	.	.	4	.	.	1	.	.	.	.	.	.	.
W16/Z7 (2)	?	2	.	.	.	.	.	.	.	.	85	.	1	67	.	.	.	1	.	.	.
W16/Z7 (3)	?	2	.	.	.	.	.	.	.	3	3	.	.	1	.	.	.	.	.	.	.
W16/Z7 (3)	3.8	0.5	1	.	.	.	.	.	1	.	1	.	.	1	1	.	.	.	.	1	.
W16/Z7 (3) bis	?	2	.	.	.	.	.	.	.	.	3	.	.	1	.	.	.	.	.	.	.
W16/Z7 (4)	?	2	.	.	.	.	.	.	.	.	6	.	.	5	.	.	.	.	.	.	.

			Waterlogged					Charred													
			Wild plants					Cereal	Wild plants												
Quadrant	Volume (l)	Smallest fraction (mm)	*Ajuga reptans* (bugle)	*Alnus* sp. (alder)	*Cornus sanguinea* (red dogwood)	*Menyanthes trifoliata* (bogbean)	*Sonchus asper* (prickly sow-thistle)	*Triticum aestivum* (bread wheat)	*Cladium mariscus* (great fen-sedge)	*Cornus sanguinea* (red dogwood)	*Corylus avellana* (hazel)	*Crataegus monogyna* (hawthorn)	*Galium aparine* (cleavers)	*Hedera helix* (ivy)	Lamiaceae (dead-nettle family)	*Lathyrus* sp./*Vicia* sp. (pea/vetch)	*Malus sylvestris* (crab appel)	*Prunus spinosa* (blackthorn)	*Quercus* sp. (oak)	Rosaceae (rose family)	*Sparganium* sp. (bur-reed)
W16/Z7 (3+4)	?	2	.	.	.	.	.	.	.	.	4	.	4	7	.	.	1	.	.	.	.
W16/Z8(4)	2.2	0.5	.	.	.	.	.	.	.	.	.	.	.	.	.	.	.	.	.	.	.
W17/Z6 (2)	?	2	.	.	.	.	.	.	.	1	1	.	.	11	.	.	.	.	.	.	.
W17/Z6 (2)	2.7	0.5	.	.	.	.	.	.	.	.	.	.	.	4	.	.	.	.	.	.	.
W17/Z9 (3)	?	2	.	.	.	.	.	.	.	.	2	.	1	.	.	.	.	.	.	.	.
W17/Z10 (4)	?	2	.	.	.	.	.	.	.	.	.	.	.	.	.	.	.	.	.	.	.
W17/Z10 (4)	2.8	0.5	4	.	.	.	.	.	.	.	.	.	.	.	.	.	.	.	1	.	.
W17/Z13 (2)	?	2	.	.	.	.	.	.	.	.	13	.	.	.	.	.	.	.	1	.	.
W18/Z2 (3)	?	2	.	.	.	.	.	.	.	2	1	.	.	6	.	.	.	2	.	.	.
W18/Z9 (2)	?	2	.	.	.	.	.	.	.	.	.	.	1	.	.	.	.	.	.	.	.
W18/Z9 (2)	3.5	1	.	.	.	.	.	.	.	.	.	.	1	.	.	.	.	.	.	.	.
W19/Z1 (4)	?	2	.	.	.	.	.	.	.	1	.	.	2	42	.	.	.	.	4	.	.
W19/Z1 (4)	2.4	0.5	.	.	.	.	.	.	.	.	.	.	.	.	.	.	.	.	.	.	.
W19/Z2 (1)	?	2	.	.	.	.	.	.	.	.	.	.	2	1	.	.	.	.	1	.	.
W19/Z2 (3)	?	2	.	.	.	.	.	.	.	2	.	.	1	.	.	.	.	.	1	.	.
W19/Z5 (4)	?	2	.	.	.	.	.	.	.	.	.	.	.	.	.	.	.	.	.	.	.
W19/Z5 (4)	4	0.5	.	.	.	.	.	.	.	.	1	.	.	.	.	.	.	.	.	.	.
W19/Z13 (4)	?	2	.	.	.	.	.	.	.	.	1	.	1	1	.	.	.	.	.	.	.
W20/Z1 (1)	?	2	.	.	.	.	.	.	.	.	.	.	.	23	.	.	1	.	.	.	.
W20/Z1 (1)	3.2	0.5	.	.	.	.	.	.	.	.	.	.	.	20	.	.	.	.	.	.	.
W20/Z2 (1)	?	2	.	.	.	.	.	.	.	.	.	.	.	.	.	.	1	.	.	.	.
W21/N2 (4)	?	2	.	.	.	.	.	.	.	.	1	.	.	.	.	.	.	.	.	.	.
W21/Z1 (4)	?	2	.	.	.	.	.	.	.	.	2	.	.	1	.	.	.	.	3	.	.
W21/Z3 (4)	?	2	.	.	.	.	.	.	.	.	.	.	.	.	.	.	.	.	.	.	.
W21/Z3 (4)	4	0.5	.	.	.	.	.	.	.	.	.	.	.	.	.	.	.	.	.	.	.
W21/Z5 (4)	?	2	.	.	.	.	.	.	.	.	3	.	.	.	.	.	.	.	.	.	.
W21/Z7 (4)	?	2	.	.	.	.	.	.	.	.	.	.	.	.	.	.	.	.	.	.	.
W21/Z9 (4)	?	2	.	.	.	.	.	.	.	1	2	.	.	.	.	.	.	.	2	.	.
W21/Z11 (3)	?	2	.	.	.	.	.	.	.	.	.	.	.	.	.	.	.	.	.	.	.
W21/Z13 (4)	?	2	.	.	.	.	.	.	.	.	.	.	.	.	.	.	.	.	.	.	.
W23/N2 (2)	?	2	.	.	.	.	1	.	.	.	10	.	.	.	.	.	.	.	.	.	.
W23/Z1 (4)	?	2	1	.	.	.	.	.	.	.	2	1	.	.	.	.	.	.	1	.	1
W23/Z3 (4)	?	2	.	.	.	.	.	.	.	.	.	.	.	.	.	.	.	.	.	.	.
W23/Z3 (4)	3	0.5	1	.	.	.	.	.	.	.	.	.	.	.	.	.	.	.	.	.	.
W25/N2 (1)	?	2	.	.	1	.	.	.	.	.	1	.	.	.	.	.	.	.	.	.	.
W25/Z1 (1)	?	2	.	.	.	.	.	.	.	.	.	.	.	.	.	.	.	.	.	.	.
W25/Z1 (1)	3	0.5	1	1	.	.	.	.	.	.	.	.	.	.	.	.	.	.	.	.	.

Table 80: Doel "Deurganckdok-sector B"; list of waterlogged and charred seeds and fruits. In grey = squares selected for detailed analysis.

The samples from Doel "Deurganckdok-sector B" yielded both charred and waterlogged seeds, with the former by far the most prevalent. The number of species found is very limited, as well as the number of seeds. Density is also very low; the 'highest' densities occur where *Corylus avellana* (hazel) and *Hedera helix* (ivy) occur together - but in both cases real numbers (minimum numbers of individuals) are lower than is shown in table 80, since fragments are included.

Discussion

Taphonomy

Charred and waterlogged seeds are thought to correspond with different phases of the site's history. In spite of the diverging ^{14}C-datings of charred seeds versus food residues on ceramics (cf 3.2.4.4.), the charred seeds, burnt bones and ceramics seem to belong together, as a result of human activity in those times when the site was at least temporarily dry enough for human occupation. It is assumed that in these dry conditions uncharred seeds must have decayed. Only charred seeds remain from the occupation phase of the site and they are treated here as one assemblage.

When the site became wetter, resulting eventually in peat overgrowing the top of the sand dune from ca. 5050±50 BP, conditions were established for anoxic, waterlogged preservation. The uncharred seeds reflect this wettening, right before peat started to accumulate on top of the sand dune and *Menyanthes trifoliata* (bogbean) is its clearest example. In the light of this, the waterlogged seeds are not taken into account in further discussion.

As sampling was representative, the low number of species and seeds and the low density reflect reality and are not an accident of sampling. This could point to limited human activity involving a limited number of species. Regarding the spatial distribution of species and seeds, in the present state of analysis no firm conclusions can be drawn. A detailed spatial analysis will only be possible when the entire sieved area is analysed[2].

Why only larger seeds are preserved remains unexplained; possibly it has something to do with deliberate selection, which meant that only species with large seeds got into the fire. The fact that most wild plants found are edible is in favour of this hypothesis. The (post-) depositional survival of only larger seeds does not seem satisfactory as an explanation, although it's true that with increasing time the survival rate of seeds diminishes.

With regard to seasonal aspects, two things are remarkable: 1) all the edible species are those that are ripe from late summer through until winter and 2) the dispersion period from *Hedera helix* is winter and spring. The fact that no species are found pertaining to other periods does not mean that man was absent in these periods. The species found only demonstrate the presence of man, not absence.

Not all the edible species are suitable for immediate consumption: heating renders them palatable, and heating and roasting is also a way to keep the fruits for a longer period. This could be an explanation for the fact that most seeds were carbonised, by accident.

2 The selection of all water-screened residues, in particular from the north-western zone, is currently ongoing. The results will be presented in detail in the second volume of the "Verrebroek and Doel Excavation Project".

Cereals and other useful plants

During the analyses one charred cereal grain of *Triticum aestivum* (bread wheat) was recovered. The grain itself is not dated, but it is certainly older than the beginning of the peat growth on top of the dune (5050±55 BP). For the coversand area of Belgium and The Netherlands only a few other early finds of this cereal have been attested: De Gouw (4200-2450 BC), Aartswoud (2850-2450 BC), Vlaardingen (2900-2500 BC) and Zeewijk (2850-2000 BC), (Van Haaster, Brinkkemper 1995). Since only one grain was recovered and rachis remains are absent, it is quite doubtful that this cereal was locally cultivated. On the other hand *Triticum aestivum* seems to be the principal crop in contemporaneous Neolithic cultures, such as the Rössen (Maastricht-Randwijck), Epi-Rössen and Michelsberg, occupying the loamy area of Belgium and adjacent countries (Bakels 1997a). In older Neolithic cultures, like the LBK, more common cereals are naked barley, emmer wheat and einkorn (van Zeist, Palfenier-Vegter 1981, 145; Bakels 1997b, 18-20, 1; Bakels 1997c; Bakels 1991, 280-281;Visser *et alii* 2001, 39), but these were not found at Doel. Their absence is the result either of limitations imposed during fieldwork or of them never having been there in the first place. After all it is plausible to assume that men did not cultivate cereals at all at Doel or just not on the dune but maybe further away on more suitable places.

EDIBLE PLANTS

In the total list of seeds and fruits found, several species could have been useful in one way or the other. Nuts from *Corylus avellana* (hazel), wild apples from *Malus sylvestris* (crab apple), sloe plums from *Prunus spinosa* (blackthorn), acorns from *Quercus* sp. (oak) and berries from *Crataegus monogyna* (hawthorn) could have been collected and eaten, whether or not first boiled or roasted. Hazelnuts are by far the best represented ones, but it has to be remembered that consuming these nuts produces a lot of waste (fragmented nutshells), which creates an overrepresentation of this species in the results of analyses. This assemblage of fruits and nuts is also mentioned at several other Final Mesolithic/Early Neolithic locations, for instance at Hardinxveld-Giessendam (Bakels, van Beurden 2001, 342-345), Wateringen 4 (Raemaekers *et alii* 1997, 157-158), Urk (Peters, Peeters 2001, 67-68) and Hekelingen (Bakels 1988, 156).

Samples of several quadrates contained a number of charred stones of *Cornus sanguinea* (red dogwood). Whether the cherries could be eaten is a matter of debate, but the stones themselves have been noted as producers of paraffin (Bakels *et alii* 2001, 399). However there are no clear indications for this kind of use at the site.

Except for hazelnut fragments, the seeds of *Hedera helix* are the most frequently attested seeds on the site of Doel "Deurganckdok-sector B". *Hedera helix* grows in woodlands and forests on the ground and in trees, so it could easily have been collected together with firewood and subsequently be charred. However, it has only very rarely been mentioned in archaeobotanical analyses from sites in Belgium and The Netherlands (Hardinxveld-Giessendam Polderweg (Bakels, van Beurden 2001) and Hekelingen (Bakels 1988)). A further question to colleagues through the archaeobotany mailing list (ARCHAEOBOTANY@JISCMAIL.AC.UK) shed no further light on this question.

The presence of *Hedera helix* thus poses a specific problem. The number of seeds, their concentration, their carbonized state and the absence of *Hedera helix* charcoal point towards the fact that the fruits of *Hedera helix* were gathered deliberately and separately. It is doubtful that a natural fire could produce concentrations of charred seeds, since the berries hanging several meters high. If *Hedera helix* was burnt by man together with wood from the trees which it grew upon, one would expect to find charcoal from *Hedera helix* as well; this is not the case (cf 3.4.3.1.).

Several uses possible of *Hedera helix* can be put forward. For its mythological, religious and folkloristic significance, inspired by, among other things, its evergreen appearance, we refer to De Cleene and Lejeune (1999). Plenty of examples of its (supposed) medicinal properties can be read in, for instance, Meirsschaut (1979). Its leaves and berries can be used for dying, producing yellows, greens and pinkish-greys (Cannon, Cannon 1994). Thrown into a fire on purpose or spilled by accident, the berries could get charred.

Another explanation for the presence of *Hedera helix* could be its use as fodder. According to Kirby (2001), referring to Rackham (1980), *Hedera helix* has been used as fodder since Neolithic times. Neolithic evidence from Switzerland is provided by Rasmussen (1989a; see also Rasmussen 1989b and Haas, Rasmussen 1993). Bonnier in his early 20th century flora (reedited by Palese, Aeschimann 1990) also mentions *Hedera helix* as fodder in times of scarcity; most animals eat the leaves fairly readily, Bonnier says. This use as fodder is at first sight a bit surprising since *Hedera helix* is poisonous, especially the berries (van Genderen, Schoonhoven 1997), but in accordance with other observations: red, roe and fallow deer eat *Hedera helix* in quantity (Kirby 2001) and sheep are also reported to feed on it (Weeda *et alii* 1999).

Therefore we would like to suggest the following hypothesis: in late winter/early spring, people at Doel "Deurganckdok-sector B" gathered young twigs of *Hedera helix*, ripped off the berries – they knew these are the most poisonous – and fed the twigs to their cattle; late winter/early spring is by its sheer nature a period of scarcity with regard to fodder. Maybe the berries were burnt to render them harmless. Although primarily one of hunter-gatherers, the Swifterbant culture goes together well with cattle on a small scale (Raemaekers 1999). At Doel, however, no absolute evidence of cattle has been found so far (cf 3.5.4.) Why only charred seeds have been found and not charred berries as well, as in Hardinxveld-Giessendam Polderweg (Bakels, van Beurden 2001), remains unclear and could perhaps be a contra-indication for the fodder hypothesis.

One other species is worthy of mention. *Cladium mariscus* (great fen-sedge) is a species of riverbank vegetation or riparian vegetation, which can be peat forming. It is useful as kindling and as thatch, but only one charred seed was detected, so it is hazardous to draw further conclusions.

Ecological interpretation

Although at first glance appearing to originate from very different habitats, almost all the species found could fit into one overall, but very local picture: a hardwood riverside woodland. Below (cf 3.4.4. and 3.6.1) this hypothesis is confronted with the results of the other disciplines and hardwood riverside woodlands are further explicated.

3.4.3.3. | Pollen analysis

Cyriel Verbruggen

One hearth-pit has been pollen analysed. From the blackish charcoal-rich layer two levels were sampled for pollen slides; both contained a large number of microscopic small uncarbonised plant remains, but no pollen grains.

3.4.4. | Conclusions

Analysis of carbonised plant remains found in different types of archaeological features at Verrebroek and Doel have yielded additional information about the palaeo-vegetation during and after the main human occupations of the sites.

At the Final Palaeolithic site of Verrebroek "Dok 2", charcoal analysis succeeded in refining the Late Glacial vegetation record by identifying *Populus* (poplar), a tree species which was not evident from pollen analysis.

At the Early Mesolithic site of Verrebroek "Dok 1", the numerous hazelnut shells (*Corylus avellana*), mainly found in surface-hearths, as well as the predominance of *Pinus sylvestris* (Scots pine) in the charcoal from hearth-pits and charcoal-pits, perfectly correspond to the Preboreal/Boreal landscape as reconstructed on the basis of pollen in organic sediments from the nearby Schelde valley.

At Doel "Deurganckdok-sector B", the preservation conditions of plant remains in relation to the Swifterbant occupation were much better than at Verrebroek, allowing a more detailed palaeo-environment reconstruction. Although seemingly originating from very different habitats, almost all the species found, both seeds and charcoal, could fit into one overall but very local picture. The picture that thus arises is in accordance with the geomorphological setting

When charcoal and charred seeds from the A-horizon, at the contact of the coversands and the peat (the so-called Swifterbant find-layer), are combined, one gets a rather 'complete' listing of the shrub and tree layer species that are found in present-day hardwood riverside woodlands. Hardwood is used here in the German sense (*Hartholz*), meaning broadleaved trees that give good timber (e.g. *Quercus* sp. (oak), *Ulmus* sp. (elm) and *Fraxinus excelsior* (ash)) and do not include softwood (*Weichholz*), such as *Alnus* sp. (alder), *Salix* sp. (willow), *Populus* sp. (poplar), etc. (Ellenberg 1988, 245). In terms of phytosociology, hardwood riverside woodlands are referred to as *Violo odoratae-Ulmetum* and *Fraxino-Ulmetum*, both belonging to the *Alno-Padion* (Wolf *et alii* 2001; Stortelder *et alii* 1999). All the shrub and tree species recorded in Doel "Deurganckdok-sector B" can grow in hardwood riverside woodlands (Wolf *et alii* 2001, 99, 173-175; van der Werf 1991), except maybe *Malus sylvestris* (crab apple). *Hedera helix* (ivy) also has its place in this environment (Wolf *et alii* 2001; van der Werf 1991). *Fraxinus excelsior, Ulmus* sp. and *Quercus robur* dominate.

The number of species growing in current hardwood riverside woodlands that are not found in Doel but could have been according to Wolf *et alii* (2001), is less than is shown in Wolf *et alii* (2001, 99). *Fagus sylvatica* (beech), *Acer platanoides* (Norway mapple) and *Populus* x *Canadensis* (hybrid black-poplar) had not yet become part of the vegetation in the 6th millennium BP (Maes 2002; Verbruggen 1996). *Acer pseudoplatanus* (sycamore) is doubtful (Maes 2002).

Hardwood riverside woodland is situated on the highest parts of a floodplain, away from the river, and on river dunes, levees and channel bars, i.e. the higher bodies within or bordering a valley. Its typical character is determined by inundation events; these occur less than 10 days a year, sometimes not even a single day a year (Wolf *et alii* 2001; other durations are mentioned by van den Balck 1998). Sedimentation is low; aeolian activity is possible. Mean highest groundwater table is lower than 1m below ground level.

The river-stretch with the hardwood riverside woodland alongside it can be within the reach of fresh water tides; the woodland itself is out of reach. For specific soil characteristics of hardwood riverside woodlands, we refer to Wolf *et alii* (2001), Van den Balck *et alii* (1998) and van der Werf (1991).

On its lower side, present day hardwood riverside woodland is flanked by softwood riverside woodland in the close vicinity of the river, with a dominance of *Salix* sp. (willow), or by a more alder carr-like type of woodland further on in the floodplain (*Alnus glutinosa*) (van der Werf 1991). However, the presence of softwood riverside woodland at Doel "Deurganckdok-sector B" is not attested by botanical macro-remains. In view of its topographical and hydrological relation with hardwood riverside woodland, one could question whether softwood riverside woodland was present in Doel but not exploited, rather than simply absent. Whether the *Alnus* sp. charcoal from the archaeological site originates from hardwood (or softwood) riverside woodland or from an alder carr-like type of woodland is not detectable.

On its upper side present-day hardwood riverside woodland is flanked by woodlands dominated by *Fagus sylvatica* and *Quercus robur* (van der Werf 1991). *Fagus sylvatica* was, however, not yet present at ca. 5500 BP. These woodlands could well have been absent in the immediate vicinity of Doel "Deurganckdok-sector B", as the site itself was already situated on top of a sand dune.
Hardwood riverside woodland-like environments have been encountered previously in archaeological research (cf Bakels 1988).

Hardwood riverside woodlands have by now virtually ceased to exist along the River Schelde (Van den Balck *et alii* 1998; Paul Van den Bremt, personal communication; Noirfalise 1984) due to the total reclamation of the valley bottom and the embankment of the river. Nevertheless, favourable conditions for its development are still present (Van den Balck *et alii* 1998). For habitats elsewhere in Belgium we refer to Noirfalise (1984).

3.5. | Archaeozoological analyses

Wim Van Neer, Anton Ervynck & An Lentacker

3.5.1. | Introduction[1]

Animal remains have been recovered from excavations at three sites, i.e., from a Final Palaeolithic site at Verrebroek "Dok 2", an Early Mesolithic site at Verrebroek "Dok 1", and an Final Mesolithic site at Doel "Deurganckdok-sector B". A fourth site, from Doel "Deurganckdok-sector C", dating back to the Michelsberg culture, did not yield any bones or shells. All finds recovered are very small due to severe fragmentation, and show a white colour as the result of incineration. That unburnt bones have not been found must be linked with the presence of sandy sediments and a low groundwater table at the site locations, allowing the percolation of (acid) rain water and resulting in an oxidised environment. Over the years, these conditions must have degraded all shells, animal (or human) bones, except those that had been burnt before, or shortly after, deposition. Heavily burnt bones go through a process of remineralisation, which ultimately makes them more resistant to chemical degradation. However, a disadvantage of the burning (for analysis) is that it also causes fragmentation and shrinking of the bones. Moreover, burnt bones are brittle which makes them prone to further fragmentation, when buried in the soil, or during (and after) excavation (cf Herrmann *et alii* 1990). In the case of the "Verrebroekdok" and "Deurganckdok" sites, it is striking that the preservation condition and the average size of the animal remains vary clearly in relation to their archaeological age, with the Final Palaeolithic finds being very small and in extremely poor condition and the Final Mesolithic bones being somewhat larger and in better condition. The only exception to this rule is the Michelsberg site which, for some reason, did not contain any animal remains at all.

3.5.2. | Verrebroek "Dok 2"

Only a small collection of 81 animal remains was recovered from this site. The material had been studied previously (Oversteegen 2000a) but contained only two identifiable fragments (of wild boar). No interpretations can thus be made about the economy or the ecology of the site.

1 *Acknowledgements:* the authors wish to thank everybody at Ghent University who has helped with the sieving and with the sorting of the sieved residues, and Luc Muylaert (Institute for the Archaeological Heritage of the Flemish Community, IAP) for the final control of the sorted material. Wim Wouters (Royal Museum for Central Africa) analysed the amphibian remains. Archaeozoological data from Neolithic sites from The Netherlands were partly collected through the Internet database 'Bone-Info' compiled by the ROB (Rijksdienst voor het Oudheidkundig Bodemonderzoek, Amersfoort) (http://wwwarchis.archis.nl:70/bone-info/index.html).

3.5.3. | Verrebroek "Dok 1"

The Mesolithic site yielded more material, of which 1190 fragments have already been studied (Oversteegen 1999). However, the interpretations made are unreliable due to the particular methodology employed during the analysis. In fact, only three bone fragments were identifiable (one from beaver, two from wild boar), while five were classified only within the broader taxonomic category of 'fish' and two were described as 'microfauna'. Of the remaining finds, roughly 50% (591 fragments) were classified in size classes, these being 'large mammal', 'medium mammal gracile', 'medium mammal robust', 'small mammal robust' and 'small mammal carnivorous'. The counts for these size classes were then replaced by counts per species on the basis of a table that lists which animals would most likely belong to the size categories (Oversteegen 1999, fig. 1). The counts for 'small mammal robust' were, for example, replaced by counts for beaver because this animal is the most likely to have been present around the site as representative of this size category (excluding, for some unknown reason wild cat and badger, species that would also fit into this group). Moreover, this is not the only strange pattern within Oversteegen's approach. According to the author (Oversteegen 1999, fig. 2 & 3), all 'large mammal' bones would belong to aurochs, excluding, for unmentioned reasons, moose, red deer, horse and wild boar. Similarly, why the possible presence of large carnivores (wolf, brown bear) is not taken into account, remains equally unclear.

The unreliability of Oversteegen's study is further illustrated by a control check of the material itself, showing that for most specimens it is impossible to be sure whether it belongs to a small carnivore or another small mammal species. Furthermore, bones from the same species can fall in different size classes because large mammals also have medium sized bones, etc. Finally, it must be considered unacceptable that the analysis discusses the use of animals within the Mesolithic sites without a single certain identification for most species, and attempts to partly reconstruct the economy of the site and its function, on the basis of only three bones of which the identification is beyond doubt. Dismissing the data and interpretations compiled by Oversteegen, the Mesolithic animal remains have been re-analysed. However, the conclusion remained that virtually all finds are unidentifiable and that no interpretations can thus be put forward about the economy or the ecology of the site. It is therefore advisable not to use or cite Oversteegen's report in future.

3.5.4. | Doel "Deurganckdok-sector B"

3.5.4.1. | Sampling and methods

The analysed burnt bone remains originate from the Swifterbant cultural layer, situated in the bioturbated A-horizon of the coversand deposits. They were found in the same stratigraphic level as the other archaeological finds, i.e. the flint artefacts, pottery fragments, charcoal fragments and burnt seeds and fruits.

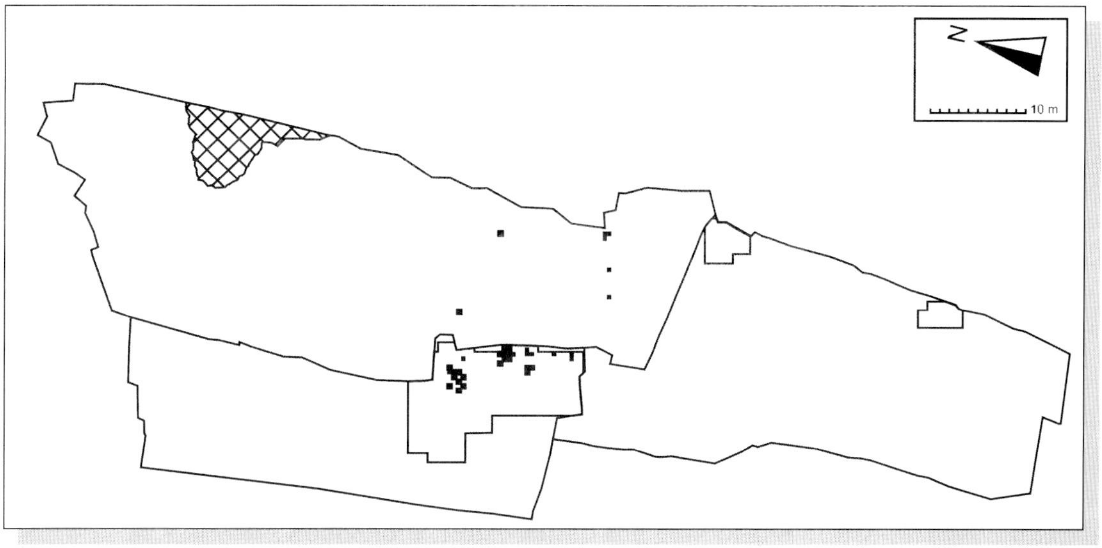

Figure 159: Doel "Deurganckdok-sector B"; distribution map of the analysed bone samples.

The animal remains studied were recovered mainly from the north-western sub-zone of the site. The material has been collected through an extensive sieving campaign, consisting of the wet sieving of large volumes of soil using a mesh size of 2mm. The sieved residues have been left to dry prior to sorting. For analysis, the residues containing the highest density of animal remains were selected, a process facilitated by the fact that the white, incinerated, bone fragments were very conspicuous amongst the matrix of dark brown plant and charcoal fragments. The residues were sorted manually at Ghent University, after which the result was controlled in the laboratory of the Institute for the Archaeological Heritage of the Flemish Community. Identifications were performed using the archaeozoological reference collections of the Royal Museum for Central Africa. In the case of the amphibian remains identifications also relied on the criteria described by Bailon (1999). Although each sample can be located spatially within the site, the material is interpreted as a whole. It was clear that the finds did not derive from well-defined structures, and there was no time available to analyse all samples (especially as the majority contained only a few finds).

3.5.4.2. | Results

Due to the already mentioned fragmented and burnt state of the animal material, most of the specimens remain unidentified. Nevertheless, in most cases, it was possible to distinguish the bone fragments of fish, amphibians and mammals. Thousands of finds belonged to the latter group but could not be attributed to species. This unidentifiable group has not been counted and is not incorporated in the inventory. In general, the mammal remains consisted mostly of the remains of large species, i.e., small fragments of the diaphysis of long bones, of epiphyses, of ribs and vertebrae, along with a small number of teeth fragments and some small compact bones that were still more or less complete. As has been said, only a few of these finds could be identified to species (tab. 81), in most cases these were small compact skeletal elements from hand or foot. In total, 27 identifications were possible, amongst which were a single bone of polecat (*Putorius putorius*), a skeletal element of an unidentified small rodent, 13 bone fragments and 5 tooth fragments of red deer (*Cervus elaphus*), and 7 bone fragments of wild boar (*Sus scrofa*). The identifications for the latter species remain somewhat doubtful because, since the Swifterbant culture is a Mesolithic-Neolithic transitional culture, bones of both domestic pig (*Sus scrofa* f. domestica) and wild boar could, in theory, have been present within the group of the *Sus* remains. For the Early Neolithic period, the distinction between the two forms of the species can only be made on the basis of their biometry (with wild boar being larger than pig), and such distinction will be most straightforward for dental material (Clason 1967). However, since the size differences between the early domestic pigs and wild boar were only slight and cross-breeding between the two forms must have occurred on a regular basis, an identification beyond doubt can rarely be achieved. In an archaeozoological study of Neolithic sites from The Netherlands, the option was not even taken to attempt any further identifications within the taxon *Sus* (Zeiler 1997). In the case of the "Deurganckdok site", the situation is even more complex due to the incineration of the bones and the associated shrinkage, which may vary between 5% to 30% (Gilchrist, Mytum 1986). Nevertheless, it was clear that, regardless of their shrinkage, all *Sus* bones found were larger than recent Belgian reference material of wild boar. Taking into account that recent wild boar in Belgium are probably smaller than Neolithic specimens of the species, it can be concluded that in all probability all *Sus* remains from "Deurganckdok" represent wild boar.

species	N
red deer (*Cervus elaphus*)	18
wild boar (*Sus scrofa*)	7
polecat (*Putorius putorius*)	1
small rodent (*Rodentia* sp.)	1
total	27

Table 81: Doel "Deurganckdok-sector B"; list of mammal remains.

Amphibians are represented by 11 finds (tab. 82), of which one belonged to common toad (*Bufo bufo*), a species that occurs in many habitats (de Witte 1948). One of the other finds comes from a frog species (*Rana* sp.), while nine bones did not allow a distinction between toad and frog.

species	N
common toad (*Bufo bufo*)	1
frog (*Rana* sp.)	1
frog or toad (*Anura* sp.)	9
total	11

Table 82: Doel "Deurganckdok-sector B"; list of amphibian remains.

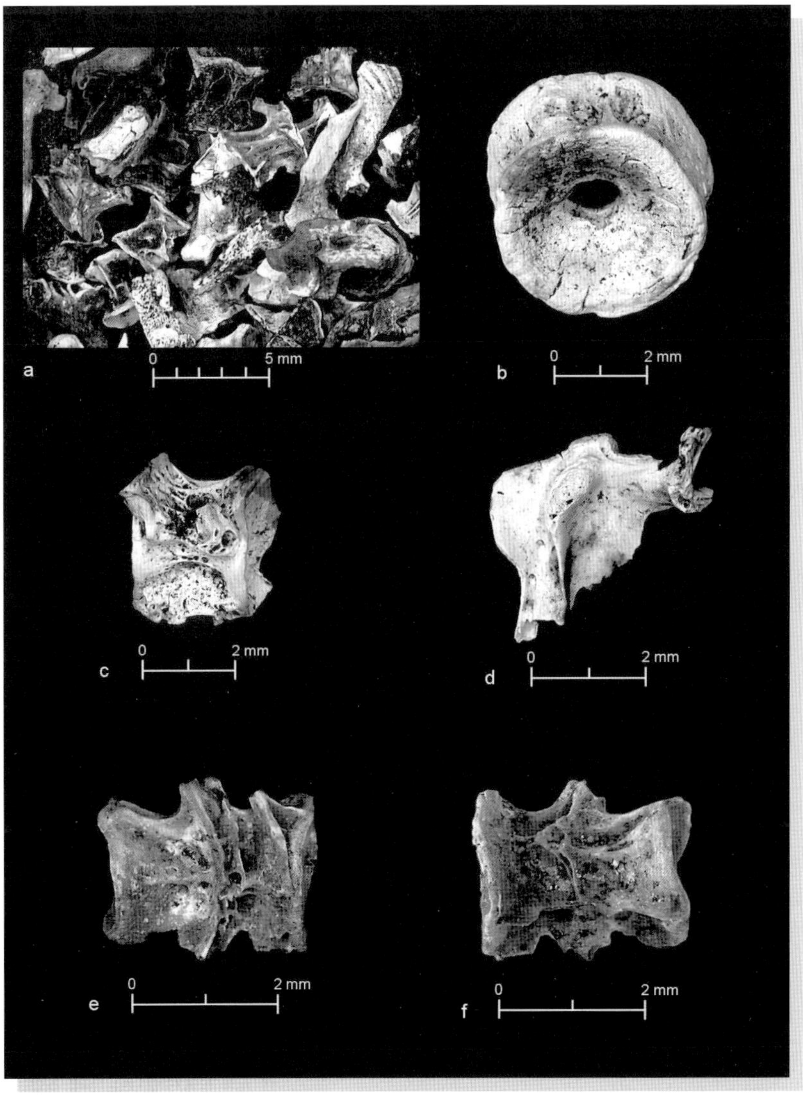

Figure 160: Doel "Deurganckdok-sector B"; (a) small sample of fish bones, (b) vertebral centre of stingray *Dasyatis pastinaca*, (c) precaudal vertebra of perch *Perca fluviatilis*, (d) basipterygium of roach *Rutilus rutilus*, (e) right lateral view of two fused cyprinid vertebrae, (f) left lateral view of the same two fused cyprinid vertebrae.

In contrast to the previous animal groups, the identifiable remains of fish were very frequent amongst the Final Mesolithic material (see fig. 160a for some examples). Of the more than 2000 fish remains, 57% could at least be classified at family level. Often, the material was even preserved in a condition that allowed the visual comparison of the finds with reference material from specimens of known standard length (SL: the distance from the tip of the snout to the base of the tail). This allowed a reconstruction of the fish sizes present within the Final Mesolithic collection, taking into account that the shrinkage will in all cases have produced an underestimation.

Virtually all identified fish bones come from freshwater species (tab. 83). The only real marine species is the stingray (*Dasyatis pastinaca*), represented by a single vertebral centre of a medium-sized animal (fig. 160b). The disc width, a common measure to evaluate the size of ray species, must have been 40 to 50cm, while the maximum width for this species can be 1.5m (Poll 1947). Stingray occurs near the coast in shallow waters on sandy sediments, which they use to partly bury themselves. It is known that the species can explore estuaries although its behaviour is not well studied within that habitat. Within the Schelde, the stingray has already been observed near Doel (Poll 1947), where at present the influence of the sea is still noticeable.

Species	N
stingray (*Dasyatis pastinaca*)	1
sturgeon (*Acipenser sturio*)	9
eel (*Anguilla anguilla*)	7
pike (*Esox lucius*)	16
silver bream (*Abramis bjoerkna*)	1
bream (*Abramis brama*)	1
silver bream or bream (*Abramis* sp.)	2
orfe / dace / chub (*Leuciscus* sp.)	2
roach (*Rutilus rutilus*)	6
rudd (*Rutilus erythrophtalmus*)	2
cyprinids, carp family (*Cyprinidae* sp.)	1149
perch (*Perca fluviatilis*)	4
total identified	1200
total unidentified	891
total	2091

Table 83: Doel "Deurganckdok-sector B"; list of fish remains.

The presence of sturgeon (*Acipenser sturio*) is proven by nine very small fragments of dermal bone plates, on which the ornamentation characteristic for the species is clearly visible. The fragments derive from the dermal scutes that protect the animal's body flanks, except for one find (a preopercular) that comes from the gill cover. Sturgeons are marine fish that reproduce in freshwater, and that therefore seasonally entered the river basins. This process, during which the fish can go very far upstream, could still be witnessed in the river Schelde until the 19th century but now sturgeon is locally extinct. It is known that within the Schelde basin the reproductive season for the species took place from the end of spring until the beginning of summer, and the fish could spawn in brackish or fresh waters (Poll 1947). The bone fragments found do not allow a size reconstruction of the animals.

All other fish remains represent freshwater species, with a strong dominance of the carp family (Cyprinidae). This is a group with many species with a very similar osteomorphology, which makes identification to species level possible for some skeletal elements only. Together with the high level of fragmentation of the Final Mesolithic material, this explains why only 14 of 1163 cyprinid remains could be classified to species. Roach (*Rutilus rutilus*) was represented by four pieces of the pelvic girdle (fig. 160d), a basioccipital (part of the cranium) and a pharyngeal plate with some of the teeth still preserved. The size reconstructions indicate an individual of 5-10cm SL, 3 of 10-15cm and 2 of 15-20cm SL. The presence of rudd (*Rutilus erythrophtalmus*) could be demonstrated by two pharyngeal plates of animals of 10-15 and 20-25cm SL. From the silver bream (*Abramis bjoerkna*) a pharyngeal plate was found, from an individual of around 10-15cm SL. A pelvic fragment could be attributed to bream (*Abramis brama*) of around 30-40cm SL, while for two similar pieces it remained unclear whether they should be classified as bream or silver bream. The only other cyprinid remains that were interesting in terms of identification are two incomplete pharyngeal plates that could not be identified to species level but that certainly belong to the genus *Leuciscus*. In our rivers, three species can be found within this genus, i.e., orfe (*Leuciscus idus*), chub (*Leuciscus cephalus*) and dace (*Leuciscus leuciscus*). Taking into account the ecological characteristics of the other fish species found at the site, it is most probable that we are dealing with orfe.

Amongst the unidentifiable cyprinid remains a clear dominance of vertebral elements could be observed, although skeletal elements of the rest of the body, including the head and the fins, are also present (tab. 84). Considering size, it is striking that the vast majority of the remains corresponds with fish between 10 and 20cm SL. Approximately 88% of the cyprinids fall within this size class, while individuals smaller than 10cm SL only form 0.2%. 11.4% comes from larger cyprinids, between 20 and 30cm SL, while the size class of 30-40cm only attains 0.3%. The cyprinid remains also show a series of pathological deformations, a phenomenon that will be discussed below.

skeletal element	NF	%
neurocranium fragment	3	0.3
pharyngeal plate	97	8.4
pharyngeal tooth	7	0.6
preopercular	1	0.1
opercular	3	0.3
epihyal	3	0.3
keratohyal	14	1.2
hypohyal	5	0.4
branchiostegal	1	0.1
articular	28	2.4
hyomandibular	1	0.1
quadrate	20	1.7
dentary	7	0.6
cleithrum	1	0.1
precaudal vertebra	507	44.1
caudal vertebra	351	30.5
precaudal or caudal vertebra	57	5.0
rib	28	2.4
basipterygium	4	0.3
pterygiophore	11	1.0
total	1149	100

Table 84: Doel "Deurganckdok-sector B"; frequency of skeletal elements within the group of the cyprinids not identified to species (NF = number of fragments, % = percentage).

Other, less frequent, freshwater species from the site include eel (*Anguilla anguilla*), pike (*Esox lucius*) and perch (*Perca fluviatilis*). Only seven eel bones were recovered: they consisted of vertebrae fragments from animals between 30 and 40cm. Pike is represented by seven vertebrae fragments, which were easily identifiable on the basis of the spongeous structure of the centres, and by seven edentate fragments of toothplates. In one case it was clear that the fragment was part of a vomer, but the other finds were so small that it was no longer possible to see whether they were parts of the vomer, the pterygoïd or the palatine. Two isolated teeth also belonged to pike. Size reconstructions were only possible for four better-preserved vertebrae, indicating medium-sized animals between 40 and 50cm SL. The teeth could have belonged to somewhat smaller individuals. Finally, from perch only two precaudal (fig. 160c) and two caudal vertebrae were found, from individuals between 15 and 20cm SL.

3.5.4.3. | Discussion

Taphonomy

Because of the preservation condition of the bones, only a few modification traces can still be observed. Cutting and chopping marks, gnawing traces of carnivores or rodents, and other modifications could have been present on the bone surfaces but are no longer visible on the fragmentary, incinerated material. The evaluation of the possible presence of more or less complete skeletons or a reconstruction of the processing of the animal carcasses is equally unattainable. As a result, there is little evidence to classify the animal remains found within a taphonomic category (*sensu* Gautier 1987). Still it can be assumed that the bones from red deer and wild boar come from animals that have been hunted for human consumption. Within this context, it needs to be mentioned again that the fact that only bones from the feet of these mammals have been identified does not indicate that only parts of the carcasses were left at the site (or were brought to it). Fragments of long bones, ribs and vertebrae have indeed been found, be it in an unidentifiable condition. Finally, the fish remains must also all represent consumption refuse. The location of the site excludes the possibility that we are dealing with the remains of natural death events, while the fact that the bones were burnt also indicates the interference of humans.

The polecat is a carnivore that was perhaps not frequently eaten within the Mesolithic period. Of course, such an assumption is based upon recent ideas on the edibility of animal species, but it can still be envisaged that this small animal was mainly hunted for its fur. This could make the polecat bone the only find belonging to the taphonomic group of the production waste. Traces of skinning or bone working are absent from the "Deurganckdok" finds but, as has been said, this could be the result of the preservation condition of the material. That fragments of antler have not been found can equally be caused by the incineration of the remains and by the fact that unburnt skeletal elements have not been preserved. Finally, the remains of a small rodent and of amphibians most probably represent intrusives.

Even though the volume of unburnt bone, that must originally have been deposited at the site but that has disappeared due to taphonomic factors, cannot be estimated, it remains striking that the number of burnt animal remains is high. The question thus arises why a large number of bones have been burnt at the site? The condition of the surviving burnt bones indicates contact with high temperatures. The white colour of the remains can indeed not be attained simply by frying, cooking or roasting food.

Experiments on fish bones, evaluating how rising temperatures affect the colour, and the macro- and microscopic appearance (Nicholson 1993), indicate that the fish remains preserved at "Deurganckdok" have been exposed to temperatures of 500 to 700 °C. For the mammal bones the same estimations are valid. This conclusion, however, does not exclude that bones have also been burnt at lower temperatures but these elements, which were not completely remineralised, have not survived within the local soil. The consumption leftovers must have been thrown into a fire, or they must have been burnt because a fire was lit on a spot where bones were deposited. However, it can also not be excluded that the consumption remains were used as fuel, a possibility that has been discussed at length within the literature (e.g., Colten 1995; Théry-Parisot 1998; Costamagno *et alii* 1999). Perhaps this explanation is more valid for the larger mammal bones than for the smaller remains. The fish bones may well just have been thrown away in a fire, unless we are dealing with specimens that one attempted to smoke (in order to preserve them), a process during which something could have gone wrong. On the other hand, it is also possible that the occupants burnt their food leftovers on purpose, in order to prevent large predators (such as brown bear) from being attracted to the site. Yet another possibility is that the burning of food leftovers was done for unknown ritual purposes; perhaps they were used as offerings to the spirits. Finally, the intrusive remains can have been present in the soil, on the spots where a fire was lit, or can represent the stomach contents of larger animals, or could have ended up amongst the other animal remains just by accident.

Hunters, fishers and gatherers

Traditionally, the Swifterbant culture is regarded as representing the last hunter-gatherer population in the coastal and sandy areas of the Low Countries. Regarding food economy, they continued a 'Mesolithic' way of life while, more inland, in the loess area, crop cultivation and animal husbandry had already started. This general view, however, is too simple and needs more nuances (van den Broeke 1991). For example, amongst the animal remains excavated at most Swifterbant sites in The Netherlands, bones from domestic mammals have been found (Raemaekers 1999, 114, tab. 3.49). This pattern could be explained by exchange mechanisms with cattle breeders living further inland, although it is perhaps more likely that the Swifterbant groups organised some animal husbandry on a small scale, while hunting, gathering and fishing remained the most important source (Raemaekers 1999, 112-115). A detailed analysis of the age structure of the domestic animal populations from the Swifterbant sites (evaluating, for example, the presence of newborns) could shed more light upon this but, unfortunately, until now, the study collections have always been too small to allow such analysis.

It is now also clear that the Swifterbant culture shows a chronological variation, leading to the subdivision into an 'Early' (4900-4600 cal. BC), a 'Middle' (4600- 3900/3800 cal. BC) and a 'Late' phase (3900/3800-3400 cal. BC) (Raemaekers 1999, 107, fig. 3.37). According to the archaeological and radiocarbon evidence (cf. 3.2.4.) the site of Doel belongs to the Early and/or first half of the Middle Swifterbant phase (4900-4450 cal. BC). An important, yet not completely solved, research question is now to establish whether the importance of domestic animals within the Swifterbant food economy remained the same during the three phases mentioned, or whether there was an increasing trend towards more and more husbandry through time. In other words: is the absence of the remains of domestic animals at the early site of "Deurganckdok" meaningful, or just the consequence of the small number of identifiable finds? The answer can hardly be found by comparison with contemporaneous sites because of the identification problems associated with the material (including the discrimination between wild boar and domestic pig) and because all these sites have only yielded small finds collections. The Early Swifterbant deposit 'L30' of the site Brandwijk (4610-4550 cal. BC according to Raemaekers 1999, 43) contained a bone of a sheep or a goat (Robeerst 1995), but this context yielded only a few animal remains and was covered by younger layers with much higher densities of bones, including those from domestic animals. Possibly, we are thus dealing with an intrusive find within the Early Swifterbant deposit. According to the first finds reports, on the older Early Swifterbant site Almere 'Hoge Vaart' (dated between 5200-4530 cal. BC, perhaps to be refined to 4900-4770 cal. BC: Raemaekers 1999, 95), the bones of cattle and perhaps also of pigs have been found (Hogestijn, Peeters 1996; Hogestijn *et alii* 1995a, 1995b, 1996; Raemaekers 1999, 95). However, the final study of this material has shown that this interpretation cannot be maintained (Laarman 2001). Phase 3 of the site Hardinxveld-Giessendam 'De Bruin', dated between 4700 and 4450 cal. BC, contained the remains of sheep, goat, pig and cattle, but the preceding phases (globally dated between 5500-4700 cal. BC) did not yield evidence for these domesticates (Oversteegen *et alii* 2001). A complicating factor for these interpretations is the fact that it is not sure at all that the sites investigated yield a sound image of the global food economy of the Early Swifterbant people. Certain sites may well have been used only during certain parts of the year, for specific purposes, making the frequencies of domestic animals difficult to evaluate. The Dutch site Bergschenhoek (4360-4050 cal. BC: Raemaekers 1999, 97) is a good example of such a site, and must have been used as a camp for hunting activities, for birds and mammals, and as a fishing site (Clason, Brinkhuizen 1993). In addition, in the case of Doel "Deurganckdok", it remains possible that we are also dealing with such a 'special activity' site at which only certain processes within the food economy took place.

It is certain that on the sand dune at "Deurganckdok" a camp was located at which the catch of hunting and fishing was brought together. People must have eaten there, and most probably must have stayed there for a while, because otherwise burnt consumption remains would not have been present at the site. The taphonomic analysis has demonstrated that the image of the use of animal products at the site is presumably strongly biased but still something can be said about the exploitation of the animal world during the Early Swifterbant period. Considering the hunting of large mammals, the interpretations of course remain very limited, due to the small finds collection, but, in any case, the species' spectrum found at "Deurganckdok" is not exceptional. A comparable Final Mesolithic site concentrating on the hunt for wild boar and red deer has been excavated at Brandwijk (The Netherlands) (Robeerst 1995).

Both Brandwijk and "Deurganckdok" must have been located within wetlands near, or in, the estuary of major rivers (see further for "Deurganckdok"). A difference between the two sites is the absence, in the latter site, of beaver or otter, species that were intensively hunted in the Final Mesolithic/Early Neolithic wetlands (Raemaekers 1999, 114, tab. 3.49; Oversteegen *et alii* 2001). This pattern, however, could be the result of the small volume of the finds collection.

Reconstructing how the fishing was organised at "Deurganckdok" is also not easy. Since fishing implements are totally absent (most probably not preserved), only the fishbones themselves can give some information. The most striking characteristic of the fish fauna is the dominance of cyprinids, together with the small body sizes of the specimens caught (even taking into account the deformation of the bone material). Apparently, the people camping at "Deurganckdok" did not go for large fish, such as (large) pike, salmon (*Salmo salar*), (large) sturgeon or catfish (*Silurus glanis*). Very large fish could have been caught individually with spears, harpoons or hooks, an activity, however, that was not highly profitable in all sorts of biotopes (because the yield depended on the fish density, the accessibility, etc.). At "Deurganckdok", one must have concentrated upon the catch of small 'white fish'. Taking into account the large variety of species and the high frequency of small specimens, it seems most plausible that the fishing methods used were not active but passive, using trapping devices that were let into the water for some length of time. Evidence for the use of nets is rather scarce for the Mesolithic and the Neolithic of north-western Europe but of fish traps more remains have been found. Dutch examples come from the Mesolithic site of Hardinxveld "De Bruin" (Louwe Kooijmans *et alii* 2001), the Swifterbant site of Bergschenhoek and from the somewhat younger site at Vlaardingen (Brinkhuizen 1986). Apparently, their general form, the materials and the manufacturing techniques used were comparable to those of traditional fish traps made until recently in the Low Countries. The fish fauna found at Vlaardingen only comprised thin-lipped grey mullet (*Liza ramada*), sturgeon and pike (Brinkhuizen 1979), but at Bergschenhoek the species' spectrum, including pike, eel, perch and cyprinids (amongst which were silver bream, bream, rudd, roach, orfe, tench and bleak) (Clason, Brinkhuizen 1993), resembled very much that of "Deurganckdok". Although details about the relative frequencies of these species, and about the body sizes of the fish caught, are not available for these Dutch sites, the resemblances between the faunas from these sites strongly suggest that the Final Mesolithic people at "Deurganckdok" also used fish traps. That the Dutch sites mentioned are younger compared to the "Deurganckdok" site, is not very important within this context, given the long, apparently unaltered, tradition of these implements. In Denmark, both Mesolithic and Neolithic remains of fish traps and fish weirs have been found (Pedersen 1995). The use of fixed fish traps has also been assumed for a number of Danish Mesolithic sites on the basis of the large variety of fish species found, although a small number of fish hooks has also been recovered (at Ertebølle and Bjørnsholm) (Bødker Enghoff 1994). The hooks, however, would only have been used for larger species.

Final Mesolithic and Neolithic sites with fish remains are known from The Netherlands, including the already mentioned sites at Bergschenhoek and Vlaardingen, but also the type location of Swifterbant (Brinkhuizen 1976), the Swifterbant site Hoge Vaart (Laarman 2001), Hazendonk and Kolhorn (Brinkhuizen 1979), site P14 in the 'Noordoostpolder' (Gehasse 1995), Urk-E4 (Oversteegen 2000b, 2001), Brandwijk (Ball 1997), and Hardinxveld-Giessendam 'De Bruin' (Beerenhout 2001). Kolhorn, situated on a sand ridge along a creek in the brackish or salt coastal area, differs from all other sites because only marine fish was present amongst the finds. Hardinxveld-Giessendam 'De Bruin' showed a strong emphasis on the catch of pike. At the site Hoge Vaart the remains of perch dominate the fish assemblage while cyprinids take the second place, followed by pike. All other sites, however, show a fish spectrum that resembles that from "Deurganckdok", i.e., a dominance of cyprinids from the bream zone (see further: ecology), and pike, eel and perch. Except for Urk-E4 and Hazendonk, the Meso-Neolithic fish faunas mentioned also yielded some marine species that enter brackish or even fresh waters. In most cases, these include flounder (*Pleuronectes flesus*) and mullets (Mugilidae sp.), species that are absent at "Deurganckdok". It is also striking that most of the Dutch sites also contained the remains of catfish, suggesting the use of hooks or spears. Regardless of these differences, the resemblance between the fish fauna from "Deurganckdok" and the Dutch sites must be the result of the use of comparable fishing techniques within a similar environment of freshwater creeks. Moreover, a chronological comparison of the assemblages of fish remains does not show any trends.

Assuming that the absence of the bones of large fish, such as catfish, is not the result of taphonomic processes, it is possible, even probable, that the fishermen at Mesolithic "Deurganckdok" had made the analysis between cost and yields for the active fishing for large specimens (costing a lot of time and effort but possibly resulting in a large yield), compared with the passive catch, with fish traps, of small cyprinids (costing considerably less time and effort but resulting in smaller yields). A combination of variables (prey density, weather conditions, local environment, time budgets available, etc.) can then have resulted in taking the option of the passive catch. For the same reasons of cost efficiency, the site inhabitants can have decided not to travel to marine waters for fishing.

Assuming that the site at "Deurganckdok" was specifically used for hunting and freshwater fishing, the question arises in which part of the year these activities were organised. Attempts to analyse the possible seasonal use of Mesolithic and Neolithic sites or to recognise the seasonal nature of specific activities at a multi-season site are always problematic (Louwe Kooijmans 1993; Zeiler 1997). Reconstructions are sometimes reliable when they are made upon the basis of finds from species that do only occur in the area in certain periods of the year (as, for example, migrating birds), or upon a detailed evaluation of the age of killed animals (assuming a single, short and fixed birth season). Seasonality estimations are less reliable when the most likely hunting season is evaluated for species that occurred during most of the year near the sites. It is often inferred that species hunted for their fur must have been killed in winter because their fur is then thickest, or, on the other hand, that some species were never hunted during the breeding season. However, whether former people always respected these rules cannot be proven. In any case, the small number of identified mammal remains from "Deurganckdok" does not allow any interpretations within this context. The absence of bird bones also seriously hampers a seasonality analysis.

Fish remains can also provide interpretation possibilities for seasonality analysis. However, analytical methods such as the observation of growth rings on vertebrae or otoliths were not possible for the collection studied, due to the preservation condition of the finds. In general, however, it is not likely that fishing was as successful in all months of the year. Catches must have yielded most during spring, when the fish spawned and migrated in large numbers into shallow waters. Weather conditions during that part of the year are also better compared to those in late autumn, winter or early spring. The presence of sturgeon at the site, which migrates into freshwater at the end of spring or during early summer, also indicates fishing during the warmer part of the year. However, it must be stressed that, even if this interpretation is valid, it does not imply that the site could not have been used during other seasons, perhaps for other activities. It must be underlined that the taphonomic history of the material prevents any view upon a range of possible activities at the site, such as hunting for fur animals, bird catching or the gathering of molluscs.

Ecology

The identified mammal bones can only give limited information about the environment around the site. Moreover, it is conceivable that species such as red deer, during prehistoric times, when the impact of people on the landscape and upon the animal world was still less pronounced, occupied larger niches, or a larger variety of biotopes, than is the case today (Louwe Kooijmans 1993). Nevertheless, it can still be meaningful that the Dutch Early Neolithic site 'Brandwijk', which, just as "Deurganckdok", is characterised by a dominance of red deer and wild boar remains, was situated in a wet landscape with a lot of open water. This can have been equally true for "Deurganckdok". Swamps or wet forests were certainly an ideal biotope for wild boar, and it is highly likely that red deer also occurred in such environments (Louwe Kooijmans 1993).

On the basis of the frequencies of the fish species found at "Deurganckdok", and taking into account their ecological characteristics, some interpretations can be put forward about the aquatic environment around the site. Taking into account that within a river, from source to estuary, a subdivision can be made between different zones, characterised by different water velocities, temperature, oxygen content, substrate, vegetation and salinity (Huet 1949), and thus also by a different fish fauna, it is clear that the fish fauna from "Deurganckdok" belongs to the 'bream zone', the zone most closely situated to the estuary of the river. This interpretation is corroborated by the dominance of cyprinids, which prefer slowly flowing or stagnant waters. In the bream zone, the water easily reaches higher temperatures, to the extent that during summer sometimes 20°C or more is attained. The transport of silt sediments from upstream and a rich plankton flora often make the water turbid. In the parts of the bream zone with clearer water, dense vegetation can develop, due to the slow velocity of the water and the rich nutrient supply. Because of this rich food supply, the bream zone is the zone with the greatest fish variety and density. Amongst the cyprinids of the bream zone species of flowing or still water can be found, but the latter group will always be more frequent. Typical species are bream, roach, perch, eel and orfe, along with two species introduced in historical times: carp (*Cyprinus carpio* f. domestica) and pike perch (*Sander lucioperca*). In parts of the bream zone with clearer water and denser vegetation, pike, roach and tench are the most important species.

The fish remains from "Deurganckdok" can thus correspond with a landscape of creeks with a lot of stagnant water, in any case a wet landscape dominated by fresh water. Within this context, it remains unclear how the presence of the stingray, the only marine species found at the site, must be explained. Is it possible that the single vertebra found comes from an animal that has been imported from the coast. Other evidence for such a connection has not been found. It is certainly remarkable that there is no material from marine species that are regularly found in brackish or even fresh water, such as flounder or thin-lipped grey mullet. On the other hand, some of the freshwater species found at "Deurganckdok" do tolerate a higher salinity. Examples are perch, and also to a certain extent pike, bream and roach. Perhaps the ecological tolerance of stingray is underestimated and the species can occur in waters where more tolerant freshwater fish also occur. At the Mesolithic site of Bjørnsholm in Denmark, stingray has also been found in association with cyprinids such as roach, rudd, bream and tench (Bødker Enghoff 1994).

A possible indication for rather stressful environmental conditions for the freshwater fish found at "Deurganckdok" is provided by the high frequency of pathological specimens (fig. 160e, f). Amongst the 351 caudal vertebrae of cyprinids 15 finds show the fusion of two subsequent vertebrae, while in two cases three caudal vertebrae were fused. Within the collection of 507 precaudal vertebrae one case of the fusion of two vertebrae has been observed. This type of pathology also occurs in recent fish, although the phenomenon is only very rarely observed in (recent) osteological reference collections. In other archaeozoological assemblages the pathology is hardly ever present, suggesting that, although precise quantifications are not yet available, the fusion of vertebrae occurs more frequently now than in the past. The site at "Deurganckdok" is certainly an exception to this rule, which implies that possibly less than optimal conditions were present for the fish fauna. Whether this was the salinity or another factor, cannot be concluded. Literature on the condition in recent fish populations points to food shortage (Sato *et alii* 1983), pollution (Muramoto 1981), mechanical or thermal stress (Backiel *et alii* 1984), infection by single-celled parasites (Lom *et alii* 1991) and a relatively high velocity of the water current (Divanach *et alii* 1997) as possible causal factors.

3.5.4.4. | Conclusions

The animal remains from "Deurganckdok" have, for the first time, shed light upon the food economy of Final Mesolithic people in Flanders' sand area. The finds illustrate the great potential for interpretations, although in this case the analysis was limited by the preservation of the remains (burning, incineration). In fact, this first approach should, in the future, be enlarged by incorporating material from spots within the landscapes, such as depressions, where better preservation conditions may have been present and where, for example, unburnt bone can be found. This could even drastically change our view on what happened in the area during the Final Mesolithic and Early Neolithic. Within the analysis of the site Almere 'Hoge Vaart', the burnt remains excavated on a sand dune provided a completely different view compared to that obtained from the contemporaneous unburnt bones recovered from a nearby creek bed (Laarman 2001). In general, it is conceivable that the fact that we mostly have information from sites on sand dunes within the wetlands of the coastal area within the Low Countries only provides a biased view on Final Mesolithic and Early Neolithic society (Louwe Kooijmans 2001c).

3.6. | General conclusions

3.6.1. | Palaeo-environment

Interdisciplinary analysis of numerous anthropogenic (mainly hearth-pits and surface-hearths) and natural features has led to a further refinement of the palaeo-environmental reconstructions obtained through pollen analysis of sequences at some of the investigated sites in the *Waasland Scheldepolders* (cf 2.4.). This was particularly the case at the Final Mesolithic site of Doel "Deurganckdok-sector B", which yielded large numbers of carbonised plant remains and bones.

Based on the radiocarbon dates on seeds and charcoal from surface-hearths, the Swifterbant occupation at this site can be correlated with the phase in which the area surrounding the sand ridge was gradually inundated as a result of increased perimarine influence (the so-called Calais-phase). The assemblage of charcoal and charred seeds recovered from surface-hearths and the A-horizon, however, reveals a picture that relates very well to a modern hardwood riverside woodland, typical of the higher bodies within or bordering a valley of a freshwater river and inundated only few days a year at most. According to the ecology of present hardwood riverside woodlands, the river stretch could have been within the reach of freshwater tides, while the woodland itself was not.

The fact that a hardwood riverside woodland, as seen from charcoal and charred seeds, is not detectable as such palynologically (cf 2.2.5.), doesn't pose a problem for interpretation in view of the spatial resolution of palynological research. By definition, one gets a mixed image composed of a local and a regional component. Moreover, the correlating sediments themselves limit the possibilities of palynological research, in contrast to e.g. peat deposits.

Another question is whether a hardwood riverside woodland combines with perimarine conditions as revealed by close by palynological analysis. The answer to this seemingly contradictory possibility could lie in the transitional position of the site of Doel "Deurganckdok", at the edge of the perimarine zone. More specifically, the palynological results show very varying values for perimarine indicator species in the clay deposits, possibly illustrating a changing environment with brackish or freshwater sedimentary conditions prevailing from time to time. Moreover, in the immediate vicinity of the coversand ridge settled by the Swifterbant people, clay deposits did not interrupt peat growth, indicating permanent freshwater conditions.

From the small number of identifiable mammal remains recovered at Doel "Deurganckdok-sector B", no further ecological reconstructions can be made. The fish remains, however, suggest the presence of a landscape dominated by stagnant or slow running freshwater. Unless the lack of species from marine or brackish water is interpreted as an economic choice, such aquatic environments must have been absent from the vicinity of the site.

3.6.2. | Mesolithic diet

Information concerning the Early Mesolithic diet is very restricted. At Verrebroek "Dok 1" the numerous carbonised nutshells point to a fairly systematic exploitation of hazelnuts. Another potential food residue is *Chenopodium album* (fat-hen), of which carbonised seeds have been found in small quantities in some hearth-pits. Indirect evidence of the processing of plants comes from microwear analysis on flint tools (Beugnier, Crombé, in press). Several flint artefacts, mainly unretouched bladelets and flakes, display a gloss created by cutting, splitting and/or scraping of plants, probably reed or hazelnut sticks/branches. However, these tools were clearly not used for harvesting plants but rather for the production and/or restoration of objects such as arrow shafts, baskets and mats. Despite the discovery of numerous burnt bone fragments, no further insight into the hunting strategy of these Early Mesolithic hunters could be obtained.

Diet reconstruction for the Final Mesolithic period has been more successful. At the Swifterbant site of Doel "Deurganckdok-sector B" many carbonised remains of potentially edible plant species have been collected. Hazelnuts, wild apples, sloe plums, acorns and hawthorn berries could all have been collected and eaten, whether or not first boiled or roasted. A single grain of *Triticum aestivum* (bread wheat), if not intrusive, most probably reflects import or exchange with contemporaneous Neolithic communities (late LBK, Rössen, Cerny, ...) rather than local production.

Furthermore, it was hypothesised that the occurrence of carbonised seeds of *Hedera helix* (ivy) in high quantities might be linked to its use as animal fodder, a common practice throughout European history. Although this hypothesis is feasible, it lacks supporting evidence in this case. From the archaeozoological analysis it could be cautiously concluded that domestic animals were most probably absent from the site, underlining the high importance of hunting as part of the organisation of the food supply. Freshwater fishing, most probably of a passive kind, using fish traps, also played a role. However, the yield from these fishing activities must be put into perspective since we are dealing with a catch of predominantly cyprinids of small dimensions. Generally, it must not be forgotten that taphonomic factors have strongly biased the archaeozoological record from the site, making it difficult to draw general economic conclusions.

In conclusion, the above palaeo-ecological evidence at the Final Mesolithic site of Doel "Deurganckdok-sector B" points to a broad-spectrum economy based on the exploitation of different kinds of natural resources. However, in the absence of other contemporaneous sites in Sandy Flanders, it is not clear whether this reflects a cultural or a functional choice. In the latter case, the subsistence strategy might be conditioned by the function of the site within the Swifterbant settlement system (residential versus non-residential sites) and/or by its seasonality.

3.6.3. | Feature function

3.6.3.1. | Hearth-pits

At Verrebroek "Dok 1" and Doel "Deurganckdok-sector B" a total of 122 hearth-pits have been excavated. At neither location are these features synchronous with the main occupation, the latter having left large quantities of settlement waste, i.e. lithic artefacts, ceramics, carbonised plant remains, etc. At Verrebroek, the hearth-pits are younger than the Early Mesolithic (Pre-boreal and Boreal) occupation, while at Doel they predate the Final Mesolithic (Swifterbant) human presence. This apparent chronological discrepancy makes discussion about the possible function(s) of these features extremely difficult.

During the last decade different scholars (Groenendijk, Peeters, Perry, ...) have tried to explain the use of these features, though no single hypothesis has proved convincing. Generally, and with the exception of charcoal, archaeological material that could possibly inform us about the use of these pits is almost totally absent. However, there is a common agreement that hearth-pits were not used simply to produce warmth/heat for the inhabitants of the site(s). Indeed, contrary to surface-hearths which radiate heat, hearth-pits retain their warmth as the fire is protected against wind (Groenendijk 1987, 98).

A first possible function which is frequently mooted is that of hazelnut roasting pits (Groenendijk 1987, 98). In the case of the hearth-pits of Verrebroek and Doel, however, this hypothesis is contradicted by the radiocarbon dates. Dates obtained on nutshells from these pits proved to be incompatible with the charcoal dates from the same features. They prove to be more compatible with the hazelnut dates obtained on samples from surface-hearths. The most plausible explanation for this anomaly is residuality. It seems likely that the few hazelnut shells found within these hearth-pits are residual material coming from nearby surface-hearths, which were destroyed during the digging of the hearth-pits. In addition, recent experiments by Score and Mithen (2000) have clearly demonstrated that deep pits are not at all suited for roasting nuts.

Mainly on the basis of ethnographical evidence, some scholars have suggested a use as cooking-pits, e.g. for drying, roasting or smoking meat or vegetable food (see Hamburg *et alii* 2001, 13). According to Groenendijk (1987, 89-99) roasting could have been done by placing the meat above the fire, into the glowing embers or into the underlying heated sand. The almost complete absence of burnt bone fragments in the Verrebroek and Doel features, however, does not support the hypothesis of meat preparation very well. The minor presence of small burnt bone fragments in a few hearth-pits at Doel can most likely be explained by intense bioturbation. On the other hand the dominance of organic over inorganic phosphate in some analysed hearth-pits at Verrebroek and Doel, as opposed to the surrounding sediment, might indicate organic residues of dead plants and/or animal material, as e.g. lipids. This is further corroborated by the presence in some hearth-pits at Verrebroek of carbonised plant remains from at least 7 different species besides hazelnut shells. Amongst these, however, there is only one possibly edible species, i.e. *Chenopodium album* (fat-hen). The other species might have been used to start fires, or they could have slipped accidentally into the filling of the pits. It is also possible that they were thrown into the fire as (settlement) waste; most of the species found are part of pioneer vegetations and they could have grown at or in the vicinity of the settlement. Furthermore, since none of these seeds have been radiocarbon dated directly, their contemporaneity with the associated features remains unproven. Referring to the dated hazelnut shells from these features, one has to take into account the possible residuality of the other plant remains too. On the other hand, the absence of comparable seeds in the surrounding surface-hearths rather supports the contemporaneity hypothesis. It should be noted that in other areas, such as the northern Netherlands, detailed botanical analysis of hearth-pits also revealed remains of potentially edible plants (Perry 1999), such as tubers of *Typha* sp. (bulrush), *Scirpus* sp. (club-rush), *Dryopteris filix-mas* (male-fern), *Beta vulgaris* ssp. *maritima* (sea beat) and *Equisetum* sp. (horsetail).

Assuming that the hearth-pits were used as cooking-pits, the absence of direct evidence of fire activity in the soil mineral material, for example reddened soil, might look strange. This absence, however, is most likely a consequence of very low iron content in the parent material both at Verrebroek and Doel. Moreover, at Doel "Deurganckdok" associated soil characteristics serve as indirect evidence of fire. Clay accumulation bands and coatings consisting of carbonised organic punctuations, both following the hearth-pit morphology, developed as a consequence of the changed physico-chemical conditions after fire activity. Presence of carbonised organic fine mass and fragments indicates both at Verrebroek and Doel a fair chance of *in situ* but incomplete burning. This fits with the presence and eventual production of charcoal in these features.

On the other hand, food processing also requires a slow burning fire. An experiment conducted by Groenendijk (1987) found that *Pinus* (pine) is an excellent fuel for producing intense, high-temperature and long glowing fires combined with low fuel consumption. According to Groenendijk this is the main reason why *Pinus* remained the most favoured fuel throughout the Mesolithic of the northern Netherlands, despite the decreased supply of this wood species in the local environment. This role for *Pinus*, however, is not supported by the data from the *Waasland Scheldepolder* area, where a clear shift in the use of firewood between the Boreal Mesolithic (Verrebroek) and the Atlantic Mesolithic (Doel) is observed. *Pinus sylvestris* (Scots pine), the dominant species at Verrebroek "Dok 1", is only attested in small amounts in the hearth-pits of Doel "Deurganckdok-sector B", while *Quercus* sp. (oak) is dominant with 97.5%. This is in accordance with the results of palynological research in Sandy Flanders: a clear increase of *Quercus* sp. in the forest vegetation from the second part of the Boreal until the beginning of the Atlantic.

Besides cooking activities, other explanations, which are even less demonstrable, have been proposed, such as the drying or smoking of non-food objects such as hides (Groenendijk 1987, 98) or the heating of flint or the production of resin (Hamburg *et alii* 2001, 13). Given the discrepancies with respect to radiocarbon dating between the hearth-pits and the main occupation phase, the weak presence or even absence of artefacts within these structures and the shift in wood type between the Boreal and the Atlantic in the study area, one may even question whether the so-called hearth-pits might not correspond to natural disturbances rather than to anthropogenic structures. This hypothesis will be further investigated and discussed in a future publication.

3.6.3.2. | Charcoal-pits

These features, only attested at Verrebroek "Dok 1", are even more difficult to interpret than the hearth-pits. The only similarity with the latter is the presence of numerous charcoal fragments and the charcoal spectrum, which is dominated by *Pinus sylvestris* (Scots pine). Deviating characteristics are the random spatial distribution and the presence of more weathered and smaller charcoal fragments. Another difference from the hearth-pits is the higher occurrence of traces of infestation (fungal or from insects), while clear cracks are absent. This points to the exclusive presence of dead wood within these features. Although at present difficult to prove, a natural origin of these features cannot be fully excluded. Further investigation is needed.

3.6.3.3. | Surface-hearths

Due to posterior soil formation processes these features are much less visible as soil marks. Their position is only traceable through spatial analysis of the burnt artefacts and ecofacts (burnt hazelnut shells, burnt bone fragments, ...). At Verrebroek "Dok 1", the surface-hearth always coincides with the highest quantities of burnt bone fragments, and frequently, but not always, with the highest quantities of carbonised hazelnut shells. Absence of large amounts of charcoal is probably due to taphonomic factors, i.e. weathering and wind activity. Additional indications of these superficial fireplaces are offered by the micromorphological and chemical analyses. An extended cross-section through a surface-hearth indicates an increasing, albeit not continuous, trend of indicators, such as the presence of carbonised organic fine mass, a Pi/Po-ratio (inorganic over organic phosphorus) close to or over 1, and an increasing total phosphate content towards the surface-hearth. Within the particular context of the studied surface-hearth, they indicate anthropogenic input of bone and/or hazelnut with subsequent burning and are thus possible indicators of *in situ* fire activity. Different types of ecofact clusters were investigated separately. The hazelnut cluster has a dominance of inorganic over organic phosphate, but no exceptionally high cation content. Probably, the porous surface and matrix of carbonised hazelnut shells stores, just like charcoal, appreciable amounts of phosphate. The burnt bone cluster has a high, mainly inorganic, phosphate content that is directly related to the presence of bone. Amorphous phosphate infilling fragments indicate bone alteration. The hazelnut/charcoal cluster has a slight dominance of organic over inorganic phosphate that fits with an input of organic material, also attested by the presence of charcoal.

Assuming that both hazelnut shells and bone fragments have been found *in situ*, these surface-hearths should probably be interpreted as domestic hearths, used for processing food, i.e. the roasting of hazelnuts and meat. However, one has to consider the possibility that both organic residues were deliberately dumped into these hearths as (additional) fuel, and/or during the process of cleaning up the living-area.

3.6.3.4. | Natural features

At Verrebroek "Dok 1", bio-galleries, tree-fall and tree-sway structures are the main observed natural disturbances that affected the remnants of the Mesolithic occupation level. The general trend for artefacts to move down into the soil is most probably related to the activity of burrowing animals such as small mammals, worms, ants, and a variety of insects. Uprooting of small trees and the grubbing activity of boar can also contribute to this movement.

The Brunified Soil horizonation is preserved in those tree-fall structures that were deep enough and of a type that enabled preservation (Langohr 1993). As they contain Mesolithic artefacts, the analysed tree-fall structures nearly all date from the period after the Mesolithic occupation, but before the development of the Podzolised Soils.

Tree-sway structures show clear traces of water-transported sediment within the soil. The morphology calls for a type of disturbance that occurs in stormy periods when soil moisture content is at or close to saturation conditions. These structures probably developed in a period when the Podzolised Soil development was already non-active, but before the burial by the Peat Soil, and are then associated with the raising, permanent groundwater table.

IV.

REFERENCES

Adderley W.P., Alberts I.L., Simpson I.A., Wess T.J., in press, Calcium-Iron-phosphate features in archaeological sediments: characterisation through microfocus X-ray scattering analyses, *Journal of Archaeological Science.*

Aitchison T.C., Ottaway B.S., Scott E.M., 1990, Statistical Treatment of Groups of Related Radiocarbon Ages, in: Mook W.G., Waterbolk H.T. (eds), *Proceedings of the second international symposium ^{14}C and Archaeology, Rixensart, Belgium* (PACT 29), 125-140.

Ameryckx J., 1960, La pédogenèse en Flandre sablonneuse, une chrono-bioséquence sur matériaux sableux, *Pédologie* X_1, 124-190.

Andersen S.T., 1984, *Forests at Løvenholm, Djursland, Denmark at present and in the past,* København, 208 pp.

Andersen B.G., Borns H.W., 1997, *The Ice Age World. An Introduction to Quaternary History and Research with Emphasis on North America and Northern Europe During the Last 2.5 Million Years,* Oslo, Scandinavian University Press.

Andersen S. H., Malmros C., 1984, Madskorpe" på Ertebøllekar fra Tybrind Vig, *Aarbøger for Nordisk Oldkyndighed og Historie,* 78–95.

Anderson H.A., Berrow M.L., Farmer V.C., Hepburn A., Russell J.D., Walker A.D., 1982, A reassessment of podzol formation processes, *Journal of Soil Science* 33, 125-136.

Anne P., 1945, Sur le dosage rapide du carbone organique des sols, *Ann. Agron.* 15, 161-172.

Arts N., 1989, Archaeology, Environment and the Social Evolution of Later Band Societies in a Lowland Area, in: Bonsall Cl. (ed), *The Mesolithic in Europe, Papers presented at the third international symposium, Edinburgh 1985,* Edinburgh, 291-312.

Atkinson T. C., Briffa K.R., Coope G.R., 1987, Seasonal temperatures in Britain during the past 22,000 years, reconstructed using beetle remains, *Nature* 325, 587-592.

Aurousseau P., 1983, Microscopy of the cambic horizon, in: Bullock P., Murphy C.P. (eds), *Soil Micromorphology. Volume 2. Soil Genesis, Proceedings of the International Working Meeting on Soil Micromorphology, August 1981*, Rothamsted, Academic Publishers, 459-465.

Augustyn B., 1977, Bijdrage tot het ontstaan van de vroegste geschiedenis van de Wase polders (van de oudste tijden tot ca. 1400), *Annalen van de Koninklijke Oudheidkundige Kring van het Land van Waas* 80, 1-95.

Augustyn B., 1999, *De veenontginning (12de – 16de eeuw). Geschiedenis van volk en land van Beveren*, Beveren.

Baales M., 2000, L'archéologie du Paléolithique final en Rhénanie du centre et du nord (Allemagne), in: Valentin B., Bodu P., Christensen M. (eds), *L'Europe centrale et septentrionale au Tardiglaciaire, Actes de la table ronde internationale de Nemours 14-15-16 mai 1997* (Mémoires du Musée de Préhistoire d'Ile-de-France 7), Nemours, 213-252.

Backiel T., Kokurewicz B., Ogorzalek A., 1984, High incidence of skeletal anomalies in carp, *Cyprinus carpio*, reared in cages in flowing water, *Aquaculture* 43, 369-380.

Bailon S., 1999, Différenciation ostéologique des anoures (Amphibia, Anura) de France, *Fiches d'ostéologie animale pour l'archéologie. Série C: Varia n°1*, Centre de Recherches Archéologiques du CNRS, APDCA, Antibes.

Baize D., 1988, *Guide des analyses courantes en pédologie*, Paris, Institut National de la Recherche Agronomique, 172 pp.

Bakels C., 1988, Hekelingen, a Neolithic site in the swamps of the Maas estuary, in: Küster H. (ed), *Der prähistorische Mensch und seine Umwelt, Festschrift für Udelgard Körber-Grohne zum 65. Geburtstag* (Landesdenkmalamt Baden-Württemberg), Stuttgart, 155-161.

Bakels C., 1991, Western Continental Europe, in: Van Zeist, Wasylikowa & Behre (eds), *Progress in Old World Palaeoethnobotany*, Rotterdam, 279-298.

Bakels C., 1997a, Broodtarwe, een specialiteit van het Zuid-Limburgse Neolithicum, *Archeologie in Limburg* 71, 1-3.

Bakels C., 1997b, De cultuurgewassen van de Nederlandse Prehistorie, 5400 v. C. – 12 v.C., in: Zeven A.C. (ed), *De introductie van onze cultuurplanten en hun begeleiders, van het Neolithicum tot 1500 AD* (Vereniging voor Landbouwgeschiedenis), Wageningen, 15-24.

Bakels C., 1997c, Le blé dans la culture de Cerny, in: Constantin C., Mordant D., Simonin D. (eds), *La culture de Cerny. Nouvelle économie, nouvelle société au Néolithique, Actes du Colloque International de Nemours, 9-11 mai 1994* (Mémoires du Musée de Préhistoire d'Ile-de-France 6), Nemours, 315-317.

Bakels C., van Beurden L.M., 2001, Archeobotanie, in: Louwe Kooijmans L.P. (ed), *Hardinxveld-Giessendam Polderweg, een mesolithisch jachtkamp in het rivierengebied (5500-5000 v. Chr.)* (Rapportage Archeologische Monumentenzorg 83), Amersfoort, 325-378.

Bakels C., van Beurden L.M., Vernimmen T.J.J., 2001, Archeobotanie, in: Louwe Kooijmans L.P. (ed), *Hardinxveld-Giessendam De Bruin, een kampplaats uit het Laat-Mesolithicum en het begin van de Swifterbant-cultuur (5500-4450 v. Chr.)* (Rapportage Archeologische Monumentenzorg 88), Amersfoort, 369-434.

Bakker M., Van Smeerdijk D.G., 1980, A Palaeoecological study of a Late Holocene section from "Het Ilperveld", Western Netherlands, *Review of Palaeobotany and Palynology* 36, 95-163.

Bal L., 1973, *Micromorphological Analysis of Soils, lower levels in the organization of organic soil materials* (Soil Survey Papers of the Netherlands, Soil Survey Institute 6), Wageningen, 175 pp.

Ball E., 1997, *De vissen van Brandwijk,* Unpublished thesis, Rijksuniversiteit Leiden.

Barkman J.J., 1992, Plant communities and synecology of bogs and heath pools in the Netherlands, in: Verhoeven J.T.A. (ed), *Fens and Bogs in the Netherlands: Vegetation, History, Nutrient Dynamics and Conservation* (Geobotany 18), Dordrecht, 173-235.

Barratt B.C., 1969, A revised classification and nomenclature of microscopic soil materials with particular reference to organic components, *Geoderma* 2, 257-271.

Bats M., 2000-2001, *Prospectie- en waarderingsonderzoek van twee steentijdsites in Zandig Vlaanderen,* Unpublished Master's thesis, Ghent University.

Bats M., Crombé Ph., Perdaen Y., Sergant J., Van Roeyen J.-P., Van Strydonck M., 2003, Nieuwe ontdekkingen in het Deurganckdok te Doel (Beveren, Oost-Vlaanderen): Vroeg- en Finaal-Mesolithicum, *Notae Prahistoricae* 23, 55-59.

Becze-Deák J., Langohr R., Verrechia E.P., 1997, Small scale secondary CaCO3 accumulations in selected sections of the European loess belt, morphological forms and potential for palaeoenvironmental reconstruction, *Geoderma* 76, 221-252.

Beerenhout B., 2001, Vissen, in: Louwe Kooijmans L.P. (ed), *Hardinxveld-Giessendam De Bruin. Een kampplaats uit het Laat-Mesolithicum en het begin van de Swifterbant-cultuur (5500-4450 v.Chr.)* (Rapportage Archeologische Monumentenzorg 88), Amersfoort, 299-325.

Begon M., Harper J.L., Townsend C.R., 1996, *Ecology. Individuals, Populations and Communities,* Oxford, Blackwell Science, 1068 pp.

Behre, K-E., 1981, The interpretation of anthropogenic indicators in pollen diagrams, *Pollen et Spores* 23, 225-245.

Bell G. H., 1973, Solubilities of normal aliphatic acids, alcohols and alkanes in water, *Chemistry and Physics of Lipids* 10, 1-10.

Beugnier V., Crombé Ph., in press, Étude fonctionnelle du matériel en silex du site Mésolithique ancien de Verrebroek (Flandres, Belgique): premiers résultats, *Bulletin de la Société Préhistorique Française.*

Bjørlykke K., 1989, *Sedimentology and Petroleum Geology*, Berlin, Springer-Verlag, 363 pp.

Boaden P.J.S., Seed R., 1985, *An Introduction to Coastal Ecology*, London, Blackie Academic & Professional, 218 pp.

Bocherens H., 1997, L'apport de la biogéochimique isotopique à la connaissance des comportements de subsistance des chasseurs cueilleurs anciens, *Etudes et Recherches Archéologiques de l'Université de Liège* 83, 235-264.

Bødker Enghoff I., 1994, Fishing in Denmark during the Ertebølle period, *International Journal of Osteoarchaeology* 4, 65-96.

Boggs S. Jr., 1987, *Principles of Sedimentology and Stratigraphy*, Columbus, Merrill Publishing Company, 784 pp.

Bohncke S., Vandenberghe J., Coope R., Reiling R., 1987, Geomorphology and palaeoecology of the Mark valley (southern Netherlands): palaeoecology, palaeohydrology and climate during the Weichselian Late Glacial, *Boreas* 16, 69-85.

Bonsall C., Cook G., Manson J. L., Sanderson D., 2002, Direct dating of Neolithic pottery; progress and prospects, *Documenta Praehistorica* 29, 47-59.

Bos H., 1998, *Aspects of the Late Glacial - Early Holocene Vegetation Development in Western Europe. Palynological and palaeobotanical investigations in Brabant (The Netherlands) and Hessen (Germany)* (LPP Contribution Series 10), Utrecht.

Bottema S., Mook-Kamps B., 2000, A note on the Allerød vegetation of Southeastern Friesland (with emphasis on the Oudehaske area), *Palaeohistoria* 39/40, 89-98.

Bouman F., Boesewinkel D., Bregman R., Devente N., Oostermeijer G., 2000, *Verspreiding van zaden*, Utrecht.

Brady N.C., Weil R.R., 1999, *The Nature and Properties of Soils*, New Jersey, 881 pp.

Brinch Petersen E., 1989, Vaenget Nord: Excavation, Documentation and Interpretation of a Mesolithic Site at Vedbaek, Denmark, in: Bonsall Cl. (ed), *The Mesolithic in Europe, Papers presented at the third international symposium, Edinburgh 1985*, Edinburgh, 325-337.

Brinkhuizen D., 1976, De visresten van Swifterbant, *Westerheem* 25, 246-252.

Brinkhuizen D., 1979, Preliminary notes on fish remains from archaeological sites in The Netherlands, *Palaeohistoria* 21, 83-90.

Brinkhuizen D., 1986, Some notes on recent and pre- and protohistoric fishing gear from northwestern Europe, *Palaeohistoria* 25, 7-53.

Brown D.A., 1984, Prospects and limits of a phytolith key for grasses in the central United States, *JAS* 11, 345-368.

Bullock P., Clayden B., 1980, The Morphological Properties of Spodosols, in: Theng B.K.G. (ed), *Soils with Variable Charge*, Lower Hutt, New Zealand Society of Soil Science 44-55, 63-65.

Bullock P., Fedoroff N., Jongerius A., Stoops G., Babel U., 1985, *Handbook for soil thin section description*, Wolverhampton, 152 pp.

Bunting, B.T., 1983, The effect of fire on humus profiles in Canadian boreal forests, in: Bullock, P., Murphy, C.D. (eds), *Soil Micromorphology*, August 1981, A B Academic Publishers, Berkhamsted, 681-688.

Buurman P., 1984, Editor's comments on the paper "The podzol concept" by L. Petersen, in: Buurman P. (ed), *Podzols*, New York, Van Nostrand Reinhold Company Inc., 8-11.

Cambridge University Press, 2003, *Cambridge International Dictionary of English*, Cambridge Dictionaries Online, [online], URL: http://dictionary.cambridge.org/.

Cannon J., Cannon M., 1994, *Dye plants and dyeing*, London.

Cauwe N., Vander Linden M., Vanmontfort B., 2001, The Middle and Late Neolithic – Le Néolithique moyen et récent, in: Cauwe N., Hauzeur A., van Berg P.-L. (eds), *Prehistory of Belgium. Special issue on the occassion of the XIVth Congress of the International Union for Prehistoric and Protohistoric Sciences* (Anthropologica et Praehistorica, Bulletin de la Société royale belge d'Anthropologie et de Préhistoire 112), Brussel, 77-89.

Chabal, L., 1988, L'étude paléoécologique de sites protohistoriques à partir des charbons de bois: la question de l'unité de mesure. Dénombrements de fragments a pesées, in: *Pact 22, Wood and Archaeology, Acts of the European Symposium held at Louvain-la-Neuve, Octobre 1987*, 189-207.

Clason A.T, 1967, Animal and man in Holland's past (*Palaeohistoria* 13), Groningen, 247 pp.

Clason A.T., Brinkhuizen D.C., 1993, Bergschenhoek, in: Clason A., Payne S., Uerpmann H.-P. (eds), *Skeletons in her cupboard. Festschrift for Juliet Clutton-Brock* (Oxbow Monograph 34), Oxford, 61-73.

Cleveringa P., De Gans W., Huybrechts W., Verbruggen C., 1988, Outline of river adjustments in small river basins in Belgium and the Netherlands since the Upper Pleniglacial, in: Lang, Schlüchter (eds), *Lake, Mire and River Environments During the last 15 000 years*, Rotterdam, Balkema, 123-132.

Cojan I., Renard M., 1997, *Sédimentologie*, Collection Enseignement des Sciences de la Terre, Paris, Masson, 418 pp.

Colten R.H., 1995, Faunal exploitation during the Middle to Late Period transition on the Northern Channel Islands, California, *Journal of California and Great Basin Anthropology* 17 (1), 93-120.

Copley M. S., Berstan R., Dudd S. N., Docherty G., Murkherjee A. J., Straker V., Payne S., Evershed R. P., 2003, Direct chemical evidence for widespread dairying in prehistoric Britain, *Proceedings of the National Academy of Sciences* USA 100, 1524-1529.

Cordemans K., Bats M., Crombé Ph., Meganck M., 2001, Extensive survey in the Antwerp harbour area: assessing the buried Mesolithic landscape at Verrebroek (East-Flanders, Belgium), *Notae Praehistoricae* 21, 79-83.

Costamagno S., Griggo C., Mourré V., 1999, Approche expérimentale d'un problème taphonomique: utilisation de combustible osseux au Paléolithique, *Préhistoire Européenne* 13, 167-194.

Crombé Ph., 1993a, Epipaleolithische en mesolithische bewoning in zandig Vlaanderen; resultaten van de opgravingscampagne 1992 op vier Oostvlaamse sites, *Notae Praehistoricae* 12, 83-94.

Crombé Ph.,1993b, Tree-fall features on Final Palaeolithic and Mesolithic sites situated on sandy soils: how to deal with it, *Helinium* 33(1), 50-66.

Crombé Ph., 1994, Recherche poursuivie sur le Mésolithique en Flandre orientale, *Notae Praehistoricae* 13, 71-78.

Crombé Ph., 1998, *The Mesolithic in Northwestern Belgium, Recent Excavations and Surveys* (British Archaeological Reports, International Series 716), Oxford.

Crombé Ph., 2002, Quelques réflexions sur la signification de la variabilité des industries lithiques mésolithique de Belgique, in: Otte M., Kozlowski J.K. (eds), *Préhistoire de la Grande Plaine du Nord de l'Europe. Les échanges entre l'Est et l'Ouest dans les sociétés préhistoriques, Actes du Colloque Chaire Francqui Interuniversitaire, Liège 26/06/2001*, (ERAUL 99), Liège, 99-114.

Crombé Ph., Cauwe N., 2001, The Mesolithic, in: Cauwe N., Hauzeur A., van Berg P.-L. (eds), *Prehistory of Belgium. Special issue on the occassion of the XIVth Congress of the International Union for Prehistoric and Protohistoric Sciences*, (Bulletin de la Société royale belge d'Anthropologie et de Préhistoire 112), Brussel, 49-62.

Crombé Ph., Meganck M., 1996, Results of an auger survey research at the Early Mesolithic site of Verrebroek "Dok" (East-Flanders, Belgium), *Notae Praehistoricae* 16, 101-115.

Crombé Ph., Van Strydonck M., 1994, Recherche poursuivie sur le site Mésolithique ancien de Verrebroek (Flandre orientale): résultats de la campagne 1994, *Notae Praehistoricae* 14, 95-102.

Crombé Ph., Verbruggen C., 2002, The Lateglacial and early Postglacial occupation of northern Belgium: the evidence from Sandy Flanders, in: Eriksen B.V. , Bratlund B. (eds), *Recent studies in the Final Palaeolithic of the European Plain, Proceedings of a UISPP Symposium, Stockholm, 14.-17. October 1999* (= Jutland Archaeological Society Publications), Stockholm, 165-180.

Crombé Ph., Groenendijk H., Van Strydonck M., 1999a, Dating the Mesolithic of the Low Countries: some practical considerations, in: Evin J., Oberlin C., Daugas J.P., Salles J.F. (eds), *Actes du colloque "C14 et Archéologie", 1998* (Mémoires de la Société Préhistorique Française 26 and Supplément de la Revue d'Archéométrie), Paris, 57-63.

Crombé, Ph., Perdaen, Y., Sergant, J., 1997, Le gisement mésolithique ancien de Verrebroek: Campagne 1997, *Notae Praehistoricae* 17, 85-92.

Crombé Ph., Perdaen Y., Sergant J., 1998, The Early Mesolithic site of Verrebroek "Dok": preliminary results of the 1998 excavation campaign, *Notae Praehistoricae* 18, 101-105.

Crombé Ph., Perdaen Y., Sergant J., 1999b, The Early Mesolithic site of Verrebroek "Dok 1": preliminary results of the 1999 excavation campaign, *Notae Praehistoricae* 19, 71-74.

Crombé Ph., Deforce K., Langohr R., Louwagie G., Perdaen Y., Sergant J., Verbruggen C., 1999c, A small Final-Palaeolithic knapping site at Verrebroek "Dok 2" (Flanders, Belgium), *Notae Praehistoricae* 19, 63-68.

Crombé Ph., Perdaen Y., Sergant J., Van Roeyen J.-P., Van Strydonck M., 2002, The Mesolithic-Neolithic transition in de sandy lowlands of Belgium: new evidence, *Antiquity* 76, 699-706.

Crombé Ph., Van Roeyen J.-P., Sergant J., Perdaen Y., Cordemans K., Van Strydonck M., 2000, Doel "Deurganckdok" (Flanders, Belgium): settlement traces from the Final Palaeolithic and the Early to Middle Neolithic, *Notae Prahistoricae* 20, 111-119.

Dauchot-Dehon M., Van Strydonck M., Heylen J., 1984, Institut Royal du Patrimoine Artistique, Radiocarbon dates X, *Radiocarbon* 26(3), 384-391.

Dauchot-Dehon M., Van Strydonck M., Heylen J., 1986, Institut Royal du Patrimoine Artistique, Radiocarbon dates XI, *Radiocarbon* 28(1), 69-77.

Davidson D.A., Dercon G., Simpson I.A., Dalsgaard K., Spek T., Plant D.A., submitted, The identification and significance of inputs to anthrosols in north-west Europe, in: *Proceedings of the Second International Conference on Soils and Archaeology, Pisa, 12.-15. May 2003*.

de Bakker H., Schelling J., 1989, *Systeem van bodemclassificatie voor Nederland, de hogere niveaus*, Wageningen, Winand Staring Centre, 209 pp.

De Bie M., Vermeersch P.M., 1998, Pleistocene-Holocene transition in Benelux, *Quaternary International* 49/50, 29-43.

De Cleene M., Lejeune M.C., 1999, *Compendium van de rituele planten in Europa. Botanisch, cultureel, gebruik*, Gent.

De Coninck F., 1980, Major mechanisms in formation of spodic horizons, *Geoderma* 24, 101-128.

De Coninck F., Righi D., 1983, Podzolisation and the spodic horizon, in: Bullock P., Murphy C.P. (eds), *Soil Micromorphology. Volume 2. Soil Genesis, Proceedings of the International Working Meeting on Soil Micromorphology, August 1981*, Rothamsted, Academic Publishers, 389-417.

De Coninck F., Greguss P., Vanhoorne R., 1966, La superposition de dépôts tourbeux datant des oscillations Alleröd et Bölling à Stabroek (Belgique), *Pedologie* XVI/3, 293-308.

Dekker L.W., Booij A.H., Vroon H.R.J., Koopman G.J., 1991, Waterhardlagen: indicatoren van een voormalig veendek, *Grondboor en Hamer* (maart), 25-30.

De Moor G., 1963, Bijdrage tot de kennis van de fysische landschapsvorming in Binnen-Vlaanderen, *Tijdschrift van de Belgische Vereniging voor Aardrijkskundige Studies* XXXII (2), 329-433.

De Moor G., Heyse I., 1978, De morfologische evolutie van de Vlaamse vallei, *De Aardrijkskunde* 4, 343-375.

Denys L., Verbruggen C., 1989, A case of drowning - the end of subatlantic peat growth and related palaeoenvironmental changes in the lower Scheldt basin (Belgium) based on diatom and pollen analysis, *Review of Palaeobotany and Palynology* 59, 7-36.

de Witte G.F., 1948, *Amphibiens et reptiles. Faune de Belgique*, Musée Royal d'Histoire Naturelle de Belgique, Bruxelles.

Dimbleby G.W., 1962, *The development of British heathlands and their soils* (Oxford Forestry Memoirs 23), Oxford, Clarendon Press, 120 pp.

Divanach P., Papandroulakis N., Anastasiadis P., Koumoundouros G., Kentouri M., 1997, Effects of water currents on the development of skeletal deformities in sea bass (*Dicentrarchus labrax*) with functional swimbladder during postlarval and nursery phase, *Aquaculture* 156, 145-155.

Dosskey M.G., Bertsch P.M., 1997, Transport of Dissolved Organic Matter through a Sandy Forest Soil, *Soil Science Society of America Journal* 61, 920-927.

Driessen P.M., Dudal R. (eds), 1991, *The Major Soils of the World. Lecture Notes on their Geography, Formation, Properties and Use,* Agricultural University Wageningen - Katholieke Universiteit Leuven, 309 pp.

Duchaufour P., 1982, *Pedology. Pedogenesis and classification,* London, George Allen & Unwin.

Dudd S. N., Evershed R. P., 1998, Direct demonstration of milk as an element of archaeological economies, *Science* 282, 1478-1481.

Dudd S.N., Evershed R.P., Gibson A.M., 1999, Evidence for varying patterns of exploitation of animal products in different prehistoric pottery traditions based on lipids preserved in surface and absorbed residues, *Journal of Archaeological Science* 26, 1473-1482.

Dufour E., Bocherens H., Mariotti A., 1999, Palaeodietary implications of isotopic variability in Eurasian Lacustrine fish, *Journal of Archaeological Science* 26, 617-627.

Ellenberg H., 1994, *Vegetation ecology of Central Europe,* fourth edition, Cambridge.

Ellis L. (ed), 2000, *Archaeological Method and Theory: An Encyclopedia,* New York, Garland Publishing Inc., 705 pp.

Enser M., 1991, Animal carcass fats and fish oils, in: Rosell J.B., Protchard J.L.R. (eds), *Analysis of oilseed, fats and fatty food,* London, Elsevier, 329-394.

Evershed R.P., Dudd S.N., Lockheart M.J., Jim S., 2001, Lipids in Archaeology, in: Brothwell D.R., Pollard A.M. (eds), *Handbook of Archaeological Sciences,* Chichester, John Wiley & Sons Ltd, 331-349.

Exaltus R., 2001, Deel 14, Micromorfologie: onderzoek aan slijpplaatmonsters van grondsporen, in: Hogestijn J.W.H., Peeters J.H.M. (eds), *De mesolithische en vroeg-neolithische vindplaats Hoge Vaart-A27 (Flevoland)* (Rapportage Archeologische Monumentenzorg 79), Amersfoort, 58 pp.

Faegri K., Iversen J., 1975, *Textbook of Pollen Analysis.* Munksgaard, Copenhagen.

Faegri K., Kaland P.E., Krzywinski K., 1989, *Textbook of Pollen Analysis by Knut Faegri and Johs. Iversen, IV edition,* Chichester.

Fagnart J.-P., Coudret P., 2000, Le Tardiglaciaire dans le Nord de la France, in: Valentin B., Bodu P., Christensen M. (eds), *L'Europe centrale et septentrionale au Tardiglaciaire, Actes de la table-ronde internationale de Nemours 14-15-16 mai 1997* (Mémoires du Musée de Préhistoire d'Ile-de-France 7), Nemours, 111-128.

FAO, 1990a, FAO - Unesco Soil Map of the World. Revised Legend, with corrections, Reprinted from 1988 with corrections, *World Resources Report* 60, Rome, FAO, Reprinted as Technical Paper 20, Wageningen, ISRIC, 1994, 140 pp.

FAO, 1990b, *Guidelines for soil description*, Rome, FAO, Soil Resources, Management and Conservation Service, Land and Water Development Division.

Farmer V.C., Russell J.D., Smith B.F.L., 1983, Extraction of inorganic forms of translocated Al, Fe and Si from a podzol Bs horizon, *Journal of Soil Science* 34, 571-576.

Firmin G., 1986, Archéobotanique et archéologie, in: Demoule J.-P., Guilaine J. (eds), *Le Néolithique de la France. Hommage à G. Bailloud*, Paris, 53-70.

Fischer A., Heinemeier J., 2003, Freshwater reservoir effect in C-14 dates of food residue on pottery, *Radiocarbon* 45, 449-466.

Fitzpatrick E.A., 1986, *An introduction to soil science*, Second edition, Harlow (England), Longman Scientific & Technical, 255 pp.

Fitzpatrick E.A., 1993, *Soil Microscopy and Micromorphology*, Chicester, John Wiley & Sons Ltd, 304 pp.

Folch J., Lees M., Stanley G.H.S., 1957, A simple method for the isolation and purification of total lipids from animal tissues, *Journal of Biological Chemistry* 226, 497-509.

Fredlund G.G., Tieszen L.T., 1994, Modern phytolith assemblages from the North American Great Plains, *Journal of Biogeography* 21, 321-335.

French H.M., 1988 (First published 1976), *The Periglacial Environment*, Harlow (England), Longman Scientific & Technical, 309 pp.

Galimov E., 1985, *The biological fractionation of isotopes*, New York: Academic Press, 16–38.

Gautier A., 1987, Taphonomic groups: How and Why, *ArchaeoZoologia* I (2), 47-52.

Gehasse E.F., 1995, *Ecologisch-archeologisch onderzoek van het Neolithicum en de Vroege Bronstijd in de Noordoostpolder met de nadruk op vindplaats P14*, Unpublished PhD thesis, Universiteit van Amsterdam.

Geological Map of Belgium (Geologische kaart van België). Kaartblad (15). Antwerpen, 2002 (elaborated in 2000), Scale 1:50000, Brussel, Koninklijk Belgisch Instituut voor Natuurwetenschappen - Belgische Geologische Dienst & Ministerie van de Vlaamse Gemeenschap - Afdeling Natuurlijke Rijkdommen en Energie.

Geological Map of Belgium (Geologische kaart van België). Kaartblad 7-1. Kapellen-Essen, 2002 (elaborated in 1999), Scale 1:50000, Brussel, Koninklijk Belgisch Instituut voor Natuurwetenschappen - Belgische Geologische Dienst & Ministerie van de Vlaamse Gemeenschap - Afdeling Natuurlijke Rijkdommen en Energie.

Gilchrist R., Mytum C., 1986, Experimental archaeology and burnt animal bone from archaeological sites, *Circaea* 4, 29-38.

Gisi U., Schenker R., Schulin R., Stadelmann F.X., Sticher H., 1997, *Bodenökologie*, Stuttgart, Georg Thieme Verlag, 350 pp.

Green W.P., Pettry D.E., Switzer R.E., 1998, Formicarious Pedons, the Initial Effect of Mound-Building Ants on Soils, *Soil Survey Horizons* 39 (2), 33-44.

Greguss P., 1959, *Holzanatomie Der Europäischen Laubhölzer und Straücher*, Akadémiai Kiadó, Budapest.

Groenendijk H.A., 1987, Mesolithic hearth-pits in the Veenkoloniën (Prov. Groningen, The Netherlands), defining a specific use of fire in the Mesolithic, *Palaeohistoria 29*, 85-102.

Groenendijk H.A., 2004, Middle Mesolithic occupation on the extensive site NP3 in the peat reclamation district of Groningen, The Netherlands, in: Crombé Ph. (ed), *Landscape-use during the Final Palaeolithic and Mesolithic in NW-Europe; the formation of extensive sites and site-complexes,* Proceedings of the XIV International Congress of the UISPP, Liège 2-8/09/01 (British Archaeological Reports, International Series), Oxford (in press).

Grosser D., 1977, *Die Hölzer Mitteleuropas. Ein mikrophotographischer Lehratlas*, Springer, Berlin.

Guillet B., 1974, Le problème de l'évolution des podzols vosgiens et ses relations avec l'histoire de la végétation, *R.F.F.* XXVI-1, 31-44.

Guillet B., 1987, L'âge des Podzols, in: Righi D., Chauvel A. (eds), *Podzols et podzolisation, Comptes rendus de la table ronde internationale «Pédologie, Pétrologie et Métallogénie de la Surface», 10-11 avril 1986*, Laboratoire de Pédologie, Université de Poitiers, 131-144.

Guillet B., Rouiller J., Soucher B., 1975, Podzolisation and clay migration in Spodosols of Eastern France, *Geoderma* 14, 223-245.

Gullentops F., Wouters L., 1996, *Delfstoffen in Vlaanderen*, Brussel, Ministerie van de Vlaamse Gemeenschap - afdeling Natuurlijke Rijkdommen en Energie, 198 pp.

Haas J.N., Pasmussen P., 1993, Zur Geschichte der Schnietel- und Laubfutterwirtschaftin der Schweiz – eine alte Landwirtschaftspraxis kurz vor dem Aussterben, *Dissertationes Botanicae* 196, 469-489.

Hamburg T., Kruijshaar C., Nientker J., Peeters J.H.M., Rast-Eicher A., 2001, Deel 13. Grondsporen: antropogene sporen en structuren, in: Hogestijn J.W.H., Peeters J.H.M. (eds), *De mesolithische en vroeg-neolithische vindplaats Hoge Vaart-A27 (Flevoland)* (Rapportage Archeologische Monumentenzorg 79), Amersfoort, 103 pp.

Hansel F. A., Copley M. S., Madureiraa L.A.S., Evershed R.P., 2004, Thermally produced omega-(o-alkylphenyl)alkanoic acids provide evidence for the processing of marine products in archaeological pottery vessels, *Tetrahedron Letters* 45, 2999-3002.

Harris S.A., French H.M., Heginbottom J.A., Johnston G.H., Ladanyi B., Sego D.C., van Everdingen R.O., 1988, *La terminologie du pergélisol et notions connexes* (Note de service technique 142), Conseil national de recherches du Canada, 154 pp.

Hartge K.H., Horn R., 1991, *Einführung in die Bodenphysik*, Stuttgart, Ferdinand Enke Verlag, 303 pp.

Havinga A.J., 1962, *Een palynologisch onderzoek van in dekzand ontwikkelde bodemprofielen (A palynological investigation of soil profiles developed in cover sand)*, PhD thesis, Wageningen, H. Veenman en zonen n.v.

Havinga A.J., 1984a, Pollen analysis of Podzols, in: Buurman P. (ed), *Podzols*, New York, Van Nostrand Reinhold Compnay Inc., 313-323.

Havinga A.J., 1984b, A 20-year experimental investigation into the differential corrosion susceptibility of pollen and spores in various soil types, *Pollen et Spores* 26 (3-4), 541-558.

Helbaek H., 1951, Ukrudtsfrø som Naeringsmiddel I førromersk Jernalder, *Kuml* 195, 65-74.

Heron C., 2001, Geochemical Prospecting, in: Brothwell D.R., Pollard A.M. (eds), *Handbook of Archaeological Sciences*, Chichester, John Wiley & Sons Ltd, 565-573.

Heron C., Evershed R. P., 1993, The analysis of organic residues and the study of pottery use, *Archaeological Method and Theory* V, 247-286.

Heron C., Evershed R. P., Goad L. J., 1991, Effects of Migration of Soil Lipids on Organic Residues Associated with Buried Potsherds, *Journal of Archaeological Science* 18, 641-659.

Herrmann B., Grupe G., Hummel S., Piepenbrink H., Schutkowski H. 1990, *Prähistorische Anthropologie. Leitfaden der Feld- und Labormethoden*, Springer Verlag, Berlin.

Heyse I., 1979, *Bijdrage tot de geomorfologische kennis van het noordwesten van Oost-Vlaanderen (België)*, Brussel, 257 pp.

Heyse I., 1983, Cryoturbation Types in Eolian Würm Late Glacial Sediments in Flanders, Belgium, *Polarforschung* 53 (2), 87-95.

Hinout J., 2002, Le Mésolithique dans le Bassin parisien. Essai de synthèse, *Préhistorie et Protohistoire en Champagne-Ardenne* 26, 15-90.

Hintermaier-Erhard G., Zech W., 1997, *Wörterbuch der Bodenkunde. Systematik, Genese, Eigenschaften, Ökologie und Verbreitung von Böden*, Stuttgart, Ferdinand Enke Verlag, 338 pp.

Hoek W., 1997, *Palaeogeography of Late Glacial Vegetations. Aspects of Late Glacial and Early Holocene vegetation, abiotic landscape and climate in the Netherlands* (Netherlands Geographical Studies 230), Amsterdam.

Hogestijn W.-J., Peeters H., 1996, De opgraving van de Mesolithische en Vroegneolithische bewoningsresten van de vindplaats "Hoge Vaart" bij Almere (Prov. Fl.): een blik op een duistere periode van de Nederlandse prehistorie, *Archeologie* 7, 80-113.

Hogestijn J.W.H., Peeters J.H.M. (eds), 2001, *De mesolithische en vroeg-neolithische vindplaats Hoge Vaart-A27 (Flevoland)* (Rapportage Archeologische Monumentenzorg 79), Amersfoort.

Hogestijn W.J., Peeters H., Schnitger W. 1996, Bewoningsresten uit het Vroeg Neolithicum bij Almere; enkele resultaten van de opgraving 'Hoge Vaart', *Cultuur Historisch Jaarboek voor Flevoland* 6, 93-115.

Hogestijn W.J., Peeters H., Schnitger W., Bulten E.E.B., 1995a, Bewoningsresten uit de Steentijd langs een voormalige loop van de Eem in zuidelijk Flevoland; de voltooiing van de A27 en de archeologie, *Cultuur Historisch Jaarboek voor Flevoland* 5, 83-102.

Hogestijn W.J., Peeters H., Schnitger W., Bulten E.E.B., 1995b, Bewoningsresten uit het Laat Mesolithicum / Vroeg Neolithicum bij Almere (Prov. Fl.): verslag van de eerste resultaten van de opgraving "A27 - Hoge Vaart", *Archeologie* 6, 66-89.

Huet M., 1949, Aperçu de la relation entre la pente et les populations piscicoles des eaux courantes, *Schweizerische Zeitschrift Für Hydrologie* 11, 332-351.

ISO 10694, 1995, *Soil quality - Determination of organic and total carbon after dry combustion (elementary analysis)*, [online], URL: http://www.iso.ch/iso/en/CatalogueListPage.CatalogueList.

ISO 13878, 1998, *Soil quality - Determination of total nitrogen content by dry combustion ("elemental analysis")*, [online], URL: http://www.iso.ch/iso/en/CatalogueListPage.CatalogueList.

Jabiol B., Brêthes A., Ponge J.-F., Toutain F., Brun J.-J., 1995, *L'humus sous toutes ses formes*, Nancy, Ecole Nationale de Génie Rural, des Eaux et des Forêts.

Jacobson G.L., Bradshaw R.H.W., 1981, The selection of sites for paleovegetational studies, *Quaternary Research* 16, 80-96.

Jacomet S., Kreuz A., 1999, *Archäobotanik: Aufgaben, Methoden und Ergebnisse vegetations- und agrargeschichtlicher Forschung*, Ulmer, Stuttgart, 368 pp.

Jaillard B., Guyon A., Maurin A.F., 1991, Structure and composition of calcified roots, and their identification in calcareous soils, *Geoderma* 50, 197-210.

Jansen J.B.H., Peeters J.H.M., 2001, Deel 6. Geochemische aspecten: verkenningen in enkele toepassingsmogelijkheden, in: Hogestijn J.W.H., Peeters J.H.M. (eds), *De mesolithische en vroeg-neolithische vindplaats Hoge Vaart-A27 (Flevoland)* (Rapportage Archeologische Monumentenzorg 79), Amersfoort, 92 pp.

Janssen C.R., 1974, *Verkenningen in de palynologie*, Utrecht.

Jongerius A., 1970, Some morphological aspects of regrouping phenomena in Dutch soils, *Geoderma* 4, 311-331.

Karte J., 1987, Pleistocene periglacial conditions and geomorphology in north central Europe, in: Boardman J. (ed), *Periglacial processes and landforms in Britain and Ireland*, Cambridge, Cambridge University Press, 67-75.

Kiden P., 1991, The Lateglacial and Holocene Evolution of the Middle and Lower River Scheldt, Belgium, in: Starkel L., Gregory K.J., Thornes J.B. (eds), *Temperate Palaeohydrology. Fluvial Processes in the Temperate Zone during the last 15 000 years*, Chichester, John Wiley and Sons Ltd, 283-299.

Kiden P., Verbruggen C., 1987, Het landschap in het Antwerpse sinds de laatste ijstijd, in: Warmenbol E. (ed), *Het ontstaan van Antwerpen. Feiten & fabels*, Antwerpen, 9-18.

Kirby K.J., 2001, The impact of deer on the ground flora of British broadleaved woodland, *Forestry* 74, 220-229.

Kjeldahl J.I., 1883, Neue Methode zur Bestimmung des Stickstoffs in organischen Körpern, *Z. Anal. Chem.* 22, 366-382.

Koch E., 1998, *Neolithic Bog Pots from Zealand, Møn,Lolland and Falster* (Nordiske Fortidsminde Series B, 16), Copenhagen.

Komárek J., Jankovská V., 2001, *Review of the Green Algal Genus Pediastrum; Implication for Pollenanalytical Research* (Bibliotheca Phycologica 108), Berlin – Stuttgart.

Koopman G.J., 1988, "Waterhard": a hard brown layer in sand below peat, The Netherlands, *Geoderma* 42, 147-157.

Kubiena W.L., 1953, *The Soils of Europe*, Madrid, 298 pp.

Kubiena W.L., 1970, *Micromorphological features of soil geography*, New Brunswick, Rutgers University Press, 254 pp.

Kuntze H., Roeschmann G., Schwerdtfeger G., 1994, *Bodenkunde*, Stuttgart, Verlag Eugen Ulmer, 424 pp.

Laarman F.J., 2001, Archeozoölogie: aard en betekenis van de dierlijke resten, in: Hogestijn J.W.H., Peeters J.H.M. (eds), *De mesolithische en vroeg-neolithische vindplaats Hoge Vaart-A27 (Flevoland)* (Rapportage Archeologische Monumentenzorg 79, deel 16), Amersfoort.

Lambinon J., De Langhe J.-E., Delvosalle L., Duvigneaud J. (eds), 1998, *Flora van België, het Groothertogdom Luxemburg, Noord-Frankrijk en de aangrenzende gebieden (Pteridofyten en Spermatofyten)*, 3th edition, Meise.

Lang G., 1994, *Quartäre Vegetationsgeschichte Europas. Methoden und Ergebnisse*, Jena – Stuttgart – New York.

Langohr R., 1993, Types of tree windthrow, their impact on the environment and their importance for the understanding of archaeological excavation data, *Helinium* XXXIII/1, 36-49.

Langohr R., Louwagie G., 2000, Tree sway turbation of soils as observed on archaeological sites in the sandy area of northern Flanders, Belgium, *Pedologie-Themata* 8, 123-129.

Langohr R., Scoppa C.O., Van Wambeke A., 1976, The use of a comparative particle size distribution index for the numerical classification of soil parent materials: application to Mollisols of the Argentinian Pampa, *Geoderma* 15, 305-312.

Lanting J.N., Aerts-Bijma A.T., van der Plicht J., 2001, Dating of cremated bones, *Radiocarbon* 43(2B), 249-254.

Lapen D.R., Wang C., 1999, Placic and Ortstein Horizon Genesis and Peatland Development, Southeastern Newfoundland, *Soil Science Society of America Journal* 63, 1472-1482.

Lee G.B., 1983, The micromorphology of peat, in: Bullock P., Murphy C.P. (eds), *Soil Micromorphology. Volume 2. Soil Genesis, Proceedings of the International Working Meeting on Soil Micromorphology, August 1981*, Rothamsted, Academic Publishers, 485 -501.

Leenders K.A.H.W., 1987, De boekweitkultuur in historisch perspektief, *Geografisch Tijdschrift* 21, 213 – 227

Leenders K.A.H.W., Verbruggen C., Van Strydonck M., 1990, De betekenis van Zundertse veenresten voor de kennis van de Holocene Westkempische en Westeuropese venen en historisch en paleo-ecologisch onderzoek, *K.N.A.G. Geografisch Tijdschrift* XXIII-4, 340-352.

Little D., Farrell E., Collins J., 1997, Land-use legacies and soil development in semi-natural ecosystems in the marginal uplands of Ireland, *Catena* 30, 83-98.

Lom J., Pike A.W., Dykova I., 1991, *Myxobolus sandrae* (Reuss, 1906), the agent of vertebral column deformities of perch *Perca fluviatilis* in northeast Scotland, *Diseases of Aquatic Organisms* 12, 49-53.

Lozet J., Mathieu C., 1997, *Dictionnaire de Science du Sol*, Paris, Technique & Documentation, 488 pp.

Louwagie G., Langohr R., Crombé Ph., Perdaen Y., Sergant J., in press, Bioturbation and Migration of Mesolithic Artefacts and Ecofacts in the Sandy Soils of the Verrebroek "Dok" Excavation (Belgium, Province of East-Flanders), in: De Bie M. (ed), *Drifted Apart: Spatial analysis of Stone Age sites in bioturbated sediments*, Proceedings of the XIV International Congress of the UISPP, Liège 2-8/09/01 (British Archaeological Reports, International Series), Oxford.

Louwe Kooijmans L.P., 1993, Wetland exploitation and upland relations of prehistoric communities in the Netherlands, in: Gardiner J. (ed), *Flatlands and wetlands: current themes in East Anglian archaeology* (East Anglian Archaeology Report 50), Norwich, 71-116.

Louwe Kooijmans L.P. (ed), 2001a, *Hardinxveld-Giessendam Polderweg. Een mesolithisch jachtkamp in het rivierengebied (5500-5000 v.Chr.)* (Rapportage Archeologische Monumentenzorg 83), Amersfoort.

Louwe Kooijmans L.P. (ed), 2001b, *Hardinxveld-Giessendam De Bruin. Een kampplaats uit het Laat-Mesolithicum en het begin van de Swifterbant-cultuur (5500-4450 v.Chr.)* (Rapportage Archeologische Monumentenzorg 88), Amersfoort.

Louwe Kooijmans L.P., 2001c, Synthese, in: Louwe Kooijmans, L.P. (ed), *Hardinxveld-Giessendam De Bruin. Een kampplaats uit het Laat-Mesolithicum en het begin van de Swifterbant-cultuur (5500-4450 v.Chr.)* (Rapportage Archeologische Monumentenzorg 88), Amersfoort, 499-528.

Louwe Kooijmans L.P., Nokkert M., 2001, Sporen en structuren, in: Louwe Kooijmans L.P. (ed), *Hardinxveld-Giessendam De Bruin. Een kampplaats uit het Laat-Mesolithicum en het begin van de Swifterbant-cultuur (5500-4450 v.Chr.)* (Rapportage Archeologische Monumentenzorg 88), Amersfoort, 75-115.

Louwe Kooijmans L.P., Hänninen K., Vermeeren C.E., 2001, Artefacten van hout, in: Louwe Kooijmans L.P. (ed), *Hardinxveld-Giessendam De Bruin. Een kampplaats uit het Laat-Mesolithicum en het begin van de Swifterbant-cultuur (5500-4450 v.Chr.)* (Rapportage Archeologische Monumentenzorg 88), Amersfoort, 435-477.

Louwye S., Declercq E., 1998, Relative water level change in the intracoastal zone of Belgium and Northern France over the last 2500 years, *Boreas* 27, 162-177.

Lowe D.R., 1975, Water escape structures in coarse-grained sediments, *Sedimentology* 22, 157-204.

Lundström U.S., van Breemen N., Bain D., 2000a, The podzolisation process. A review, *Geoderma* 94, 91-107.

Lundström U.S., van Breemen N., Jongmans A.G., 1995, Evidence for microbial decomposition of organic acids during podzolisation, *European Journal of Soil Science* 46, 489-496.

Lundström U.S., van Breemen N., Bain D.C., van Hees P.A.W., Giesler R., Gustafsson J.P., Ilvesniemi H., Karltun E., Melkerud P.-A., Olsson M., Riise G., Wahlberg O., Bergelin A., Bishop K., Finlay R., Jongmans A.G., Magnusson T., Mannerkoski H. Nordgren A., Nyberg L., Starr M., Tau Strand L., 2000b, Advances in understanding the podzolisation process resulting from a multidisciplinary study of three coniferous forest soils in the Nordic Countries, *Geoderma* 94, 335-353.

Mackay J.R., 1999, Periglacial Features Developed on the Exposed Lake Bottoms of Seven Lakes that Drained Rapidly after 1950, Tuktoyaktuk Peninsula Area, Western Arctic Coast, Canada, *Permafrost and Periglacial Processes* 10, 39-63.

Macphail R.I., 1998, A Reply to Carter and Davidson's "An Evaluation of the Contribution of Soil Micromorphology to the Study of Ancient Arable Agriculture", *Geoarchaeology* 13, 549-564.

Macphail R., Cruise G., Engelmark R., Linderholm J., 2000, Chapter 8. Integrating Soil Micromorphology and Rapid Chemical Survey Methods: new developments in reconstructing past rural settlement and landscape organisation, in: Roskams S. (ed), *Interpreting Stratigraphy. Site evaluation, recording procedures and stratigraphic analysis, papers presented to the Interpreting Stratigraphy Conferences 1993-1997* (BAR, International Series 910), Oxford, 71-80.

Maes B., 2002, Bomen en struiken in Nederland. Inheems, autochtoon, exoot en archeofiet, *Gorteria* 28, 1-20.

Maréchal R., Maarleveld G.C., 1955, L'extension des phénomènes périglaciaires en Belgique et aux Pays-Bas, *Mededelingen van de Geologische Stichting, Nieuwe Serie* 8, 77-86.

Maréchal R., Tavernier R., 1974, *Pedologie. Atlas van België. Commentaar bij de bladen 11 A en 11 B. Uittreksels van de bodemkaart. Bodemassociaties*, Nationaal Comité voor Geografie, Commissie voor de Nationale Atlas.

Mathieu C., Pieltain F., 1997, *Analyse physique des sols. Méthodes choisies*, Paris, 275 pp.

McDonnell, J.G., 2001, Pyrotechnology, in: Brothwell D.R., Pollard A.M. (eds), *Handbook of Archaeological Sciences*, Chichester, John Wiley & Sons Ltd, 493-505.

McKinley J.I., Bond, J.M., 2001, Cremated Bone, in: Brothwell D.R., Pollard A.M. (eds), *Handbook of Archaeological Sciences*, Chichester, John Wiley & Sons Ltd, 281-292.

Mehra O.P., Jackson M.L., 1960, Iron oxide removal from soils and clays by a dithionite-citrate system buffered with sodium bicarbonate, *Proc. 7th Nat. Conf. on Clays and Clay Minerals*, New York, Permagon Press, 317-327.

Meirsschaut G., 1979, *Herbarius oft Cruydt-boek van Rembertus Dodonaeus*, Kruishoutem, (facsimile van Anoniem, 1644, *Cruydt-boeck Remberti Dodonaei*, Antwerpen).

Miklauz R., 1908, Beiträge zur Kenntnis der Humussubstanzen, *Z. Moorkult. Torfverwert.* 6, 285-327.

Minnaert G., Verbruggen C., 1986, Palynologisch onderzoek van een veenprofiel uit het Doeldok te Doel, *Buitengewone uitgaven van de Koninklijke Oudheidkundige Kring van het Land van Waas* 19 (*Bijdragen van de Archeologische Dienst Waasland* I), 201-208.

Mithen S., Finlay N., 2000, Staosnaig, Colonsay: Excavations 1989-1995, in: Mithen S. (ed), *Hunter-gatherer landscape archaeology. The Southern Hebrides Mesolithic Project 1988-98* (McDonald Institute Monographs), Cambridge, 359-441.

Mithen S., Finlay N., Carruthers W., Carter S., Ashmore P., 2001, Plant Use in the Mesolithic: Evidence from Stoasnaig, Isle of Colonsay, Scotland, *Journal of Archaeological Science* 28, 223-234.

Mokma D.L., Buurman P., 1982, *Podzols and podzolisation in temperate regions*, Wageningen, International Soil Museum (ISM Monograph 1), 126 pp.

Moore P.D., 1986, Hydrological changes in mires, in: Berglund B.E. (ed), *Handbook of Holocene Palaeoecology and Palaeohydrology*, Chichester, John Wiley & Sons, 91-107.

Moore P.D., 1989, The ecology of peat-forming processes: a review, *International Journal of Coal Geology* 12, 89-103.

Moore P.D., Webb J.A., Collinson M.E., 1991, *Pollen analysis*, Cambridge.

Morras H., 1973, L'identification des matériaux carbonatés par coloration différentielle, *Technical note* 1, Gent, Rijksuniversiteit Gent, Bodemmineralogie en mikromorfologie, 9 pp.

Morton J.D., Schwarcz H.P., 2004, Palaeodietary implications from stable isotopic analysis of residues on prehistoric Ontario ceramics, *Journal of Archaeological Science* 31, 503-517.

Mottram H.R., Dudd S.N., Lawrence G.J., Stott A.W., Evershed R.P., 1999, New chromatographic, mass spectrometric and stable isotope approaches to the classification of degraded animal fats preserved in archaeological pottery, *Journal of Chromatography A* 833, 209-221.

Moulin B., 1991, *Hauterive-Champréveyres 3. La dynamique sédimentaire et lacustre durant le Tardiglaciaire et le Postglaciaire* (Archéologie Neuchâteloise 9), Saint-Blaise, Editions du Ruau, 142 pp.

Mückenhausen E., 1982, *Die Bodenkunde und ihre geologischen, geomorphologischen, mineralogischen und petrologischen Grundlagen*, Frankfurt am Main, DLG-Verlag, 579 pp.

Munaut A.V., 1967, *Recherches paléo-écologiques en Basse et Moyenne Belgique* (Acta Geographica Lovaniensia 6), Louvain.

Munaut A.V., Paulissen E., 1973, Evolution et paléo-écologie de la vallée de la Petite Nèthe au cours du post-Würm (Belgique), *Extrait des Annales de la Société Géologique de Belgique* 96, 301-346.

Munsell, 2000, *Munsell Soil Color Charts* (Revised washable edition), New York, GretagMacbeth.

Muramoto S., 1981, Variations of some elements in cadmium-induced malformed fish, *Bulletin of Environmental Contamination and Toxicology* 27, 193-200.

Mys M., Gullentops F., Janssens P., Wuytack M.J., Stinissen H., 1983, De holocene evolutie van de alluviale vlakte van de Beneden-Schelde, *Tijdschrift van de Belgische Vereniging voor Aardrijkskundige Studies* 1, 7-33.

Newell R.R., 1981, Mesolithic Dwelling Structures: Fact and Fantasy, in: Gramsch B. (ed), *Mesolithikum in Europa. 2. Internationales Symposium Potsdam, 3. bis 8. April 1978* (Veröffentlichungen des Museums für Ur- und Frühgeschichte Potsdam 14/15), 235-284.

Nicholson R.A., 1993, A morphological investigation of burnt animal bone and an evaluation of its utility in archaeology, *Journal of Archaeological Science* 20, 411-428.

Noirfalise A., 1984, *Forêts et stations forestières en Belgique*, Gembloux.

Odgaard B.V., 1994, *The Holocene vegetation history of northern West Jutland, Denmark* (Opera Botanica 123), Copenhagen, 171 pp.

Otte M., Vandermoere N., Heyse I., Léotard J.-M., 1984, Maldegem et le Paléolithique récent du nord-ouest européen, *Helinium* 24, 105-126.

Oudemans T.F.M., Boon, J.J., 1991, Molecular archaeology: analysis of charred (food) remains from prehistoric pottery by PY-GC-MS, *Journal of Applied and Analytical Pyrolysis* 20, 197-227.

Oversteegen J.F.S., 1999, *De mesolithische opgraving Verrebroek 'Dok' campagne 1996: Verslag van het onderzoek naar de resten van dierlijk botmateriaal* (Archaeo-Zoo Rapporten 2 (1)), Amsterdam.

Oversteegen J.F.S., 2000a, *Verrebroek 'Dok' 2* (Archaeo-Zoo Rapporten), Amsterdam.

Oversteegen J.F.S., 2000b, *Analyse van het dierlijke botmateriaal van de vroeg-neolitische vindplaats Urk-E4*, (BIAX-rapport 46B), Leiden.

Oversteegen J.F.S., 2001, Archeozoölogie, in: Hogestijn J.W.H., Peeters J.H.M. (eds), *De opgraving van de mesolithische en neolithische vindplaats Urk-E4 (Domineesweg, gemeente Urk)* (Rapportage Archeologische Monumentenzorg 93), Amersfoort, 43-48.

Oversteegen J.F.S., van Wijngaarden-Bakker L.H., Maliepaard R., van Kolfschoten Th., 2001, Zoogdieren, vogels, reptielen, in: Louwe Kooijmans L.P. (ed), *Hardinxveld-Giessendam De Bruin. Een kampplaats uit het Laat-Mesolithicum en het begin van de Swifterbant-cultuur (5500-4450 v.Chr.)* (Rapportage Archeologische Monumentenzorg 88), Amersfoort, 209-297.

Paepe R., 1968, Les sols fossiles Pléistocènes de la Belgique, *Pedologie* XVIII/2, 176-188.

Paepe R., Vanhoorne R., 1967, The stratigraphy and palaeobotany of the Late Pleistocene in Belgium, *Mémoires pour servir à l'explication des cartes géologiques et minières de la Belgique* 8, Bruxelles.

Palese R., Aeschimann D., 1990, *La grande flore de Gaston Bonnier*, volume 3, Paris.

Pals J.P., van Geel B., 1976, Rye Cultivation and the Presence of Cornflower (Centaurea Cyanus L.), Berichten van de Rijksdienst voor Oudheidkundig Bodemonderzoek 26, 199-204.

Pals J.P., van Geel B., Delfos A., 1980, Palaeoecological studies in the Klokkeweel bog near Hoogkarspel (Noord-Holland), *Review of Palaeobotany and Palynology* 30, 371-418.

Passi S., Cataudella S., Marco P. D., Simone F. D., Rastrelli L., 2002, Fatty Acid Composition and Antioxidant Levels in Muscle Tissue of Different Mediterranean Marine Species of Fish and Shellfish, *Journal of Agricultural and Food Chemistry* 50, 7314-7322.

Paulissen E., Munaut A.V., 1969, Un horizon blanchâtre d'âge Bølling à Opgrimbie, *Acta Geographica Lovaniensia 7*, 65-91.

Pedersen L., 1995, 7000 years of fishing: stationary fishing structures in the Mesolithic and afterwards, in: Fischer A. (ed), *Man and sea in the Mesolithic. Coastal settlement above and below present sea level* (Oxbow Monograph 53), Oxford, 75-86.

Peeters J.H.M., Hanreats E., Hogestijn J.W.H., Jansma E., 2001, Deel 12: dateringen: ^{14}C-analyse en dendrochronologie, in: Hogestijn J.W.H., Peeters J.H.M. (eds), *De mesolithische en vroeg-neolithische vindplaats Hoge Vaart-A27 (Flevoland)* (Rapportage Archeologische Monumentenzorg 79), Amersfoort.

Perlès C., 1977, *Préhistoire du feu*, Paris, Masson, 180 pp.

Perry D., 1999, Vegetative Tissues from Mesolithic Sites in the Northern Netherlands, *Current Anthropology* 40(2), 231-237.

Peters F.J.C., Peeters J.H.M., 2001, *De opgraving van de mesolithische en neolithische vindplaats Urk-E4 (Domineesweg, gemeente Urk)* (Rapportage Archeologische Monumentenzorg 93), Amersfoort.

Petersen L., 1976, *Podzols and Podzolisation*, Copenhagen, 293 pp.

Pissart A., 1987a, *Geomorphologie périglaciaire, Belgische Franqui Leerstoel 1987*, Liège, Edition Laboratoire de Géomorphologie et de Géologie du Quaternaire, 135 pp.

Pissart A., 1987b, Weichselian periglacial structures and their environmental significance: Belgium, the Netherlands and northern France, in: Boardman J. (ed), *Periglacial processes and landforms in Britain and Ireland*, Cambridge, Cambridge University Press, 77-85.

Poll M., 1947, *Faune de Belgique. Poissons marins*, Musée Royal d'Histoire Naturelle de Belgique, Bruxelles.

Poncelet L., Martin H., 1947, Esquisse climatographique de la Belgique, *Institut Royal Météorologique de Belgique - Mémoires* XXVII, Bruxelles.

Pons L.J., 1987, *Bodem, land en landschap*, College, Landbouwuniversiteit Wageningen, Regionale Bodemkunde, 26 maart 1987.

Pons L.J., 1992, Holocene peat formation in the lower parts of the Netherlands, in: Verhoeven J.T.A. (ed), *Fens and Bogs in the Netherlands: Vegetation, History, Nutrient Dynamics and Conservation* (Geobotany 18), Dordrecht, 7-79.

Price T.D., Whallon R., Chappell Jr. & S., 1974, Mesolithic Sites near Havelte, Province of Drenthe (Netherlands), *Palaeohistoria* 16, 7-61.

Punt W., Blackmore S., Clarke G.C.S., Hoen P. (eds), 1984-1995, *The Northwest European Pollen Flora* I-VII, Amsterdam.

Rackham O., 1980, *Ancient woodland*, London.

Raemaekers D.C.M., 1999, *The Articulation of a "New Neolithic". The meaning of the Swifterbant Culture for the process of neolithisation in the western part of the North European Plain (4900-3400 BC)* (Archaeological Studies Leiden University 3), Leiden.

Raemaekers D.C.M., 2001, Aardewerk en verbrande klei, in: Louwe Kooijmans L.P. (ed), *Hardinxveld-Giessendam De Bruin. Een kampplaats uit het Laat-Mesolithicum en het begin van de Swifterbant-cultuur (5500-4450 v.Chr.)* (Rapportage Archeologische Monumentenzorg 88), 117-152.

Raemaekers D.C.M., Bakels C.C., Beerenhout B., Van Gijn A.L., Hänninen K., Molenaar S., Paalman D., Verbruggen M., Vermeeren C., 1997, Wateringen 4: a settlement of the Middle Neolithic Hazendonk 3 Group in the Dutch Coastal Area, *Analecta Praehistoria Leidensia* 29, 143-191.

Rasmussen P., 1989a, Leaf-foddering of livestock in the Neolithic: archaeobotanical evidence from Weier, Switzerland, *Journal of Danish Archaeology* 8, 51-71.

Rasmussen P., 1989b, Leaf foddering in the earliest Neolithic agriculture. Evidence from Switzerland and Denmark, *Acta Archaeologica* 60 (1990), 71-86.

Reimer P. J., Hughen K.A., Guilderson T.P., McCormac G., Baillie M.G.L., Bard E., Barratt P., Beck J.W., Buck C.E., Damon P.E., Friedrich M., Kromer B., Bronk Ramsey C., Reimer R.W., Remmele S., Southon J.R., Stuiver M., van der Plicht J., 2002, Preliminary Report of the First Workshop of the IntCal04 Radiocarbon Calibration/Comparison Working Group, *Radiocarbon* 44(3), 653-662.

Robeerst J.M.M., 1995, *De Neolithische fauna van de donk 'Het Kerkhof' bij Brandwijk, Alblasserwaard,* Unpublished report, Universiteit Leiden.

Romans J.C.C., Robertson L., 1983, The general effects of early agriculture on the soil profile, in: Maxwell G.S. (ed), *The impact of aerial reconnaissance on archaeology* (CBA Research Report 49), London, 136-141.

Rosen A.M., 1992, Preliminary identification of silica skeletons from the Near Eastern archaeological sites: an anatomical approach, in: Ray G.Jr., Mulholland S.C. (eds), *Phytolith Systematics. Emerging Issues* (Advances in Archaeological and Museum Science 1), 129-148.

Rovner I., 2001, Cultural behavior and botanical history: phytolith analysis in small places and narrow intervals, in: Meunier, J. D., Colin, F. (eds), *Phytoliths: applications in earth sciences and human history,* Balkema Publishers, 119-127.

Rowell D.L., 1997, *Soil Science: Methods and Applications,* Essex (England), Addison Wesley Longman Limited, 350 pp.

Runhaar J., Groen C.L.G., van der Meijden R., Stevers R.A.M., 1987, Een nieuwe indeling in ecologische groepen binnen de Nederlandse flora, *Gorteria* 13, 277-359.

Ruthenberg K., 1991, Untersuchung von Teer und Pech: Archäologische Fragestellungen - archäometrische Möglichkeiten, *Acta praehistorica et archaeologica* 23, 97-102.

Ruthenberg K., 1997, Historical Development and Comparison of Analytical Methods for the Identification of Tar and Pitch, in: Brzeziński W., Piotrowski W. (eds), *Proceedings of the First International Symposium on Wood Tar and Pitch,* Warsaw, State Archaeological Museum, 173-179.

Sato M., Kondo T., Yoshinaka R., Ikeda S., 1983, Effect of water temperature on the skeletal deformity in ascorbic acid-deficient rainbow trout, *Bulletin of the Japanese Society of Scientific Fisheries* 49, 443-446.

Schachtschabel P., Blume H.-P., Brümmer G., Hartge K.H., Schwertmann U., 1998, *Scheffer/Schachtschabel. Lehrbuch der Bodenkunde*, Stuttgart, Ferdinand Enke Verlag, 494 pp.

Schaetzl R., Johnson D., Burns S., Small T., 1989, Tree uprooting: a review of terminology, process and environmental implications, *Canadian Journal of Forestry Research* 19, 1-11.

Scheffer F., Ulrich B., 1960, *Humus und Humusdüngung. Band I. Morphologie, Biologie, Chemie und Dynamik des Humus*, Stuttgart, Ferdinand Enke Verlag.

Schwan J., 1990, Noncryogenic deformations in Loch Lomond Stadial to Early Flandrian coversands in North Lincolnshire, England, *Geologie en Mijnbouw* 69, 173-178.

Schweingruber F.H., 1990a, *Anatomy of European Woods. An atlas for the identification of European trees, shrubs and dwarf schrubs*, Bern.

Schweingruber F.H., 1990b, *Microscopic Wood Anatomy. Structural variability of stems and twigs in recent and subfossil woods from Central Europe*, Swiss Federal Institute for Forest, Snow and Landscape Research, Birmensdorf.

Score D., Mithen S., 2000, The Experimental Roasting of Hazelnuts, in: Mithen S. (ed), *Hunter-gatherer landscape archaeology. The Southern Hebrides Mesolithic Project 1988-98* (McDonald Institute Monographs), Cambridge, 507-512.

Sergant J., 2004, *De aantrekkingskracht van een zandrug. Ruimtelijke analyse van een vroegmesolithische site te Verrebroek-Dok. Deel 1: Tekst*, Unpublished PhD thesis Ghent University, 343 pp.

Simmons I.G., 1996, *The Environmental Impact of Later Mesolithic Cultures. The Creation of Moorland Landscape in England and Wales*, Edinburgh, Edingburgh University Press, 260 pp.

Simpson I.A., Perdikaris S., Cook G., Campbell J.L., Teesdale W.J., 2000, Cultural Sediment Analyses and Transitions in Early Fishing Activity at Langenesværet, Vesterålen, Northern Norway, *Geoarchaeology* 15, 743-763.

Slager S., van de Wettering H.T.J., Geenen S.S., 1976, Anwendungsmöglichkeiten der Bodenmikromorfologie bei der Auswertung von Befunden der Ausgrabungen zum Neolithikum auf der Aldenhovener Platte, *Bonner Jahrbücher* 176, 313-317.

Smit J.L., 1995, De grootste boreaal-mesolithische nederzetting van Nederland, in: Constandse-Westermann T.S., Niekus M.J.L.Th., Smit J.L. (eds), *Bundel Mesolithicumdag Veendam*, Veendam, 7-18.

Soil Map of Belgium (Bodemkaart van België). Lillo 14E, 1960, Scale 1:20000, Brussel, Militair Geografisch Instituut.

Soil Map of Belgium (Bodemkaart van België). Beveren-Waas 27E, 1961, Scale 1:20000, Brussel, Militair Geografisch Instituut.

Soil Science Society of America, 1997, *Glossary of Soil Science Terms*, Wisconsin, USA, Soil Science Society of America.

Spek Th., Bisdom E.B.A., van Smeerdijk D.G., 2001a, Deel 7. Bodemkunde en landschapsecologie I: veranderingen in bodem en landschap, in: Hogestijn J.W.H., Peeters J.H.M. (eds), *De mesolithische en vroeg-neolithische vindplaats Hoge Vaart-A27 (Flevoland)* (Rapportage Archeologische Monumentenzorg 79), Amersfoort, 143 pp.

Spek Th., Bisdom E.B.A., van Smeerdijk D.G., 2001b, Deel 8. Bodemkunde en landschapsecologie II: aanvullend onderzoek naar landschapsvormende processen, in: Hogestijn J.W.H., Peeters J.H.M. (eds), *De mesolithische en vroeg-neolithische vindplaats Hoge Vaart-A27 (Flevoland)* (Rapportage Archeologische Monumentenzorg 79), Amersfoort, 128 pp.

Stace C., 1997/2001, *New Flora of the British Isles*, Cambridge, 2nd edition.

Stapert D., Veenstra H.J., 1988, The section at Usselo; brief description, grain-size distributions, and some remarks on the archaeology, *Palaeohistoria* 30, 1-28.

Stoklund B., 1999, Turf manuring on the Danish island of Læsø, *Danish Journal of Geography, Special Issue* 1, 209-214.

Stoops G., 1983, SEM and light microscopic observations of minerals in bog-ores of the Belgian Campine, *Geoderma* 30, 179-186.

Stoops G., 1998, Key to the ISSS "Handbook for Soil Thin Section Description", *Natuurwetenschappelijk Tijdschrift* 78, 193-203.

Stoops G., Eswaran H., 1985, Morphological characteristics of wet soils, in: *Wetland Soils: Characterization, Classification and Utilization, Proceedings Workshop 26/03 - 05/04/1984*, Manila, Philippines, International Rice Research Institute, 177-189.

Stortelder A.H.F., Schaminée J.H.J., Hommel P.W.F.M.,1999, *De vegetatie van Nederland. Deel 5. Plantengemeenschappen van ruigten, struwelen en bossen*, Uppsala – Leiden.

Stuiver M., van der Plicht J. (eds), 1998, INTCAL98: calibration issue, *Radiocarbon* 40(3), 1041-1159.

Stützer A., 1998, Early stages of podzolisation in young aeolian sediments, western Jutland, *Catena* 32, 115-129.

Swanson D.K., Ping C.-L., Michaelson G.J., 1999, Diapirism in Soils due to Thaw of Ice-Rich Material near the Permafrost Table, *Permafrost and Periglacial Processes* 10, 349-367.

Talbot M.R., Allen P.A., 1996, Lakes, in: Reading (ed), *Sedimentary Environments: Processes, Facies and Stratigraphy*, 83-124.

Tamm O., 1922, Om bestämning av de oorganiska komponenterna i markens gelkomplex, *Medd. Stat. Skogsförsöksanst.* 19, 385-404.

Tavernier R., Ameryckx J., 1957, Le postpodzol en Flandre sablonneuse, *Pédologie* VII, 86-96.

Tavernier R., Smith G., 1957, The concept of Braunerde (Brown Forest Soil) in Europe and the United States, *Advances in Agronomy* IX, 217-289.

Théry-Parisot I., 1998, *Economie du combustible et paléoécologie en contexte glaciaire et périglaciaire, Paléolithique moyen et supérieur du sud de la France*, Thèse d'Université de Paris I, Paris.

Théry-Parisot, 2001, *Economie des combustibles au Paléolithique. Expérimentation, taphonomie, anthracologie* (Dossier de documentation archéologique 20), Paris, 195 pp.

Twiss, P.C., 1992, Predicted world distribution of C3 and C4 grass phytoliths, in: Rapp, G. Jr., Mulholland, S.C. (eds), *Phytolith Systematics. Emerging Issues* (Advances in Archaeological and Museum Science 1), 113-128.

USDA, 1992, *Keys to Soil Taxonomy* (SMSS Technical Monograph 19), Soil Survey Staff, Soil Conservation Service, United States Department of Agriculture, Blacksburg, Virginia, Pocahontas Press Inc.

van Breemen N., Buurman P., 1998, *Soil Formation* , Dordrecht, Kluwer Academic Publishers, 377 pp.

Van den Balck E., Hoffmann M., Meire P., 1998, *Vegetatie van alluviale vlakten langs West-Europese rivieren en de invloed van abiotische factoren en beheer: een literatuuroverzicht*, Rapport IN 98.20, Brussel.

van den Broeke P.W., 1991, De introductie van de landbouw: lössboeren en onverstoorbare kustbewoners, in: Bloemers J.H.F., van Dorp T. (eds), *Pre- & protohistorie van de Lage Landen*, Open Universiteit, De Haan, 213-225.

Van Der Hammen T., Wijmstra T.A. (eds), 1971, The Upper Quaternary of the Dinkel valley (Twente, Eastern Overijssel, The Netherlands), *Mededelingen Rijks Geologische Dienst, Nieuwe Serie* 22, 55-214.

Van der Sluijs P., Maarleveld G.C., 1963, Dekzandruggen uit de Jonge Dryastijd in Zeeuws-Vlaanderen (Cover-sand ridges from the Late Dryas time in Zealand Flanders), *Boor en Spade* XIII, 21-26.

Van der Werf S., 1991, *Bosgemeenschappen* (Natuurbeheer in Nederland 5), Wageningen.

van Geel B., 1976, *A palaeoecological study of holocene peat bog sections, based on the analysis of pollen, spores and macro and microscopic remains of fungi, algae, cormophytes and animals* (Academisch proefschrift Universiteit van Amsterdam), Amsterdam.

van Geel B., 2001, Non-pollen palynomorphs, in: Smol J.P., Birks H.J.B., Last W.M. (eds), *Tracking Environmental Change Using Lake Sediments, Terrestrial, Algal and Silicaceous Indicators vol. 3*, Dordrecht, 99-119.

van Geel B., Bohncke S.J.P., Dee H., 1981, A palaeoecological study of an upper Late Glacial and Holocene sequence from "De Borchert", The Netherlands, *Review of Palaeobotany and Palynology* 31, 367-448.

van Geel B., Coope G.R., Van Der Hammen Th., 1989, Palaeoecology and stratigraphy of the Late Glacial type section at Usselo (The Netherlands), *Review of Palaeobotany and Palynology* 60, 25-129.

van Geel B., Klink A.G., Pals J.P., Wiegers J., 1986, An Upper Eemiaan lake deposit from Twente, eastern Netherlands, *Review of Palaeobotany and Palynology* 47, 31-61.

van Geel B., Buurman J., Brinkkemper O., Schelvis J., Aptroot A., van Reenen G., Hakbijl T., 2003, Environmental reconstruction of a Roman Period settlement site in Uitgeest (The Netherlands), with special reference to coprophilous fungi, *Journal of Archaeological Science* 30 (7), 873-883.

van Genderen H., Schoonhoven L.M., 1997, *Chemisch-ecologische flora van Nederland en België. Een inleiding over aard en ecologische betekenis van secundaire plantenstoffen* (met medewerking van Fuchs A.), Utrecht.

Van Haaster H., 1997, De introductie van cultuurgewassen in de Nederlanden tijdens de Middeleeuwen, in: Zeven A.C. (ed) (met bijdragen van Bakels C.C., Van Haaster H., Pals J.-P. & Zeven A.C.), *De introductie van onze cultuurplanten en hun begeleiders, van het Neolithicum tot 1500 AD*, Wageningen, 53-104.

Van Haaster H., Brinkkemper O., 1995, RADAR, a Relation Archaeobotanical Database for Advanced Research, *Vegetation History and Archaeobotany* 4, 117-125.

Vanhoorne R., Verbruggen C., 1975, Problèmes de subdivision du Tardiglaciaire dans la région sablonneuse du Nord de la Flandre en Belgique, *Pollen et Spores* XVII/4, 525-543.

Van Neer W., Ervynck A., Lentacker A., Crombé Ph., Sergant J., Perdaen Y., Van Strydonck M., Van Roeyen J.-P., 2001, Dierenresten uit een vroege Swifterbant-nederzetting te Doel-Deurganckdok (Vlaanderen, België): jachtwild, maar vooral veel vis, *Notae Praehistoricae* 21, 85-96.

Van Ranst E., 1991, *Soil Genesis. Concepts of Soil Development. Formation of Diagnostic Horizons and Materials*, Lecture notes, Gent, State University Ghent, International Training Centre for Post-Graduate Soil Scientists.

Van Ranst E., 1995, *Clay mineralogy. Crystal structures, identification, analysis, and chemistry of clay minerals and clays*, Lecture notes, State University Ghent, International Training Centre for Post-Graduate Soil Scientists, 287 pp.

Van Ranst E., De Coninck F., 1983, Evolution of glauconite in imperfectly drained sandy soils of the Belgian Campine, *Zeitschrift für Pflanzenernährung und Bodenkunde* 146, 415-426.

Van Ranst E., Verloo M., Demeyer A., Pauwels J.M., 1999, *Manual for the Soil Chemistry and Fertility Laboratory*, Gent, Ghent University, 243 pp.

Van Roeyen J.-P., 1990, *Mesolithische bewoning in de Wase Scheldepolders. Opgravingen en prospecties (1981-1983)*, Unpublished Master's thesis, Ghent University.

Van Roeyen J.-P., Minnaert G., Van Strydonck M., Verbruggen C., 1991, Melsele-Hof ten Damme: prehistorische bewoning, landschappelijke ontwikkeling en kronologisch kader, *Notae Praehistoricae* 11, 41-49.

Van Roeyen J.-P., Verbruggen C., Klinck B., Meersschaert L., 2001, Het Deurganckdok te Doel (Beveren, O.-Vl.). Paleolandschappelijk en archeologisch onderzoek, *Annalen van de Koninklijke Oudheidkundige Kring van het Land van Waas* 104 (Bijdragen van de Archeologische Dienst Waasland 5), Sint-Niklaas, 437-484.

Van Strydonck M., van der Borg K., 1990-91, The construction of a preparation line for AMS-targets at the Royal Institute for Cultural Heritage Brussels, *Bulletin Koninklijk Instituut voor het Kunstpatrimonium* 23, 228-234.

Van Strydonck M., Boudin M., Hoefkens M., in preparation, Enkele beschouwingen in verband met het dateren van gecremeerde beenderen.

Van Strydonck M., Crombé Ph., Maes A., 2001, The site of Verrebroek "Dok" and its contribution to the absolute dating of the Mesolithic in the Low Countries, in: Carmi I., Boaretto E. (eds), *Proceedings of the 17th International Radiocarbon Conference, Judean Hills, Israel, June 18-23* (Radiocarbon 43), 997-1005.

Van Strydonck M., Forest L., Landrie M., Hendrix V., van der Borg Kl., de Jong A.F.M., 1995, *Royal Institute for Cultural Heritage, Radiocarbon Dates XV*, Brussels, 48 pp.

Van Strydonck M., Landrie M., Boudin M., Grootes P.M., Nadeau M.-J., Sparks R., Keppens E., 2002, *Royal Institute for Cultural Heritage, Radiocarbon Dates XVIII*, Brussels, 70 pp.

Van Strydonck M., Nelson D.E., Crombé Ph., Bronk Ramsey C., Scott E.M., Van der Plicht J., Hedges R.E.M., 1999, Rapport du groupe de travail: les limites de méthode du carbone 14 appliquée à l'archéologie: What's in a ^{14}C date / Que'est ce qu'il y a dans une date ^{14}C, in: Evin J., Oberlin C., Daugas J.P., Salles J.F. (eds), *Actes du colloque "C14 et Archéologie", 1998* (Mémoires de la Société Préhistorique Française 26 and Supplémént de la Revue d'Archéométrie), Paris, 433-448.

Van Vliet-Lanoë B., 1985, Frost Effects in Soils, in: Boardman J. (ed), *Soils and Quaternary Landscape Evolution*, Chichester, John Wiley & Sons Ltd., 117-156.

Van Vliet-Lanoë B., 1988, The significance of cryoturbation phenomena in environmental reconstruction, *Journal of Quaternary Science* 3(1), 85-96.

Van Vliet-Lanoë B., 1991, Differential Frost Heave, Load Casting and Convection: Convering Mechanisms; a Discussion of the Origin of Cryoturbations, *Permafrost and Periglacial Processes* 2, 123-139.

Van Vliet-Lanoë B., 1996, Relations entre la contraction thermique des sols en Europe du Nord-Ouest et la dynamique de l'inlandsis weichélien, *C.R. Acad. Sci. Paris* 322 (IIa), 461-468.

Van Vliet-Lanoë B., 1998, Frost and soils: implications for paleosols, paleoclimates and stratigraphy, *Catena* 34, 157-183.

Van Vliet-Lanoë B., Coutard J.P., Pissart A., 1984, Structures caused by repeated freezing and thawing in various loamy sediments: a comparison of active, fossil, and experimental data, *Earth Surface Processes and Landforms* 10, 553-565.

Van Vliet-Lanoë B., Seppälä M., Käyhkö J., 1993, Dune dynamics and cryoturbation features controlled by Holocene water level change, Hietatievat, Finnish Lapland, *Geologie en Mijnbouw* 72, 211-224.

Van Zeist W., Palfenier-Vegter R.M., 1981, Seeds and fruits from the Swifterbant S3 site, *Palaeohistoria* 23, 105-168.

Vaughan D., Russell J.D., Ord B.G., Anderson H.A., Fraser A.R., 1989, Characterization of a Scottish Dopplerite, *Geoderma* 43, 325-335.

Verbruggen C., 1971, *Postglaciale landschapsgeschiedenis van Zandig Vlaanderen*, Unpublished PhD thesis, Ghent University, 440 pp.

Verbruggen C., 1977, (met medewerking van W. Van Dongen), De geokronologie van het Postpleniglaciaal in Zandig-Vlaanderen op basis van pollenanalyse en ^{14}C-onderzoek, *Natuurwetenschappelijk Tijdschrift* 58, 233-256.

Verbruggen C., 1979, Vegetational and palaeoecological history of the Late Glacial period in Sandy Flanders (Belgium), *Acta Universitatis Ouluensis* 82, *Geol.* 3, 133-142.

Verbruggen C., 1999, Quaternary palaeobotanical evolution of Northern Belgium, *Geologica Belgica* 2/3-4, 99-110.

Verbruggen C., Denys L., 1995, Early tidal influence on the Lower Schelde, Belgium, in: Gullentops F. (ed), *Wetlands in Flanders, Contributions to Palaeohydrology of the temperate Zone in the Last 15 000 years* (Aardkundige Mededelingen 6), Leuven, 167-169.

Verbruggen C., Denys L., Kiden P., 1991, Paleo-ecologische en geomorfologische evolutie van Laag- en Midden-België tijdens het Laat-Kwartair (Palaeoecological and geomorphological evolution in Lower- and Middle Belgium during the Late-Quaternary), *De Aardrijkskunde* 3, 357-376.

Verbruggen C., Denys L., Kiden P. 1996, Belgium, in: Berglund B.E., Birks H.J.B., Ralska-Jasiewiczowa M., Wright H.E. (eds), *Palaeoecological events during the last 15000 years: regional syntheses of palaeoecological studies of lakes and mires in Europe*, Chichester, John Wiley & Sons, 553-574.

Verhart L.B.M., 2000, *Times fade away. The neolithization of the southern Netherlands in an anthropological and geographical perspective* (Archaeological Studies Leiden University 6), Leiden.

Verloo M., 1992, *Agrochemie (Agrochemistry)*, Lecture notes, Gent, Faculteit van de Landbouwwetenschappen (Faculty of Agronomy), 147 pp.

Vermeersch P.M., 1984, Du Paléolithique final au Mésolithique dans le nord de la Belgique, in: Cahen D., Haesaerts P. (eds), *Peuples chasseurs de la Belgique préhistorique dans leur cadre naturel*, Bruxelles, 181-193.

Vernet J.-L., 1973, Etude sur l'histoire de la végétation du sud-est de la France au Quaternaire d'après les charbons de bois principalement, *Paléobiologie continentale* IV(1), 1-90.

Visser D., Whitton C., Brinkkemper O., Hogestijn J.W.H., 2001, Deel 11: Archeobotanie: de analyse van botanische macroresten, in: Hogestijn J.W.H., Peeters J.H.M. (eds), *De mesolithische en vroeg-neolithische vindplaats Hoge Vaart-A27 (Flevoland)* (Rapportage Archeologische Monumentenzorg 79), Amersfoort, 1-83.

Vollbrecht J., 2003, Mesolithic settlement structures in Reichwalde – Preliminary observations on Mesolithic sites, in: Larsson L., Kindgren H., Knutsson K., Loeffler D., Akerlund A. (eds), *Mesolithic on the Move: Papers presented at the Sixth International Conference on the Mesolithic in Europe, Stockholm 2000* (Oxbow Books Ltd.), Oxford, 269-277.

Waller M., 1994, *The Fenland Project, Number 9: Flandrian Environmental Change in Fenland* (East Anglian Archaeology 70), Cambridge.

Wang C., McKeague J.A., Kodama H., 1986, Pedogenic Imogolite and Soil Environments: Case Study of Spodosols in Quebec, Canada, *Soil Science Society of America Journal* 50, 711-718.

Washburn A.L., 1979, *Geocryology. A survey of periglacial processes and environments*, London, Edward Arnold, 406 pp.

Waterbolk H.T., 1999, De mens in het preboreale, boreale en atlantische bos, *Paleo-Aktueel* 10, 68-73.

Weeda E.J., Westra R., Westra Ch. en Westra T., 1985, *Nederlandse oecologische flora, wilde planten en hun relaties 1*, Haarlem.

Weeda E.J., Westra R., Westra Ch., Westra T., 1999, *Nederlandse oecologische flora. Wilde planten en hun relaties 2*, herdruk, s.l.

Whittow J., 1984, *The Penguin Dictionary of Physical Geography*, London, Pinguin Books Ltd, 591 pp.

Williams J.T., 1975, Cultivated Plants - Past, Present and Future, in: Steele F., Bourne A. (eds), *The Man/Food Equation, Proceedings of a Symposium held at the Royal Institution London, September 1973*, London, Academic Press, 97-113.

Wolf R.J.A.M., Stortelder A.H.F., de Waal R.W., van Dort K.W., Hennekens S.M., Hommel P.W.F.M., Schaminée J.H.J., Vrielink J.G., 2001, *Ooibossen* (Bosecosystemen van Nederland 2, Natuurhistorisch Bibliotheek 68), Utrecht.

Woodman P., 1985, *Excavations at Mount Sandel 1973-77* (Northern Ireland Archaeological Monographs 2), Belfast.

WRB (FAO, ISRIC, ISSS), 1998, *World Reference Base for Soil Resources* (World Soil Resources Reports 84), Rome, FAO, 88 pp.

Zeiler J.T., 1997, *Hunting, fowling and stock-breeding at neolithic sites in the western and central Netherlands,* Unpublished PhD thesis, Rijksuniversiteit Groningen.

Zvelebil M., 1994, Plant Use in the Mesolithic and its Role in the Transition to Farming, *Proceedings of the Prehistoric Society* 60, 35-74.

List of authors

Major contributions

Jan Bastiaens
Flemish Heritage Institute, Phoenix building, Koning Albert-II-laan 19, box 5, B-1210 Brussels, Belgium. jan.bastiaens@lin.vlaanderen.be

Philippe Crombé
Ghent University, Department of Archaeology and Ancient History of Europe, Blandijnberg 2, B-9000 Ghent, Belgium. philippe.crombe@UGent.be

Koen Deforce
Flemish Heritage Institute, Phoenix building, Koning Albert-II-laan 19, box 5, B-1210 Brussels, Belgium. koen.deforce@lin.vlaanderen.be

Anton Ervynck
Flemish Heritage Institute, Phoenix building, Koning Albert-II-laan 19, box 5, B-1210 Brussels, Belgium. anton.ervynck@lin.vlaanderen.be
Ghent University, Department of Archaeology and Ancient History of Europe, Blandijnberg 2, B-9000 Ghent, Belgium.

Vanessa Gelorini
Flemish Heritage Institute, Phoenix building, Koning Albert-II-laan 19, box 5, B-1210 Brussels, Belgium.
Coupure 942, B-9000 Ghent, Belgium. vanessa.gelorini@pandora.be

Bart Klinck
Ghent University, Department of Geography, Krijgslaan 281/S8, B-9000 Ghent, Belgium.
Archeologische Dienst Waasland, Regentiestraat 63, B-9100 Sint-Niklaas, Belgium. b_klinck@hotmail.com

Roger Langohr
Ghent University, Department of Geology and Soil Science, Krijgslaan 281/S8, B-9000 Ghent, Belgium. roger.langohr@UGent.be

An Lentacker
Flemish Heritage Institute, Phoenix building, Koning Albert-II-laan 19, box 5, B-1210 Brussels, Belgium. an.lentacker@ename974.org

Geertrui Louwagie
Ghent University, Department of Geology and Soil Science, Krijgslaan 281/S8, B-9000 Ghent, Belgium. geertrui.louwagie@UGent.be or geertrui_louwagie@hotmail.com

Lieselotte Meersschaert
Archeologische Dienst Waasland, Regentiestraat 63, B-9100 Sint-Niklaas, Belgium. lmeersschaert@hotmail.com

Yves Perdaen
Ghent University, Department of Archaeology and Ancient History of Europe, Blandijnberg 2, B-9000 Ghent, Belgium. yves.perdaen@UGent.be

Joris Sergant
Ghent University, Department of Archaeology and Ancient History of Europe, Blandijnberg 2, B-9000 Ghent, Belgium. joris.sergant@UGent.be

Wim Van Neer
Royal Museum of Central Africa, B-3080 Tervuren, Belgium. vanneer@africamuseum.be

Mark Van Strydonck
Royal Institute for Cultural Heritage, Jubelpark 1, B-1000 Brussels, Belgium. mark.vanstrydonck@kikirpa.be

Cyriel Verbruggen
Ghent University, Department of Geography, Krijgslaan 281/S8, B-9000 Ghent, Belgium. cyriel.verbruggen@UGent.be

Other contributions

Machteld Bats
Ghent University, Department of Archaeology and Ancient History of Europe, Blandijnberg 2, B-9000 Ghent, Belgium. machteld.bats@UGent.be

Karl Cordemans
Ghent University, Department of Archaeology and Ancient History of Europe, Blandijnberg 2, B-9000 Ghent, Belgium.
Vlaamse Landmaatschappij, Gulden-Vlieslaan 72, B-1060 Brussels, Belgium. karl.Cordemans@vlm.be

Oliver E. Craig
University of Newcastle upon Tyne, School of Historical Studies, Newcastle upon Tyne, NE1 7RU, United Kingdom. o.e.craig@ncl.ac.uk

Nathalie Van der Putten
Ghent University, Department of Geography, Krijgslaan 281/S8, B-9000 Ghent, Belgium. nathalie.vanderputten@UGent.be

Luc Vrydaghs
Ghent University, Laboratory of Palaeontology, Krijgslaan 281/S8, B-9000 Ghent, Belgium. luc_vrydaghs@yahoo.co.uk.

The authors wish to express their gratitude to J. O'Driscoll (University Ghent, English Department) for revising their use of English in the original manuscripts (except chapters 2.1. 3.3. and 3.5.).